Digital Twins in Industrial Production and Smart Manufacturing

Digital Twins in Industrial Production and Smart Manufacturing

An Understanding of Principles, Enhancers, and Obstacles

Edited by

Rajesh Kumar Dhanaraj
Symbiosis International (Deemed University), Pune, India

Balamurugan Balusamy
Shiv Nadar Institute of Eminence, Delhi, India

Prithi Samuel
SRM Institute of Science and Technology, Chennai, India

Ali Kashif Bashir
Manchester Metropolitan University, UK

Seifedine Kadry
Noroff University College, Kristiansand, Norway

IEEE*Press*

WILEY

Published by John Wiley & Sons, Inc., Hoboken, New Jersey.
Published simultaneously in Canada.

For general information on our other products and services or for technical support, please contact our Customer Care Department within the United States at (800) 762-2974, outside the United States at (317) 572-3993 or fax (317) 572-4002.

Wiley also publishes its books in a variety of electronic formats. Some content that appears in print may not be available in electronic formats. For more information about Wiley products, visit our web site at www.wiley.com.

Library of Congress Cataloging-in-Publication Data Applied for:

Hardback: 9781394195305

Cover Design: Wiley
Cover Image: © AF-studio/Getty Images

Set in 9.5/12.5pt STIXTwoText by Straive, Chennai, India

Contents

About the Editors

Dr. Rajesh Kumar Dhanaraj is a distinguished professor at Symbiosis International (Deemed University) in Pune, India. Before joining Symbiosis International University, he served as a professor at the School of Computing Science & Engineering at Galgotias University in Greater Noida, India. His academic and research achievements have earned him a place among the top 2% of scientists globally, a recognition bestowed upon him by Elsevier and Stanford University. He earned his BE degree in Computer Science and Engineering from Anna University Chennai, India, in 2007. Subsequently, he obtained his MTech degree from Anna University, Coimbatore, India, in 2010. His relentless pursuit of knowledge culminated in a PhD in computer science from Anna University, Chennai, India, in 2017. He has authored and edited over 60 books on various cutting-edge technologies and holds 23 patents. Furthermore, he has contributed over 150 articles and papers to esteemed refereed journals and international conferences, in addition to providing chapters for several influential books.

Dr. Dhanaraj has shared his insights with the academic community by delivering numerous tech talks on disruptive technologies. He has forged meaningful partnerships with esteemed professors from top QS-ranked universities around the world, fostering a global network of academic excellence. His research interests encompass Machine Learning, Cyber-Physical Systems, and Wireless Sensor Networks. Dr. Dhanaraj's expertise in these areas has led to numerous research talks on Applied AI and Cyber Physical Systems at various esteemed institutions.

Dr. Dhanaraj has earned the distinction of being a senior member of the Institute of Electrical and Electronics Engineers (IEEE). He is also a member of the Computer Science Teacher Association (CSTA) and the International Association of Engineers (IAENG). Dr. Dhanaraj's commitment to academic excellence extends to his role as an associate editor and guest editor for renowned journals, including Elsevier Computers and Electrical Engineering, Human-centric Computing and Information Sciences, Emerald - International Journal of Pervasive Computing and Communications, and Hindawi - Mobile Information Systems.

His expertise has earned him a position as an expert advisory panel member of Texas Instruments Inc., USA.

Google Scholar: https://scholar.google.com/citations?hl=th&user=8t9sO-QAAAAJ

Scopus: https://www.scopus.com/authid/detail.uri?authorId=56884774100

Orchid ID: https://orcid.org/0000-0002-2038-7359

LinkedIn: https://www.linkedin.com/in/dr-rajesh-kumar-dhanaraj-89578423

Dr. Balamurugan Balusamy is currently working as an associate dean student in Shiv Nadar Institute of Eminence, Delhi, India. Prior to this assignment he was a professor, School of Computing Sciences and Engineering and Director International Relations at Galgotias University, Greater Noida, India. His contributions focus on Engineering Education, Block chain and Data Sciences. His Academic degrees and twelve years of experience working as a faculty in a global University like VIT University, Vellore, has made him more receptive and prominent in his domain. He does have 200 plus high impact factor papers in Springer, Elsevier, and IEEE. He has done more than 80 Edited and authored books and collaborated with eminent professors across the world from top QS ranked university. Professor Balamurugan Balusamy has served up to the position of associate professor in his stint of 12 years of experience with VIT University, Vellore. He had completed his bachelors, masters, and PhD degrees from top premier institutions from India. His passion is teaching and adapts different design thinking principles while delivering his lectures. He has published 80+ books on various technologies and visited 15 plus countries for his technical course. He has several top-notch conferences in his resume and has published over 200 of quality journal, conference and book chapters combined. He serves in the advisory committee for several startups and forums and does consultancy work for industry on Industrial IOT. He has given over 195 talks in various events and symposium.

Dr. Prithi Samuel is currently working as an assistant professor in the Department of Computational Intelligence at SRM Institute of Science and Technology, Kattankulathur Campus, Chennai. She has completed her PhD in Information and Communication Engineering from Anna University, Chennai. She has got over 16 years of teaching experience in reputed engineering colleges in Coimbatore and Chennai. She is a pioneer researcher in the areas of Automata Theory, Deep learning, Machine Learning, Computational Intelligence Techniques, and Natural Language Processing. She has published papers in leading International Journals and International Conferences and published books and book chapters in Wiley, IEEE Press, Taylor and Francis, Springer, and Elsevier. She is an active IEEE, ACM, and ISTE Member.

Dr. Ali Kashif Bashir is a reader of Networks and Security at the Manchester Metropolitan University, UK. He is also affiliated with the University of Electronic Science and Technology of China (UESTC), China, National University of Science and Technology, Islamabad (NUST), Pakistan, and University of Guelph, Canada. He is managing several research and industrial projects as PI and Co-I, accumulatively of around 7 million pounds. He also reviews funding proposals for EPSRC, UK; Commonwealth, UK; National Science and Engineering Research Council (NSERC), Canada; MITACS, Canada; Irish Research Council (IRC), Ireland; and Qatar National Research Fund (QNRF), Qatar. Since 2016, he is serving IEEE Technology, Policy and Ethics as EIC, and AE of several IEEE, Springer, IET, and MDPI Journals. He has delivered more than 30 invited talks across the globe, organized 40+ guest editorials, and chaired around 35 conferences and workshops.

Dr. Seifedine Kadry has a bachelor's degree in 1999 from Lebanese University, an MS degree in 2002 from Reims University (France) and EPFL (Lausanne), PhD in 2007 from Blaise Pascal University (France), an HDR degree in 2017 from Rouen University (France). His research currently focuses on Data Science, medical image recognition using AI, education using technology, and applied mathematics. He is an IET fellow and IETE fellow, member of European Academy of Sciences and Arts. He is a full professor of data science at Noroff University College, Norway.

List of Contributors

P. Abinaya
Rathinam College of Arts and Science
Coimbatore
India

Azween Abdullah
Faculty of Applied Science and
Technology
Perdana University
Kuala Lumpur
Malaysia

B. Akoramurthy
Department of CSE
National Institute of Technology
Puducherry
Karaikal
Tamil Nadu
India

K. Aravinda
New Horizon College of Engineering
Bellandur Post
Bengaluru
India

K. Arun Kumar
Rathinam College of Arts and Science
Coimbatore
India

V. Asha
New Horizon College of Engineering
Bellandur Post
Bengaluru
India

Balamurugan Balusamy
Shiv Nadar Institute of Eminence
Delhi
India

Ali Kashif Bashir
Manchester Metropolitan University
Manchester
UK

Subrata Chowdhury
Department of CSE
Sreenivasa Institute of Technology and
Management Studies
Chittoor
Andhra Pradesh
India

Rajesh Kumar Dhanaraj
Symbiosis Institute of Computer
Studies and Research (SICSR)
Symbiosis International (Deemed
University)
Pune
India

V. Dhanashree
Department of Electronics and
Instrumentation Engineering
Sri Sairam Engineering college
Chennai
Tamilnadu
India

Sumit Singh Dhanda
School of Computer Science &
Engineering
IILM University
Greater Noida
UP
India

Jagjit Singh Dhatterwal
Department of Artificial Intelligence &
Data Science
Koneru Lakshmaiah Education
Foundation
Vaddeswaram
Andhra Pradesh
India

K. Dhivya
Department of CSE
National Institute of Technology
Puducherry
Karaikal
Tamil Nadu
India

Ahmed A. Elngar
Faculty of Computers and Artificial
Intelligence
Beni-Suef University
Beni-Suef City
Egypt

Eugene Berna
Department of Artificial Intelligence
and Machine Learning
Bannari Amman Institute of
Technology
Erode
Tamil Nadu
India

P. Ganeshkumar
Department of Computer Science
Al Imam Mohammad Ibn Saud
Islamic University (IMSIU)
Riyadh
Saudi Arabia

R. Gayathri
Department of Electronics and
Communication Engineering
Rajalakshmi Engineering College
Chennai
India

Erik Geslin
Faculty of Interactive Media – Games
CNAP Lab
Noroff University College
Kristiansand
Norway

S. Gnanavel
Department of Computing
Technologies
SRM Institute of Science and
Technology
Chennai
India

R. Gopal
Department of Computer Science and
Engineering
University of Buraimi
Al Buraimi
Oman

G.S. Gopika
Department of CSE
Sathyabama Institute of Science and
Technology
Chennai
India

Ramya Govindaraj
School of Computer Science
Engineering and Information Systems
(SCORE)
Vellore Institute of Technology
Vellore
Tamil Nadu
India

Simon Grima
University of Malta
Msida
Malta

Kariyappa Janani
Manipal institute of Technology
Manipal
India

J. Jeyalakshmi
Department of Computer Science and
Engineering
Amrita School of Computing
Amrita Vishwa Vidhyapeetham
Chennai
India

S. Jeyalakshmi
Vels Institute of Science Technology &
Advanced Studies
Chennai
India

Seifedine Kadry
Noroff University College
Kristiansand
Norway

Kuldeep Singh Kaswan
Department of Computer Science &
Engineering
Galgotias University
Greater Noida
India

Rajendra Narasimman Karthika
Department of Information
Technology
Saveetha Engineering College
Chennai
Tamil Nadu
India

Ponmurugan Karuppiah
King Saud University
Riyadh
Saudi Arabia

Balasubramanian P. Kavin
Department of Data Science and
Business Systems
SRM Institute of Science and
Technology
Kattankulathur
Chengalpattu
Tamil Nadu
India

Surbhi Bhatia Khan
Department of Data Science
School of Science
Engineering and Environment
University of Salford
Manchester
United Kingdom

Eashaan Manohar
Department of Computer Science and
Engineering
Rajalakshmi Engineering College
Chennai
Tamil Nadu
India

Pravin Satyanarayan Metkewar
School of Computer Science and
Engineering
Dr Vishwanath Karad MIT World
Peace University
Kothrud
Pune
Maharashtra
India

Ramamoorthy Poorvadevi
Department of Computer Science and
Engineering
Mahavidyalaya
Sri Chandrasekharendra Saraswathi
Viswa Mahavidyalaya
Kancheepuram
India

A. Prasanth
Department of Computer Science and
Engineering
Vel Tech Rangarajan Dr. Sagunthala
R&D Institute of Science and
Technology
Chennai
Tamil Nadu
India

Sheena Christabel Pravin
School of Electronics Engineering
VIT
Chennai
India

S. Prem Kumar
Rathinam College of Arts and Science
Coimbatore
India

R.S. Rajasree
Department of AI & ML
New Horizon College of Engineering
Bangalore
India

Sriramulu Ramamoorthy
Department of Computing
Technologies
SRM Institute of Science and
Technology
Kattankulathur Campus
Chennai
India

R. Ramya
SRM Institute of Science and
Technology
Katankulathur
Tamil Nadu
India

B. Saiganesh
Department of Electronics and
Communication Engineering
Rajalakshmi Engineering College
Chennai
Tamil Nadu
India

Diego Saldivar
Faculty of Interactive Media – Games
Noroff University College
Kristiansand
Norway

Prithi Samuel
Department of
Computational Intelligence
School of Computing
SRM Institute of Science and
Technology
Kattankulathur Campus
Chennai
India

J. Saranya
Department of Electronics and
Communication Engineering
Rajalakshmi Engineering College
Chennai
India

V.S. Selvakumar
Department of Electronics and
Communication Engineering
Rajalakshmi Engineering College
Chennai
India

Gan H. Seng
School of AI and Advanced Computing
XJTLU Entrepreneur College
(Taicang)
Xi'an Jiaotong – Liverpool University
Suzhou
Jiangsu
P.R. China

Achyut Shankar
University of Warwick
Coventry
UK

M. Shanmugathai
Department of English
Sri Sairam Engineering College
Chennai
Tamilnadu
India

B. Shuriya
PPG Institute of Technology
Coimbatore
India

P. Sivaprakash
Rathinam College of Arts and Science
Coimbatore
India

M. Sree Krishna
Department of CSE
Sathyabama Institute of Science and
Technology
Chennai
India

Gautam Srivastava
Department of Computer Science and
Math
Lebanese American University
Beirut
Lebanon

S. Subha
Department of Electronics and
Instrumentation Engineering
Sri Sairam Engineering College
Chennai
Tamilnadu
India

S. Suganthi
Department of Electronics and
Communication Engineering
Rajalakshmi Engineering College
Chennai
India

B. Surendiran
Department of CSE
National Institute of Technology
Puducherry
Karaikal
Tamil Nadu
India

S. Sree Varagi
Department of Electronics and
Instrumentation Engineering
Sri Sairam Engineering College
Chennai
Tamilnadu
India

Govindaraj Venkatesan
Department of Civil Engineering
Saveetha Engineering College
Chennai
Tamil Nadu
India

K. Vijay
Department of Computer Science and
Engineering
Rajalakshmi Engineering College
Chennai
Tamil Nadu
India

Vikneswaran Vijean
Faculty of Electronic Engineering &
Technology
Universiti Malaysia Perlis
Kampus Alam UniMAP
Arau
Perlis
Malaysia

M. Yogeshwari
Vels Institute of Science Technology &
Advanced Studies
Chennai
India

1

Journey to Digital Twin Technology in Industrial Production: Evolution, Challenges, and Trends

Prithi Samuel¹, Rajesh Kumar Dhanaraj², Balamurugan Balusamy³, Ali Kashif Bashir⁴, and Seifedine Kadry⁵

¹Department of Computational Intelligence, School of Computing, SRM Institute of Science and Technology, Kattankulathur Campus, Chennai, India
²Symbiosis Institute of Computer Studies and Research (SICSR), Symbiosis International (Deemed University), Pune, India
³Shiv Nadar Institute of Eminence, Delhi, India
⁴Manchester Metropolitan University, Manchester, UK
⁵Noroff University College, Kristiansand, Norway

1.1 Introduction

Modern technologies have significantly changed the industrial landscape, resulting in higher production, sustainability, and efficiency. The technology known as the "Digital Twin" (DT) is one such groundbreaking breakthrough that has completely changed industrial manufacturing. The "Digital Twin" concept has caused a stir in many fields because of the benefits it offers in terms of optimization, prediction, and reliability that can be achieved by constructing a digital clone of a physical asset.

The phrase "digital twin" was first used around the turn of the millennium in the sectors of aerospace and automotive engineering. With the advent of Internet of Things (IoT), data analytics, and AI, the true potential of DT is only now being realized [1]. By combining real-time sensor data with cloud computing and machine learning (ML) capabilities, more powerful DTs were made available, which could accurately reproduce complex industrial systems.

Despite its revolutionary promise, the deployment of DT technology has not been without challenges. One of the main problems has been the difficulty of integrating and sharing data between systems. Massive amounts of data are generated from numerous sources by industrial systems, and it has proven to be difficult to combine these data into a cohesive DT model [2, 3]. The transmission of sensitive data from physical assets to their virtual counterparts while maintaining data security and privacy is still a top concern for businesses.

Making high-fidelity DTs for massive industrial systems is a complex process, which presents another difficulty. Using sophisticated modeling techniques and computer capacity, which might be resource-intensive, is necessary to accurately represent a physical asset. In order to use DTs effectively, enterprises have had to spend in upskilling their workers because there is a learning curve connected with the technology.

In 2003, during a seminar on whole product lifecycle management, University of Michigan, Professor Michael Grieves first introduced the concept of DTs. The term "digital mirror" is also used to explain this innovation. The meaning of this technology has evolved since then due to the presentations of other academics who have offered their own definitions of it [9–14]. According to the Encyclopedia of Production Engineering, "the Digital Twin is a representation of an active unique 'product' which can be a real device, object, machine, service, intangible asset, or a system consisting of a product and its related services" [12, 19, 20]. The term "simulation model" is used to describe a computer representation of a physical system that can be used for research, education, and deliberation in the present (as in Figure 1.1), which is collected by sensors in the field and used to control physical equipment [15, 16, 21].

The disparity between the rising need for the best medical care and its timely, affordable supply continues to be a problem everywhere in the world. In circumstances where the healthcare system lacks the capacity to effectively offer, community healthcare practitioners enhance access to essential healthcare services [4]. For the duration of the Ebola outbreak, the authors in [5, 16] looked at the value of community-based health systems to maintain the care of children's illnesses. They advised that enhancing drug supply chain management and spreading awareness of infection prevention during times of crisis.

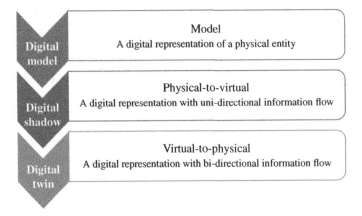

Figure 1.1 Level of integration.

1.2 Systematic Review Analysis

This study aims to summarize the key components of DT–based predictive maintenance models. This work can thus serve as a catalyst for additional primary research in this area in the future. To compile all pertinent primary research, we conduct a semi-automated systematic literature review (SLR). Prior to conducting the SLR, a review process was created based on the recommendations from [3], who state that a predefined, strictly followed protocol decreases researcher bias and promotes rigor and reproducibility. The adopted review methodology is displayed in Figure 1.2.

In accordance with the third protocol criterion, publications are categorized according to their application sectors, as shown in Figure 1.3. The most pertinent

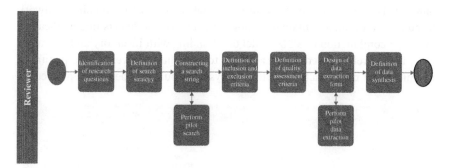

Figure 1.2 Systematic review protocol.

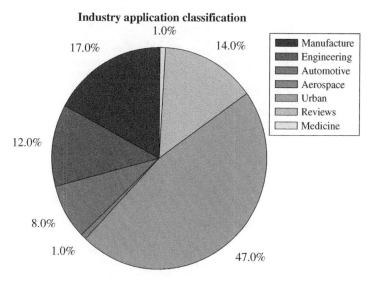

Figure 1.3 Application of industry by classification.

fields include urban environments, manufacturing, reviewing, engineering, automotive, and medicine, as shown in Figure 1.3. Automotive-specific manufacturing publications were employed in the section of the manufacturing domain that dealt with the automotive industry for this investigation. A few publications on goods and operations logistics from the urban and engineering areas were also divided. These five fields—smart cities and urban areas, freight logistics, health care, engineering, and automotive.

1.2.1 Logistics Applications of Digital Twins

Applications of DTs in logistics span the entire value chain, from container fleet management and cargo tracking to system design. IoT sensors on individual containers, for example, show where they are and monitor for contamination or damage. This information feeds into a virtual replica of the container network to guarantee optimal container deployment. Figure 1.4 depicts one application of DTs, a combination of a 3D model of a building with inventory and operating data, in a warehouse setting. In addition to providing an overarching picture of inventory and infrastructure health, the system would also be able to predict the

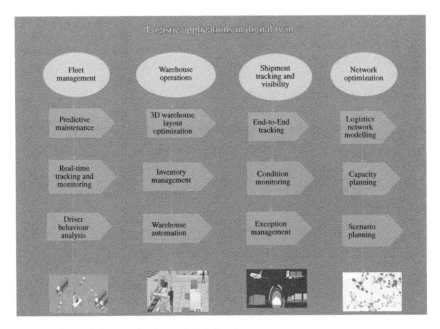

Figure 1.4 Logistics applications of digital twins.

future and take autonomous action regarding deliveries and stock levels. Major logistical nodes or international logistics networks fall under the same rule.

Markus Kückelhaus, VP of Innovation and Trend Research at DHL Customer Solutions & Innovation, has observed that "digital twins are becoming an increasingly valuable resource for businesses." He then adds, "Powered by IoT, cloud computing, AI, and advanced visualization tools." While these technologies are essential, it might be challenging for a DT buildup. So, to fully achieve the potential, there must be tight interaction among all partners throughout the value chain.

DHL tackles implementation issues including cyber security worries in the trend report, but emphasizes that the business rationale for deploying digital objects are getting stronger. As related technology becomes more reliable and affordable, businesses across a wide range of sectors may find DTs indispensable for managing complex systems of assets in real time and enhancing the efficiency of their processes. The research takes into account the resources and adjustments needed to properly implement DTs in logistics.

1.2.2 A Manufacturing Process

Figure 1.5 is a model that is helped along in its unique expression by five components: physical world sensors, actuators, integration, data, and analytics.

- In the manufacturing process, sensors send forth signals that the twin can decode to learn about the real-world process's operations and surroundings.
- Sensor data, such as operating and environmental variables, can be coupled with enterprise data in addition to the design requirements, the bill of materials (BOMs), and the 10 enterprise systems. Technical drawings, logs of consumer complaints, and connections to other data streams are just some examples of the information that may be gathered.
- Integration: Sensors can exchange data between the digital and physical worlds via integration technologies.
- The DT employs analytics tools to conduct research via computer modeling and graphical presentation of collected data.
- DTs are represented by the "digital" side of Figure 1.1. A DT is a duplicate of a physical object that can be used to compare and contrast how well actual performance matches expectations.
- Actuators: When a real-world physical action is called for, the DT will trigger the appropriate actuators.

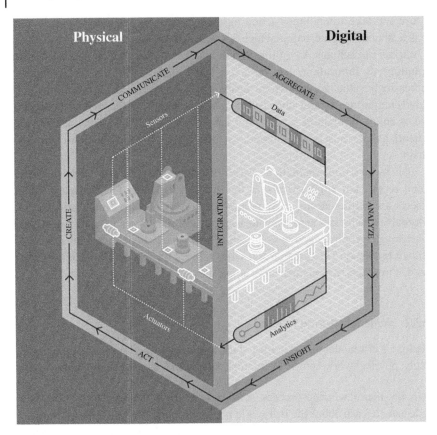

Figure 1.5 Manufacturing process in DT.

1.3 · Development of Industry

Smart manufacturing and the ideas of Industry 4.0 underpin the entire manufacturing sector today. During the First Industrial Revolution (also known as Industry 1.0), steam engines were widely used to replace human labor with mechanized output. During the Second Industrial Revolution, mass production techniques like assembly lines were perfected. During the Third Industrial Revolution, automation and computerization of production became commonplace [1, 2]. Industry 4.0 and paves the way for more flexible and efficient manufacturing processes (Figure 1.6). Using data to better operations, forecast when machines will break down, and quickly adapt to changing consumer demands are the primary purposes of this revolution.

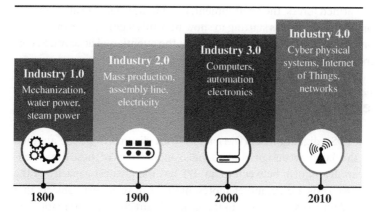

Figure 1.6 History through industry development.

Today, DTs are most commonly employed in engineering and manufacturing for simulating real-world processes and creating accurate digital replicas of physical objects. In O&S management and supply chain management, DTs are useful for a wide variety of purposes, such as tracking operations, maintaining transportation, providing remote support, visualizing assets, and making adjustments to designs. Figure 1.7 illustrates the breadth of industries that could profit from this technology. Transportation, aviation, construction, agriculture, mining, utilities, retail,

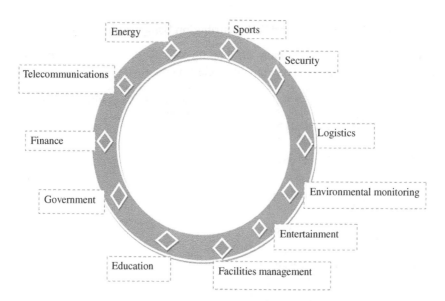

Figure 1.7 Industries using digital twin technology.

healthcare, the armed forces, the environment, and public safety are all examples of such sectors. The literature is starting to show how this technology can be used in various commercial contexts, and it has caught the attention of academics, corporate leaders, and industry experts all over the world [10, 17].

1.4 Assessments of Models

Digital shadow (DS), digital model (DM), and DT are distinct from one another. By comparing the data flow designs and intended applications of these three technologies, we can distinguish between them. DT has the greatest capacity of the three (Figure 1.8). DMs are computer representations of real-world objects that may be simulated and analyzed. Due to the lack of bidirectional, automatic data flow between the model and the outside world, digital representations can only ever be approximations. The world in DS is modeled digitally from scanned laser data, but there is no feedback loop between the model and the real world.

DM is useful for concept development and industrial design. DTs are useful tools for assessing real-time manufacturing, and DSs are effective technologies for tracking production [2]. These three technologies are effective instruments for real-time decision-making, remote monitoring of manufacturing processes, and continuous process analysis in industry. This leads to shorter manufacturing times, lesser waste, and higher revenue levels.

1.4.1 Workload Balance and Task Distribution Using Digital Twins

It is a method for proportionally dividing up assembly tasks between human and mechanical workers in real time [9]. Selecting the part, transferring it to the

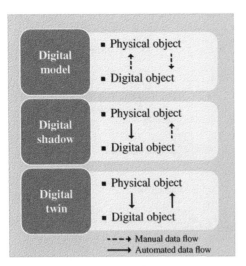

Figure 1.8 Comparisons of three models.

assembly area, and mounting it in the correct orientation are just a few examples of the many steps involved in each activity. Many factors, including the geometry and physics of the parts, the quality of the assembly, the level of risk involved, and the expertise of both human and robotic laborers, determine whether or not a certain job is robot-friendly.

1.4.2 An Outline for a Digital Shadow of Production

The roadmap has two dimensions because of its matrix structure. The first dimension talks about the different roles of the DS's complexity levels (Figure 1.9). The stages of development are described in the second dimension. There are already ideas or perhaps initial implementations for the first level. An empty initial stage indicates that there are no operational (realized) notions. There are four stages in the roadmap. Each stage functions as a logical unit. When a complexity level reaches its fourth or, more precisely, third stage, it is realized. This roadmap matrix's components all make reference to digital data or information. If the fourth stage of development of the fourth complexity level is attained, the maximum stage of expansion, or rather the optimal state, is obtained.

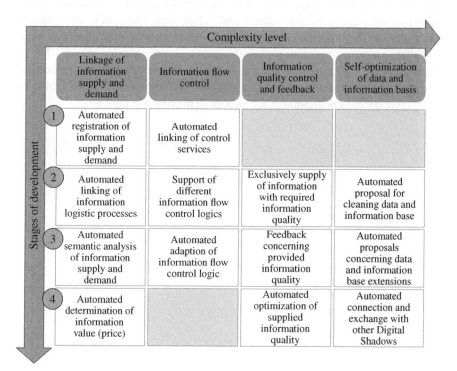

	Complexity level			
Stages of development	Linkage of information supply and demand	Information flow control	Information quality control and feedback	Self-optimization of data and information basis
1	Automated registration of information supply and demand	Automated linking of control services		
2	Automated linking of information logistic processes	Support of different information flow control logics	Exclusively supply of information with required information quality	Automated proposal for cleaning data and information base
3	Automated semantic analysis of information supply and demand	Automated adaption of information flow control logic	Feedback concerning provided information quality	Automated proposals concerning data and information base extensions
4	Automated determination of information value (price)		Automated optimization of supplied information quality	Automated connection and exchange with other Digital Shadows

Figure 1.9 Roadmap for digital shadow.

1.5 Technologies of DT

1.5.1 DT in Internet of Things (IoT)

In many sectors, the IoT and DT technologies have been the driving forces behind the digital transformation [2]. Businesses now have new potential to streamline operations, improve decision-making, and increase overall efficiency thanks to the power of IoT combined with DT capabilities. A system, process, equipment, or object's digital representation of its physical counterpart Figure 1.10. It includes real-time data integration, sensory data, and sophisticated analytics in addition to a straightforward 3D model. In essence, it replicates the physical object and its behavior, allowing businesses to monitor, assess, and manage the physical asset remotely through its digital equivalent.

1.5.2 How Does It Operate?

An IoT DT is built on the foundation of continuous data exchange between a real-world object and its digital counterpart. Environmental and operational parameters such as temperature, pressure, humidity, and mobility are monitored in real time by built-in sensors and IoT devices. The data is subsequently submitted to the DT for analysis, which can be performed in a variety of ways. The DT integrates incoming data with cutting-edge algorithms, AI, and ML to mimic the physical asset's behavior in real time. Businesses can learn important details about performance, health, and possible problems from the DT of an asset whenever changes or anomalies are noticed in the physical object.

Figure 1.10 Technologies of DT.

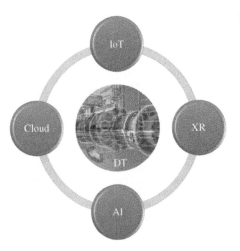

1.5.3 DT in Cloud Computing

Due to their complementary nature, businesses can benefit from utilizing both cloud computing and digital twin technologies. To improve its scalability, accessibility, and collaborative virtualization, a "Digital Twin" is a digital clone of a physical asset or system that is controlled and operated in the cloud. Cloud computing and digital twins benefit businesses because they complement one another. To generate scalable, readily available, and collaborative digital representations of real-world assets and systems, "Digital Twin" is used in the context of cloud computing to describe the deployment and use of DT models and data within cloud-based infrastructures [22, 23].

1.5.4 DT in Artificial Intelligence (AI)

A significant synergy is created when the idea of a digital twin and AI are merged, enhancing the capabilities of both technologies. In order to increase analytics, predictive abilities, and autonomous decision-making, "digital twins" are created when AI algorithms and models are applied to digital representations of real assets or systems. I symbolise a significant advance in how industries are being transformed by digital technology. Organizations can achieve previously unheard-of levels of efficiency, predictability, and autonomy thanks to the synergy between these two technologies. As AI and DT technologies advance and become more widely used, their integration is anticipated to revolutionize a number of industries by fostering creativity, efficiency, and long-term growth [24, 25].

1.5.5 DT in Extended Reality (XR)

The terms virtual reality (VR), augmented reality (AR) as well as mixed reality (MR) are all subsets of "extended reality" (XR). Interactive and immersive "digital twins" are made possible by merging XR technology with digital representations of real-world items, systems, and settings. It represents an innovative strategy that has great significance for many different sectors. Organizations may increase training, optimize operations, and spur creativity by utilizing XR's immersive capabilities to engage with virtual representations of real assets and settings. It is expected that the integration of XR and DT technologies will change the ways in which we engage with the real and virtual worlds, opening up novel possibilities for immersive experiences.

1.6 A Digital Twin System with a Modular Layout

Important procedures in industry, such as remote control and virtual machining, can be enhanced with the use of DT technology. To meet the needs of a wide

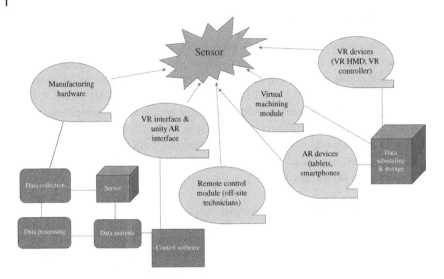

Figure 1.11 The digital twin system's modular layout.

variety of industrial processes, the authors of this study employ a modular design approach to the DT system [4] (see Figure 1.11).

- Server functions as a centralized device that communicates all data, including digital and control signals, among many distinct modules. The sent data is bundled as a stream and is sent through the socket, a dependable transmission technique. In order to enable bidirectional data flow and communication between modules, we do things like decode and repackage stream data and activate the heartbeat mechanism.
- The manufacturing module includes both the physical equipment of production and the administrative software necessary to keep them operational. One or more physical manufacturing units constitute the manufacturing hardware of a DT system. The actual industrial equipment cannot be mapped from cyberspace to the physical world without communicating with the corresponding control software.
- Data collection, analysis, storage, and distribution are all handled by separate submodules within the DT system's data processing module. The information is initially received by the data collecting module as a socket stream. After decoding, the information is converted into digital form and transferred along with control signals to a submodule of data analysis. The database stores and organizes all of these signals and data.
- The remote 3D human computer interaction (HCI) in the remote-control module enables remote technicians to access the monitoring and remote-control

services. The Unity 3D-based VR operator interface, several VR headsets, and the DT framework are all examples. A virtual reality headgear includes a display and can be used with a pair of controllers. A closed-loop control system is formed by the control software's input digital data and the interface's output control signals.

- The virtual machining module offers a service for on-site workers that avoids the need for physical testing by simulating the machining process digitally, thus enhancing the realism of the actual manufacturing environment. This component grants entry to the AR interface in Unity3D, the DT mode, and the AR hardware.

1.6.1 Deep Transfer Learning for Fault Diagnosis with a Digital Twin

Because it provides a new angle from which to see problems, the DT is a crucial component of smart manufacturing. Because of the dynamic, ever-evolving nature of the manufacturing process, it is impractical to assume that the distributions of the training data and the test data are the same in order to construct a reliable diagnostic model (see Figure 1.12 for an illustration of this point).

We provide a two-stage defect detection approach [16] that utilizes a digital twin and deep transfer learning (DFDD) to identify faults during both the design and maintenance phases. Running an ultrahigh fidelity model in a virtual environment is one technique to find issues with a deep neural network (DNN)–based diagnosis model that were overlooked during the design phase. The acquired diagnosis

Figure 1.12 Fault diagnosis system.

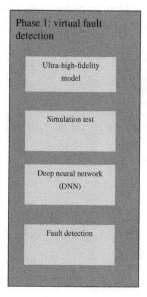

model can be used in the actual world with the help of deep transfer learning, allowing for constant monitoring and prompt repairs. A correct diagnosis can then be made with minimal investment of time and energy. As an example of the usefulness of DFDD in detecting manufacturing flaws, we use a car body side assembly line. These findings prove that our approach is superior and effective.

1.7 Automated Guided Vehicle Scheduling Based on Digital Twins

Figure 1.13 depicts the workshop's general layout. Only the automated guided vehicle (AGV) scheduling is taken into account in this case, and the machine's allocation has already been finished by default. Therefore, a different machine is used for each workpiece procedure in the processing [6]. Workpiece blanks are initially stored in the S-shaped material storage room. The AGV then delivers the blanks to the designated machine Mx once an order has been placed. AGVs transport the completed parts to the area F storage facility after many processes [5, 9].

- AGV collisions and deadlocks are not taken into account.
- The AGV follows the shortest path driving principle.
- It is only capable of carrying one workpiece at a time.
- It can ignore loading and unloading times; it can charge all AGVs simultaneously.
- It cannot perform charging tasks while in motion.
- It drives at a steady speed and has no upper limit on the amount of time it can store raw materials or finished goods.

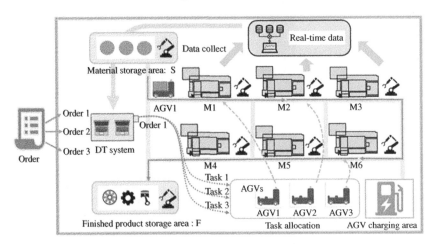

Figure 1.13 The layout of manufacturing.

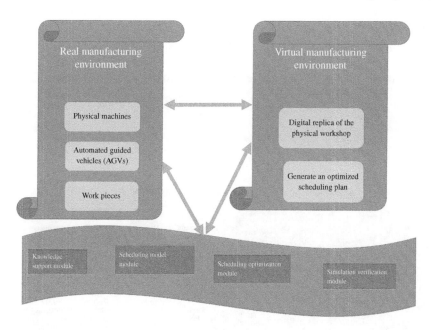

Figure 1.14 Framework for DTDAS.

1.7.1 Fundamental Concepts DTDAS

DT-based dynamic AGV scheduling (DTDAS) is composed of a physical workshop and its virtual analogue. The two halves' communication is seen in Figure 1.14. Information from the physical plant is used to update the digital scheduling model dynamically with respect to real time. The real-world workshop then follows the plan that was generated in the virtual workshop. Knowledge-supporting, scheduling, optimizing, and verifying simulations are the four components that make up the virtual setting [5, 7]. The knowledge support module's scheduling model makes predictions about when AGVs will be available based on the module's dynamic energy consumption model. Transportation responsibilities and costs are factored into the scheduler as well. The schedule optimization approach will be used to resolve this time-keeping issue. The jobs are assigned to the appropriate machines by the execution system after the scheduling plan has been gathered and verified through simulation [8].

1.8 Creating a Digital Twin

Although the ideas behind DTs are generally applicable, the conceptual architecture of DTs (Figure 1.15) can be seen as a detailed or "under the hood"

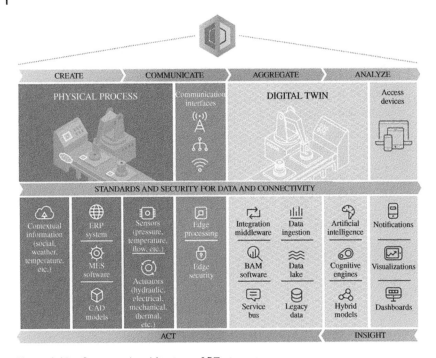

Figure 1.15 Conceptual architecture of DT.

examination of the enabling components that comprise the digital twin model of the manufacturing process (Figure 1.5). You will have a better grasp of the full system in no time if you follow these six guidelines.

Gathering information on the physical process and its surroundings is essential for reaching this objective. A number of sensors are installed in the physical process during the construction phase. There are two broad classes to be applied to the sensor data. There are two types of data that can affect the performance of a physical asset: (i) Internal data can include things like tensile strength, displacement, torque, and color uniformity; (ii) in the case of a productive asset (which may comprise numerous wireless intrusion prevention system (WIPs), external data may include temperature, barometric pressure, and moisture level.

Physical process and digital platform are in continual dialogue, allowing for seamless, real-time, bidirectional integration and interaction.

- Connecting sensors and process historians, processing signals and data locally, and then uploading the results to the platform is what the edge interface does.
- Communicative interfaces: These are used to transmit information between the sensor function and the integration function.

- Filling in the crevices: As sensor and communication technology improves, so do people's security concerns.

o **Aggregate:** The aggregate stage can enable the processing and analytics-ready data intake into a data repository. Either on-site or in the cloud data processing and aggregation are options.

o **Analyze:** Visualization is used in the analysis process to analyze and comprehend data. Advanced analytics platforms and technology allow data scientists and analysts to build iterative models that provide insights, ideas, and support for decision-making.

o **Insight:** Performance gaps between the digital twin model and its physical world duplicate are made clear by interactive dashboards and visualizations known as "insight steps."

o **Act:** By providing back actionable data into the underlying physical asset and digital process, a digital twin can boost their performance.

1.9 Digital Twins in Industry 5.0

In collaborative operations, 5IR strives to integrate these cognitive computer capabilities with insightful human thought. Therefore, it is possible that 5IR may

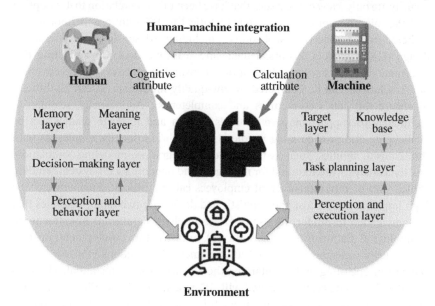

Figure 1.16 Architecture of human–machine integration.

result in a shift in norms as well as a fundamental transformation of industry and production processes. According to the author [18, 23], 3D digital visualization technology can gather, monitor, and analyze information on industrial equipment in real time. Factors such as temperature, humidity, rotation rate, vibration, and on/off switch were among those measured and recorded. In a typical setup, alarms can be categorized and managed in such a way that the alert prompt is instantly triggered and the appropriate 3D IoT virtual manufacturing equipment is located. Various sound alarms were also present so that the supervisors could quickly identify any operational dangers in Figure 1.16. In addition, it could direct fault treatment in 3D dynamic mode. The manager terminal received the analysis results and displayed them to the managers. Then, a single 3D page displayed many pieces of information that had been rendered using 3D IoT-3D digital simulation technology. Accordingly, each maintenance worker might oversee several machines, enable the information island, and increase productivity.

1.10 Implementation Obstacles

Despite DT's promising future, the technology now struggles with many of the same issues as AI and IoT do. Data standardization, data management, and data security are only a few of the issues that have been raised in relation to its adoption and the migration of existing systems. Critical to DT solutions is information gathered from a wide variety of sources, including IoT sensors, computer-aided design (CAD) models, historical data, and real-time data. Consolidating this information into a single model is a time-consuming process. The effectiveness of DT solutions can be hampered by low-quality input data. It can be difficult to ensure the data's correctness and completeness. Additionally, scaling DT technology to support massive volumes of data and intricate models can be challenging and provide a big hurdle [1, 13, 14]. Additionally, employees with the requisite abilities, such as those in data integration, modeling, simulation, and data analysis, are required for managing and operating a DT system. Finding skilled staff or retraining current employees can be difficult. Furthermore, the high implementation cost and sophisticated design are two major obstacles that are anticipated to hinder the use of DT technology. Significant investments in technological platforms (sensors, software), infrastructure development, upkeep, and security measures are needed to implement DT solutions [15, 16]. Finally, the cost of operating and maintaining the DT network is substantial. The high initial and ongoing costs associated with DT systems might be difficult for small and medium-sized businesses to justify. It is predicted that DT's high fixed costs and complex infrastructure will impede its wider implementation.

1.11   Conclusion

The transition to DT technology in industrial production has been a transformational process, one that has been both difficult and full of opportunity. Technology advancements, increased stakeholder collaboration, and a growing emphasis on data security have all contributed to DTs' potential to become an essential tool for industries seeking to achieve previously unheard-of levels of efficiency, sustainability, and innovation in the future's dynamic and competitive environment.

The DT could help in producing actual value, finding new revenue streams, and solving critical strategic difficulties. Companies may now begin the process of creating a digital twin with less up-front expense and a faster time to value than ever before thanks to the capabilities, scalability, agility, and decreased cost of current technology. There are many places in the product life cycle where DTs might be useful, and they may even help us solve some of science's biggest mysteries. Maybe the real question is not if you should get started, but how to get the most out of it and remain ahead of the curve as quickly as possible.

DT assists businesses in creating more dynamic, developed supply chain models that enable them to anticipate possible problems before they materialize and to take appropriate action. Businesses use DT solutions to optimize transportation routes and predict the environmental impact of using different sources of raw materials. Manufacturers benefited substantially from the DTs' efficient monitoring, diagnostic, and prescriptive analytic capabilities, which were vital to their process changes in order to increase corporate agility and competitiveness. Despite the unstable economic atmosphere brought on by the COVID-19 epidemic, DT solutions assisted businesses in making informed judgments.

Progress is being made in the study of DTs, but challenges remain. The inability to coordinate the physical and digital worlds is one possible root cause of such issues, along with a dearth of high-fidelity simulation models at several sizes. This study presents a human–robot assembly system controlled by DTs, which can be used for human resource consulting (HRC) in variant-oriented assembly environments like the manufacturing industry. The results demonstrate the method's potential for automating operations while yet permitting individualized building.

DT solutions are important to the success of Industry 4.0 because of their capacity to improve productivity, creativity, and environmental friendliness across a wide range of manufacturing processes. To be more specific, DT solutions aid in guiding supply networks to become more reliable for companies.

By sharing their experiences, businesses from different sectors would be better able to learn from one another and adopt best practices in DT implementation. They will help pinpoint any technological hurdles that stand in the way of widespread use and effective application.

Further proof of DTs' usefulness could mean more money spent on research and development as well as a heightened focus on integrating DTs into existing procedures. The relentless progress of DT technology will ultimately assist businesses across a wide variety of industries by enhancing operational efficiency, elevating decision-making, and igniting innovation.

References

1 Attaran, M., Attaran, S., and Celik, B.G. (2023). The impact of digital twins on the evolution of intelligent manufacturing and Industry 4.0. *Advances in Computational Intelligence* 3: 11. https://doi.org/10.1007/s43674-023-00058-y.

2 Fuller, A., Fan, Z., Day, C., and Barlow, C. (2020). Digital twin: enabling technologies, challenges and open research. *IEEE Access* 8: 108952–108971.

3 Dinter, R.V., Tekinerdogan, B., and Catal, C. (2022). Predictive maintenance using digital twins: a systematic literature review. *Information and Software Technology* 151: 107008.

4 Geng, R., Li, M., Hu, Z. et al. (2022). Digital Twin in smart manufacturing: remote control and virtual machining using VR and AR technologies. *Structural and Multidisciplinary Optimization* 65: 321.

5 Han, W., Xu, J., Sun, Z. et al. (2022). Digital Twin-based automated guided vehicle scheduling: a solution for its charging problems. *Applied Sciences* 12: 3354.

6 Huang, Z., Shen, Y., Li, J. et al. (2021). A survey on AI-driven digital twins in industry 4.0: smart manufacturing and advanced robotics. *Sensors* 21: 1–35.

7 Ibrahim, M., Rassõlkin, A., Vaimann, T., and Kallaste, A. (2022). Overview on digital twin for autonomous electrical vehicles propulsion drive system. *Sustainability* 14: 601.

8 Ikimi, O. (2020). Digital twin technology and its impact on manufacturing and businesses. February 7. Online: https://www.allaboutcircuits.com/news/digital-twin-technology-and-its-impact-on-manufacturing-and-business/.

9 ISODIS (2020). *Automation Systems and Integration—Digital Twin Framework for Manufacturing*, 23247-1–23247-4. Geneva: International Organization for Standardization.

10 Kaarlela, T., Pieska, S., and Pitkaaho, T. (2020). Digital twin and virtual reality for safety training. *11th IEEE International Conference on Cognitive Infocommunications – CogInfoCom* (23–25 September). Online: https://ieeexplore.ieee.org/stamp/stamp.jsp?tp=&arnumber=9237812.

11 Kenett, R.S., Swarz, R., and Zonnenshain, A. (2020). *Systems Engineering in the Fourth Industrial Revolution: Big Data, Novel Technologies, and Modern Systems Engineering*. Wiley.

12 Kumar, D., Singh, R.K., Mishra, R., and Vlachos, I. (2023). Big data analytics in supply chain decarbonisation: a systematic literature review and future research directions. *International Journal of Production Research* https://doi .org/10.1080/00207543.2023.2179346.

13 Lv, Z. and Xie, S. (2021). Artificial intelligence in the digital twins: state of the art, challenges, and future research topics. *DigitalTwin* https://doi.org/10 .12688/digitaltwin.17524.1.

14 Martínez-Olvera, C. (2022). Towards the development of a digital twin for a sustainable mass customization 4.0 environment: a literature review of relevant concepts. *Automation* 3: 197–222.

15 Vijay, K. and Jayashree, K. (2023). Dynamic Request Redirection and Workload Balancing of Cloud in Business Applications. *Recent Trends in Computational Intelligence and Its Application: Proceedings of the 1st International Conference on Recent Trends in Information Technology and its Application (ICRTITA, 22).* CRC Press. p. 416.

16 Xu, Y., Sun, Y., Liu, X., and Zheng, Y. (2019). A digital-twin-assisted fault diagnosis using deep transfer learning. *IEEE Access* 7: 19990–19999. https://doi.org/ 10.1109/ACCESS.2018.2890566.

17 Attaran, M. and Celik, B.G. (2023). Digital Twin: benefits, use cases, challenges, and opportunities. *Decision Analytics Journal* 6: 100165. ISSN 2772-6622, https://doi.org/10.1016/j.dajour.2023.100165.

18 Vijay, K., Sabarish Abishek, W.R., Sabarish, V.U., and Sanjeev Krishnan, R. (2023). Private Cloud Storage using Raspberry PI via Virtual Network Computing. An Analysis, 2023 International Conference on Computer Communication and Informatics (ICCCI), Coimbatore, India. pp. 1-7. https://doi.org/10.1109/ ICCCI56745.2023.10128489.

19 Samuel, P., Jayashree, K., Babu, R., and Vijay, K. (2023). Artificial intelligence, machine learning, and IoT architecture to support smart governance. In: *AI, IoT, and Blockchain Breakthroughs in E-Governance* (ed. K. Saini, A. Mummoorthy, R. Chandrika, and N. Gowri Ganesh), 95–113. IGI Global https://doi.org/10.4018/978-1-6684-7697-0.ch007.

20 Babu, R. and Jayashree, K. (2015). A survey on the role of IoT and cloud in health care. *International Journal of Scientific, Engineering and Technology Research (IJSETR)* 04 (12): 2217–2219. ISSN 2319-8885.

21 Park, K.T., Lee, J., Kim, H.J., and Noh, S.D. (2020). Digital twin-based cyber physical production system architectural framework for personalized production. *International Journal of Advanced Manufacturing Technology* 106: 1787–1810.

22 Jaeyalakshmi, M. and Kumar, P. (2016). Task scheduling using meta-heuristic optimization techniques in cloud environment. *IJ Intelligent Systems and Applications* 5.

23 Lv, Z. (2023). Digital Twins in Industry 5.0. *Research* 6: Article 0071. https://doi.org/10.34133/research.007.

24 Sorna Shanthi, D., Vijay, P., Samblesswin, S. et al. (2023). RATSEL: a game-based evaluating tool for the dyslexic using augmented reality gamification. In: *Information and Communication Technology for Competitive Strategies (ICTCS 2021)*, Lecture Notes in Networks and Systems, vol. 400 (ed. A. Joshi, M. Mahmud, and R.G. Ragel). Singapore: Springer https://doi.org/10.1007/978-981-19-0095-2_69.

25 Ravikumar, S., Kumar, K.A., and Koteeswaran, S. (2018). Dismemberment of metaphors with grid scratch via kernel k-means. *Journal of Computational and Theoretical Nanoscience* 15 (11–12): 3533–3537. https://doi.org/10.1166/jctn .2018.7657.

2

The State-of-the-Art Digital Twin Components in Industrial IoT Production: Theoretical and Practical Applications

Kariyappa Janani[1], Sriramulu Ramamoorthy[2], Ramamoorthy Poorvadevi[3], and R. Gopal[4]

[1]*Manipal institute of Technology, Manipal, India*
[2]*Department of Computing Technologies, SRM Institute of Science and Technology, Kattankulathur Campus, Chennai, India*
[3]*Department of Computer Science and Engineering, Mahavidyalaya, Sri Chandrasekharendra Saraswathi Viswa Mahavidyalaya, Kancheepuram, India*
[4]*Department of Computer Science and Engineering, University of Buraimi, Al Buraimi, Oman*

2.1 Introduction

Rapid innovation in digital twin (DT) technology is reshaping the IIoT sector. DTs were developed to save money on maintenance and replacements by making a perfect digital copy of a real object and its surroundings. As a result of having access to this digital representation, engineers may spot problems before they occur, thus cutting down on unplanned downtime and expensive repairs. The twin display of a real property which may be utilized into action predicts and reacts to changes in its behavior, much like its tangible counterpart. At the University of Michigan, Dr. Michael Grieves initially introduced DT technology [1]. DT technology could only exist as a game-changing tool with the groundwork laid by Dr. Grieves's work. In the years 2019 since its invention, this technology has found applications in everything from the simulation and prediction of the behavior of intricate systems regarding the monitoring of a patient's vital signs in a medical context.

An IoT/IIoT application involved with DT technology architecture including devices, equipment, networks, clouds, and the closed loop produces customized goods for end consumers Figure 2.1. It comprises Perception, Network, Processing, and Application layers. Sensors, RFID scanners, security cameras, GPS modules, and so on, comprise the perception layer. In an industrial context, these devices may be used with AGVs, drive systems, automated systems, and so on. Engineers in the IIoT sector now have an effective tool in DT technology for keeping tabs on, interacting with, and comprehending the behavior of physical assets

Digital Twins in Industrial Production and Smart Manufacturing:
An Understanding of Principles, Enhancers, and Obstacles, First Edition. Edited by Rajesh Kumar Dhanaraj,
Balamurugan Balusamy, Prithi Samuel, Ali Kashif Bashir, and Seifedine Kadry.

Control and optimization

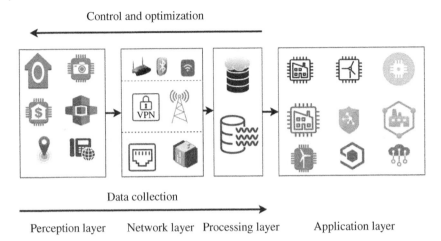

Data collection

Perception layer   Network layer   Processing layer   Application layer

Figure 2.1   General digital twin industrial IoT architecture.

in the real world. Engineers can prevent problems from occurring and adjust by first developing a digital representation of the actual asset. As a result, the asset's performance is maximized, and the need for maintenance and replacement parts is minimized. The advent of DTs greatly expands the potential applications of the IIoT sector. Engineering monitoring and control of physical assets are being transformed by DTs, which offer more efficiency, higher performance, and enhanced safety [2].

The IIoT refers to the monitoring and controlling of physical processes by utilizing internet-connected devices and sensors in industrial settings such as factories, warehouses, and utilities. When paired with DT technology, IIoT has the potential to alter the way industrial operations are carried out thoroughly. This will make it possible to achieve new levels of efficiency, safety, and sustainability. The following are some instances of how DT can be developed within the industry of IIoT:

- Asset Management: Monitoring and improving the performance of manufacturing assets such as equipment and machinery may be accomplished with the aid of DT. This type of management is referred to as asset management. Operators can spot abnormalities and proactively fix problems by tracking key performance indicators (KPIs), such as the amount of energy used, the temperature, and vibration. This helps reduce downtime and increases the longevity of assets. Modeling and simulating industrial processes like manufacturing and DT can be used for supply chain management, which can also be utilized for process optimization. Operators can find possibilities for optimization and efficiency improvement by running a variety of various situations and configurations through the testing process [3].

- Predictive Maintenance: DT can be used to predict when maintenance is required for industrial equipment. This can reduce the likelihood of unanticipated downtime and make it possible to do proactive repairs. Using sensor data to monitor equipment functioning and discover trends enables operators to take preventative measures before equipment breakdowns occur. DTs can watch and optimize product quality in real time, an essential part of quality control. Operators can spot problems and take corrective action by performing data analysis on sensor readings and comparing those readings to previously defined quality benchmarks.

- Safety Monitoring: DTs can be utilized to perform safety monitoring of industrial processes to identify and mitigate any potential risks. Operators can take measures to reduce the likelihood of accidents and enhance the safety of their workplaces by evaluating sensor data and spotting unusual patterns of behavior. Because DT technology can enable new levels of efficiency, productivity, and security in the IIoT industry, it has become an essential instrument for industrial operators who wish to optimize their operations and maintain their competitive edge in the 21st century [4].

The DT technology is susceptible to security breaches, just like any other type of technology that deals with sensitive data or directs essential business processes. In the business of the IIoT, where DT is increasingly utilized to monitor and optimize industrial processes, cybersecurity risks can have significant repercussions. These effects can range from data breaches to physical and bodily injury. The following is a list of potential forms of cybercrime that could be committed against DTs operating in the physical world.

- Breach of Data: If the data about the DT is not adequately protected, it may be susceptible to being stolen, manipulated, or accessed illegally. This can lead to sensitive industrial data being compromised, including production techniques, trade secrets, and personal information about customers and employees [5].

- Malware and Ransomware: Attacks using malware and ransomware can potentially compromise DT systems, either wreaking havoc on industrial processes or holding critical data hostage in exchange for a ransom payment. Denial of Service (DoS) assaults can flood DT systems with so much traffic that they are inaccessible to legitimate users and wreak havoc on industrial activities.

- Attacks of a Physical Nature: If DTs can manage physical equipment, they may be susceptible to attacks of a material kind, such as sabotage or tampering, which can result in injury or property damage. Risks from internal dangers, such as personnel with malevolent intent or inadvertent mistakes, can also be a risk to the security of the DT. These dangers can be reduced by implementing appropriate access controls and training.

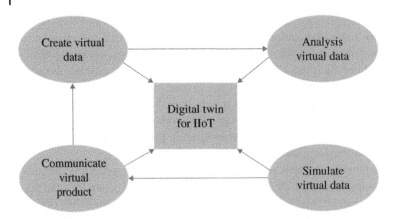

Figure 2.2 Digital twin-based framework for IIoT applications.

Industrial operators must take a holistic approach to DT security to handle security concerns. This may entail taking precautions such as encrypting data, implementing access controls, segmenting networks, monitoring potential threats, and training employees [6]. In the IIoT, operators can assist in ensuring the safety, dependability, and security of their DT systems by following these measures.

Figure 2.2 provides a digital visual depiction. Consider that the amount of ways DT may be implemented expands by the day. Under these circumstances, it is abundantly evident that experts in the field of science still need to accept the characteristics of an original DT; A corporation could be persuaded to state that it has implemented the most recent technology, some of which are capable of asset monitoring and are referred to as DTs.

This approach prevents the simplification of the complexities and potential brief advantages of a DT system that is fully functioning [7]. The characteristics described below provide a good indication of what defines a DT actively involved in enhancing knowledge of the system and serves as a foundation for future academic initiatives. The findings of previous research provide evidence that substantiates these traits. Current academic research in this field has concentrated chiefly on finding ways to integrate DTs into systems that run in real time [8]. The following is a list of some of the qualities or features that DTs possess:

- The term "digital twin" refers to a model of a product, service, or method used in the actual world.
- To accurately represent the physical things or processes they are meant to represent, DTs are continually updated with new information in real time.
- Deep learning may create additional enhancements and assertions to a system using DT, which are utilized for seeing and analyzing real things or actions.

2.1.1 Application Using Digital Twin

The DT is a powerful technique rapidly embraced across various industries. It offers tremendous promise for enhanced workplace productivity, safety, and accuracy. DT may be included in apps to boost their overall performance in various industries, including retail, aerospace engineering and aeronautics, manufacturing, healthcare, energy, automotive, smart cities, maritime, education, and construction Figure 2.3. This is made possible by the capabilities of DTs to simulate, measure, and analyze a physical environment. In the retail industry, the technology known as DT can be used to analyze consumer activity and improve the efficiency of product placement within stores. This gives merchants the ability to boost sales while also enhancing their customer shopping experiences. In the fields of aerospace engineering and aeronautics, a DT can simulate a variety of climatic conditions. This provides engineers with data to help them build safer aircraft and improves overall aviation safety. The performance of a production system can be optimized by using a DT in the manufacturing sector [9]. This includes everything from the efficiency of energy consumption to the maintenance of machines. In medicine, DTs can mimic patient outcomes, which can then be used

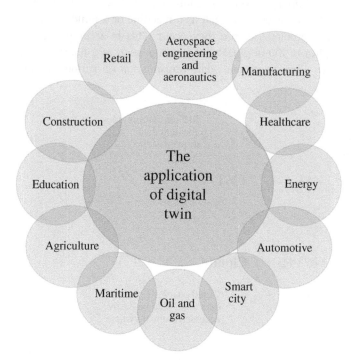

Figure 2.3 Digital twin applications.

to inform treatment decisions. In the fields of energy and automobile, DTs can be used to enhance efficiency while simultaneously reducing emissions.

In the meanwhile, they can be utilized in smart cities to improve urban planning through improvements in sensor technology and data collection. In the maritime industry, DT can assist in evaluating the performance of ships. In the education sector, they can be utilized in the construction of virtual classrooms and the simulation of learning environments. Lastly, they can be utilized in the building industry as part of the economy to improve the design and functionality of buildings. DT presents a massive opportunity for growth across a wide range of business sectors, promising enhanced productivity, safety, and precision. DTs can mimic, measure, and analyze a physical environment. Because of this, they have the potential to transform the design, performance, and maintenance of processes, goods, and services [10].

2.1.2 Technologies Involved in Digital Twin

The IoT and artificial intelligence (AI) can be utilized quite effectively in DT technology, which captures, analyses, and models physical items digitally Figure 2.4. The DT technology allows for continuous following and analysis of the performance of a physical asset, such as a machine or a building. It can help identify potential problems before they occur. This is accomplished by creating a virtual replica of the original asset in issue, such as a machine or a building. Because they allow continuous data transmission on various metrics

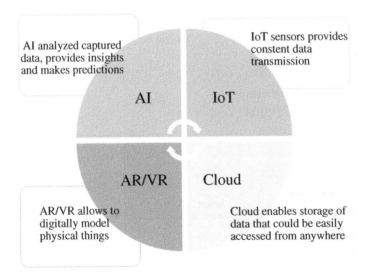

Figure 2.4 Technologies involved in digital twin.

like temperature, pressure, and vibration, IoT sensors play an essential part in the ecosystem of DTs. Following this, the data is entered into an analytics system that is powered by AI [11].

This system can recognize patterns and abnormalities, as well as generate forecasts regarding probable breakdowns or the need for maintenance. The combination of augmented reality and virtual reality (AR/VR) technology can produce a virtual image of a physical thing that is both lifelike and interactive. This opens the door to remote monitoring and virtual training for maintenance and repair. In general, the technology of DTs is a potent instrument for IIoT, as it enables more efficient and effective maintenance, enhanced asset management, and improved decision-making according to data collected in real-time insights [12].

The use of digital technologies to enhance and perfect the workings of different kinds of systems and procedures is referred to as "digital processing and evaluation," and it might refer to an entire industry, a smart city, or a network. Six different levels of digital transformation are involved when intelligent manufacturing is being discussed. The initial layer is known as the zero layer and is also known as the manufacturing area. This layer refers to the actual production space that is used during the manufacturing process. It requires the installation of some different sensors and monitoring systems to gather information about the production process and transmit it to higher layers so that it may be analyzed. The second layer is known as the infrastructure, and it refers to the information technology that is used to support the manufacturing process.

Figure 2.5 this layer of the intelligent manufacturing system is responsible for deploying various computing and storage resources, in addition to networking technologies that make it possible for different system layers to communicate with one another [13].

The third layer is operations, which refers to the administration and supervision of the production process. This layer oversees the actual making of the product. It entails implementing various control systems and algorithms which, when combined, make it possible to evaluate and alter the production method in real time. The integration of a variety of intelligent technologies, including machine learning (ML), AI, and big data analytics, is the focus of the fourth layer, which is referred to as the intelligent manufactory [14]. This layer gives the system the ability to learn from the data it has collected and improve the manufacturing process as it is arising at the present time. The DT is the fifth layer, the digital

Figure 2.5 Digital twin layer architecture.

replica of the production in the real-world process. Because this layer is there, the system can mimic various scenarios and improve the manufacturing process without actual physical experiments. In general, the computerized methods and assessment of the industry, smart city, or interconnected systems entail the incorporation of a variety of digital technologies to maximize the effectiveness

Table 2.1 Difference between industrial simulation and digital twin.

Method	Industrial simulation	Digital twin
Definition	To simulate an industrial process or system, a model that is based on software is used	A digital copy of a real-time system or operation that is linked to the original in real time so that the virtual copy may monitor and control the behavior of the actual system
Purpose	Utilized for teaching, improving, and testing various industrial processes and systems	Utilized for monitoring, controlling, and improving the performance of physical systems and procedures in actual time
Inputs	Math and sensor-based models	Real-time sensor, equipment, and other data
Outputs	Data generated via simulations and forecasts of how the system would behave in various scenarios	Information gathered in real time on how a physical system is functioning, together with forecasts and suggestions for improving performance
Applications	Utilized in planning, evaluation, and improving industrial processes and systems	Real-time control and monitoring and optimization of physical systems and processes are used in various industries, including the aerospace, industrial, and energy sectors
Key features	Focuses on modeling or simulating the response of a system or process to changes in the operating environment	When anything is connected to a physical system in real time, it enables monitoring and controlling the system's behavior
Benefits	Makes it possible to optimize and conduct tests on industrial processes and systems within a setting that is both safe and under control	It offers real-time insights and suggestions for enhancing the effectiveness, safety, and efficiency of physical processes and systems
Limitations	Restricted because of the precision and intricacy of the mathematical models used in the simulation	Required to develop a reliable digital twin, a substantial sum of money must be spent on sensors, data infrastructure, and software engineering

of the systems and procedures that are being used. When applied to intelligent manufacturing, the six-layered method makes it possible for the system to become more effective, efficient, and adaptable to shifting market conditions.

Table 2.1 by "IIoT simulation," we mean a virtual copy of something tangible in an industrial system, such as a factory or power grid. By simulating the system, its performance may be tested and improved in a risk-free and regulated setting without impacting live operations. Simulation of IIoT systems allows for the optimization of resources, the minimization of waste, and the anticipation of issues. DTs, however, are digital copies of real-world things like machines and buildings. Using sensor readings and other facts, a DT of the physical model may be built and analyzed for behavior and performance. This may increase efficiency, performance, and the ability to foresee maintenance needs [15].

2.1.3 Survey Organization

Section 1: Introduction to Digital Twin with IIoT Overview. Section 2: Literature Survey of Digital Twin with IIoT Recent Papers. Section 3: Challenges in Digital Twins. Section 4: Integration of Digital Twins with IIoT Applications. Section 5: IIoT-DT with Blockchain Approach. Section 6: Security Risks of Digital Twins. Section 7: Countermeasures. Section 8: Conclusions.

2.2 Literature Survey

Nowadays, AR and DT are attracting study from academia and business [16]. In the present people-centric development, AR embraces the possibility to incorporate players into the forthcoming human–cyber–physical systems (HCPS) that use DT. Numerous review publications have explored the merits of AR and DT. However, they are mainly domain-specific. To bridge the distance between them, this study performs a state-of-the-art AR-assisted DT literature (until 17 July, 2022) across diverse industrial sectors, encompassing 118 chosen publications. Following the engineering lifecycle, AR-assisted DT application scenarios and functionalities are presented. The manufacturing process, design of services, and human-machine interface are all current hot topics. Finally, AR's virtual, hybrid, and cognitive twin benefits are examined. Lastly, AR assisted DT problems and future perspectives for the human-centered transformation of future industry, including product design, robotics, cyber–physical interface, and ergonomics. Uses of DT technology have been widening exponentially, and as a result, the method in which organizations function is undergoing a radical transformation [17]. In the recent past, DTs have been used to leverage critical commercial applications; going forward, the technology will expand to other use cases, applications, and industries. This proposed research work determine how DT might enhance intelligent automation in many industries. This article explains the idea, discusses the growth and evolution of DT, reviews the essential tools that enable DTs, analyses the trends and challenges associated with DT, and investigates the applied areas of DT in various industries. In this comprehensive research, we suggest the identification of shortcomings and connections between existing uses of the DT in the marine field and applications in other industrial domains [18]. In addition, the technique that was used in this study may be utilized in future research to offer a balanced and unbiased summary of the research that has been done on the subject. The study brought to light the fact that the literature seldom touches on the design and decommissioning stages, suggesting that more research must be done on these areas, particularly with the design of future ships. As an industry, construction, smart cities, and healthcare digitize, DTs will be extensively employed [19]. Transportation, energy, and military systems employ IoT and require decision assistance; this DT will help. Several forms of DTs might communicate with each other and with their physical assets. Misusing these technologies might delay and complicate decision-making, which could have serious consequences. Safely implementing DTs on diverse infrastructures is crucial. Many attack avenues exist in such infrastructures. Security and safety are in danger from most attacks. Nonetheless, this report summarizes several approaches to address these dangers. Metaverse, which combines VR goggles and other gear with a digital counterpart, is growing in popularity. Metaverse security should be examined, Digital Twin-enabled edge network (DITEN) is a framework that integrates DT into wireless edge networks to enable immediate data analysis and asset optimization [20]. Federated learning was employed to generate DTs by utilizing the operational histories of devices for DT modeling. These digitized duplicates were then used as templates for future doppelgangers. Data privacy has been significantly improved due to eliminating the need to communicate raw data in federated learning. We then decomposed the challenge and used the Deep Neural Network (DNN) to allocate a communication resource, minimizing inter-action expenses in distributed learning. The proposed technique was validated by numerical findings obtained from a standard real-world dataset, which indicated that it has the potential to enhance effective communication while simultaneously lowering the total energy cost. Proposed architecture for a DITEN for 6G to sense the MEC environment most effectively [21]. When a user travels from one set of edge servers to another and requires computing job offloading, the DITEN project formalized the mobility offloading problem to reduce the offloading delay while adhering to the limitation of cumulative migration cost. Deep reinforcement learning, also known as a system of actors and critics, solves the reduced version of the original formal issue of multipurpose optimization under uncertainty. The original problem was simplified with the help of Lyapunov. In the end, we conducted simulations to evaluate the performances of the various schemes and compare them to one another. According to the numerical findings, our suggested method can cut down on the time that computational jobs take to complete and the rate at which they fail, all while maintaining a low overall system cost. In the proposed model, we investigated adaptive federated learning for DT-driven IIoT [19]. The suggested system can dynamically alter the number of times per day that distributed learning is aggregated in response to changes in the channel state because it uses DT, which can skillfully record the network's deviation as it changes over time. The suggested technique exceeds the tested system in training accuracy, convergent motion effect, and consumption of energy, according to numerical findings. Sun et al. developed air-ground systems with shared learning based on analysis from DTs; we have created two different incentive mechanisms [22]. These mechanisms are respectively referred to as the static and dynamic incentives. To document the present state of the network nodes and support the process of making choices for federated learning, a computerized air-ground network will be built. Developed incentive systems offer the ability to reward high-performing clients in a highly mobile and heterogeneous network environment. The experiment's findings suggest that incentive programs may enhance the global model's precision and use of energy, enabling reliable collaborative learning in air-ground systems. The proposed method makes it easier to map in both directions between a physical structure and its digital counterpart [23]. It also makes it easier for structures, machines, and people to interact with one another, which is a necessary step toward developing an intelligent monitoring system that operates currently. To lessen the number of data that needed to be processed and the strain placed on the digital model's computing resources, a fog computing layer was added before the cloud layer. Using case studies of structural damage detection in bridges utilizing DL algorithms, it was shown that Connected Digital Twin System Health Monitoring (cDTSHM) is feasible, achieving an accuracy of 92% in the process. To guarantee the successful operation of the factory-simulated environment, we must provide an information exchange distribution strategy by altering the updated training program of the client and server systems and allowing certain pieces of manufacturing machinery to participate in federated training [24].

We aim to find the best solution to the contacts overhead allocation issue that crops up during DT architecture federated learning. In this study, the actual experimental setup that will be used to evaluate the efficiency of the suggested approach is established. The findings indicate that the strategy we have presented has the possibilities for improved communication results of the training model while simultaneously lowering the communication overhead. Our recommended solution is more suited for the infrastructure setup of the DT system of complicated manufacturing scenarios. This study demonstrated an effective fault detection system that correctly identifies early defects in industrial equipment by integrating DT and AI [25]. The system was able to do this via the integration of both technologies. In the proposed architecture, the DT model of the industrial machine was built by first producing the virtual reality testing model and then updating continuously measured data from the property of value. This was done to complete the construction of the model. After that, many machine learning techniques were included in the proposed framework for defect identification and diagnosis to perform data categorization. A case study on the identification of a failure in a triplex pump was used together with many other datasets that were created for learning, and the recommended hybrid framework was evaluated using this method.

The findings of the experiments showed that the hybrid GA-ML technique elasticities produced superior outcomes compared to employing ML methods independently. It was discovered that the proposed hybrid framework successfully achieved a high detection accuracy of 95% while using the hybrid GASVM classifier. We describe an innovative architecture for creating DTs of household appliances, which includes placing the twins in the fog layer and interfacing them with one another and the cloud via blockchain [26]. Three accelerators transcend cloud system restrictions and handle social and cognitive IoT applications on household devices; the distributed ledger, fog, and digital copies enable this. Safe twin-appliance association, virtual figments, ironclad rights to ownership, and DT democratization are essential. BCs secure each layer's connection to household items, their owners, digital wallets, and the enterprises that make them in the architecture. We provide further information on several examples of applications that can operate on this architecture, such as selling data on appliances, buying components, and basing prices on use. In this proposed work, a method for synchronization and skew estimation for innovative clocks with DTs was developed to mitigate the impact of complex surroundings on the different types of oscillators used in industrial IoT systems [27]. The suggested technique can eliminate not required and wasteful packet exchange thanks to a better knowledge of the clock's behavior in various operating settings. The DT-enabled synchronization method that was presented successfully achieved much-improved clock accuracy while also reducing the number of packets necessary for network functioning. Under rigorous network conditions, however, the performance benefits were much more noticeable, indicating that the suggested technique is less subject to the pensioner data verification (PDV) and dynamic operational parameters to maintain a higher clock precision.

Difficulties in the digital twin: The difficulties that have recently arisen regarding incorporating DTs into IR 5.0 are discussed in the next section. From the perspective of development and research, the difficulties encountered in this area are also related to other technological domains, including blockchain, IoT semantic models, data science, and intelligent learning.

2.2.1 An Overview of IT Network Architecture

The network's existing architecture does not combine data analysis and interaction across the IoT into the same model. The brains of the operation and the IoT must communicate for the DT to perform correctly. This calls for a network infrastructure. The excellent simulation created by the DT will be made even better by these intelligent networks. DT is not feasible to get the intended results regarding data simulation or visualization IT network architecture needs to be put in place.

2.2.2 Data Interpretation

The second difficulty associated with the DT is the massive quantity of information that must be collected to construct the setting for the simulation. The information has to be digitized for this purpose, and the data stream must have a high-quality sequence. Inconsistent simulation results caused by analog data or signals, including noise, would result from this. The DT would also underperform. For this reason, the reliability and uniformity of the information, as well as the quantity of IoT nodes, are crucial elements in the deployment of DT. It is vital to enter the appropriate variety and quantity of data at the appropriate periods to optimize the results of data simulations or visualization.

2.2.3 Privacy and Security of Data

Although the whole functioning of these innovative businesses is dependent on the information created by the system that is going to be utilized in the simulation study, data security, and privacy pose significant hurdles to the DT linked with industries. As a first order of business, the system produces a mountain of data, which is perceptive. Even a slight alteration might compromise a system's safety to these numbers. Integrating DTs with cutting-edge technology is necessary to prevent these security breaches. Data flow between IoT nodes and processors will also be made possible by this integrated security mechanism. Because of the trust and privacy concerns inside many businesses, advancing data protection and safety in DT is essential.

2.2.4 Trustworthiness

Regarding DT-based intelligent devices, trust in the organization's and the users' data has always gone hand in hand. As a result, establishing trust is the next significant obstacle for DTs to overcome. It is crucial for the company's reputation and product quality that customers positively receive the final product. Integration with a highly secure architecture and continuous improvement at each stage of development is essential for DT technology to provide high-quality data. Data security may be strengthened in several ways, one of which is by developing DT systems based on smart contracts to address issues of data privacy and trust.

2.2.5 Predictions

It might be challenging to bring the DT into an organization and have it work in real time [6]. If an enterprise wants to meet its customers' needs, it must implement a sophisticated IoT infrastructure. The only way for researchers to get there is to build a robust DT infrastructure from the ground up. Because of current technology developments, businesses no longer need to implement DTs into their existing infrastructure. As a result, meeting the needs of businesses is a difficulty that the DT must also overcome. Similar difficulties have been encountered by the IoT and distributed ledger technology before they entered actual market connectivity. Challenges for DTs, as seen by businesses, are solely on user privacy and data security.

2.2.6 Standardized Simulation

For every kind of technical advancement, researchers are required to first provide a standardized model for the purpose of its implementation in the real world. The modeling of intelligent infrastructure presents the DT of physical activity with the next set of obstacles that must be overcome. Companies are required to adhere to the standard model for the integration of their system, beginning with the gathering of data and the modeling of DTs. This integration of the system might take the form of a simulation- or physically-based model that is driven by data. Standardized modeling's essential function ensures the correct execution of dynamic semantics integration of the DT into the organization.

2.2.7 Domain Model

Domain modeling presents the last obstacle for DTs. A multidomain infras-tructure is required to grow DTs [7]. Domain re-engineering is required for a successful DT since only accurate data can be passed via its many iterations. The correct design of the DT is ensured by domain modeling that incorporates disciplines like data science, ML, and the IoT. Hence, future domain modeling for the DT model will improve stage virtualization.

Several DTs represent the synchronization of real-time industry updates with acquiring property resources. This includes data on the status of equipment as well as its history. With the performance enhancement using a data-driven strategy, the DTs that have been established effectively communicate with one another. This is accomplished by forecasting the inherent dangers inside remote production systems. The DT works together by implementing information analytics to assess a methodology that considers data, comprehends the system's products by applying collective expertise, and provides continuous defect monitoring and remediation. Humans may be assisted in making appropriate judgments about the semantic-based IoT design patterns of distributed production systems by better understanding the predicted prospective dangers. Nonetheless, the DT concept is still in its very early stages, and there are many challenges to conquer before it can be implemented. These challenges are characterized by the following characteristics [8]:

- Data exchange: To define the model of DT and the data encryption, it is necessary first to establish a framework and a system for exchanging information among businesses.
- Access control: In a decentralized production system, DTs aid a single entity. Thus, a cost-effective security system to gather real-time data across multiple organizations is necessary for authenticating digital data in the system.
- If there is even a single mistake in the data, the whole centralized system will crash, and it will take much longer to get the right conclusion. Although decentralization necessitates global data from companies, it also offers a consensus method for the decision-making process.
- Blockchain-based system for making decisions: An incomplete or inaccurate dataset may cause the centralized system to crash, extending the time needed to reach the right conclusion. Decisions in a decentralized system are made using a consensus procedure, which necessitates access to information from all over the world.
- Possibility and dependability: A sector must adopt several DTs to denote the many components that make up a system, such as things, elements, sensors, nodes, people, and systems. Because of this, required to give feasible and dependability in DT inside the production unit, this distributed architecture may be accomplished via several DTs.

Using blockchain technology with DT offers several benefits, including reliable traceability, scalability, information sharing, decentralized ledgers, end-to-end connections, and security [28]. With blockchain, several DTs can work together in a structured and detailed manner, applying group information to regulate and data on the assembly of the product should be monitored. If producing digital twins, specific processes may be automated with the help of a smart contract, which improves the reliability and efficacy of data interchange, provides data that can be trusted, enables minimal information processing, and enables tracking events for interested parties. In addition, most of the attention paid during earlier research focused on the use of blockchain for virtual twins. It ignores alternative ways to create constructive DTs using organizational business data and safety procedures. Blockchain-based digital copies generating operational information in global intelligent manufacturing facilities provide several obstacles that must be explored to discover, analyses, and resolve any issues.

2.3 Digital Twin Integrated IIoT Application

An analysis and summary of three theoretical applications simultaneously, seven manufacturing-phase applications, and quality of service applications are provided to reflect the many kinds of DT applications that may be found across the many cycle stages (Figure 2.6). The literature that is now accessible has very little information regarding the application of DTs throughout the years of retirement. Distinct parts of these applications are analyzed separately.

2.3.1 Design-phase Applications

By fusing the data model and the physical system of the product and iteratively optimizing both, DTs may shorten the design cycle and cut down on rework costs. There are typically four stages to the design process: Four stages are included in the design process: (i) defining the task, (ii) creating an idea, (iii) creating an embodiment, and (iv) designing the details. The case of a local shoemaker was used by Chen et al. [29]. The integrity of data and traceability of products and the availability of knowledge were all improved since the cobbler was in charge of every step of the shoemaking process from start to finish. Under this basic "cobbler model," the cobbler was aware of the needs of his clientele and the

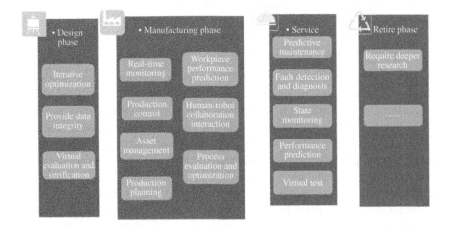

Figure 2.6 IIoT application with DT lifecycle phase.

limitations imposed by their budgets. He was well-versed in the processes and materials needed to produce various products. With the increasing complexity and diversity of goods, the DT might one day replace "the cobbler's mind." According to Feng et al., the DT gives designers access to the whole digital footprint of items throughout the design process [30]. It "engineered" the transformation of massive datasets into actionable intelligence. Designers may use the data right away at various stages of the design process to make better choices. In the study, DT was envisioned to aid in the definition of tasks, the development of concepts, and the execution of virtual verification. DTs made it possible to verify whether the product met the specified design and user needs. There was a proposal for a complete design and manufacturing engineering's DT model of the process to help engineers improve their designs, how DTs might aid in the virtual validation of hardware. We discussed the disconnect between engineering design automation technologies and DT. A solid-fuel rocket engine a shift from the more commonplace experience- and semi-experience-based design mode to one that is instead informed by data and models [31].

2.3.1.1 Optimization Through Iteration
Constant iteration in the direction of the product specification throughout the design process, from ideation to implementation, is the hallmark of good design. The DT of a product may be used to trace its historical footprints and to monitor its progress toward optimization to achieve iterative optimization. A DT-based approach may increase the precision and effectiveness of product material selection, represent the development of dynamic parameters, and enable between static setup and dynamic implementation designs to be optimized iteratively. Compared to conventional simulation, DTs allow for accurate performance predictions and identify possible issues pointing to where optimization should be focused.

2.3.1.2 Provide Integrity of Data
The "cobbler model" fits poorly with the paradigm of collaborative design that is prevalent today, as was described above. Processes formerly carried out by a single individual are now broken up into multiple sections; this causes knowledge to be held in isolated silos and eventually fragments both product and procedure information. As a result, a great deal of valuable information conducive to effective decision-making is spread among many stakeholders. The following generation often faces issues encountered by the generation before them, although these difficulties may have been avoided by using the information gained from the prior generation [29]. Product lifecycle management (PLM) invented the phrase "digital twin." The virtual twin evaluates and collects real space information to support design decisions. The virtual twin enables digital processes, which means that comprehensive data may be delivered without exposing confidential or proprietary

information, which is beneficial for various viewpoints and stakeholders. Because of this, the creation of a security policy for the DT is required.

2.3.1.3 Virtual Assessment and Verification

Evaluation is performed to bring about a reduction in the number of contradictions that exist between the actual behavior and the intended behavior. The conventional focus has been confirming and verifying prerequisites and removing difficulties and failures. Still, the DT model also presents a chance to detect and eradicate the unpredicted undesirable. A model with a significant amount of correctness of both the product and the surroundings in which it will be used is provided by the DT approach, another benefit of this method compared to the conventional assessment and verification method. As a result, an interaction on several scales and across multiple disciplines between the item and its potential in the future operational environment is achieved. During the time of design known as "design," the "digital twin" assists in "virtual prototyping" and evaluates various product iterations in a variety of "various applications."

2.3.2 An Overview of Manufacturing Applications

Manufacturing is often understood as an advanced manufacturing method that turns raw materials into finished products. This transformation takes place throughout the manufacturing process. Manufacturing, on the other hand, is moving away from primary processes and toward intelligent processes. This is because there are now higher expectations placed on product characteristics and a greater need for quick market reaction. The physical and virtual worlds must interact in a closed-loop fashion for modern production. DT is about realizing physical-virtual contact and interaction.

Grieves conceived of three different applications for the DT. First, we can visualize the original manufacturing operations, check the digital model against the real thing, then work with others to get a complete picture of what is happening in the factory. As mechanical and electrical technologies have evolved, manufacturing has become more automated. Automated behavior is produced in response to predetermined, well-planned sequences of activities. In today's world, manufacturing components can communicate with their environment to gain awareness of their context. As a result, they can make intelligent choices on their own without being explicitly programmed. The DT is essential to making this vision a reality.

During the manufacturing phase, many brand-new ideas, frameworks, and new concepts are being suggested. This study demonstrated how a DT might turn a cyber-physical manufacturing system becomes an independent entity. one. investigated a revolutionary idea known as the DT shop floor (DTSF). It involved the real shop level, digital shop level, shop floor service system, and the digital ground level. Wright's DT solved customized manufacturing and dispersed manufacturing-related difficulties [32].

The virtual twin tracks current information, monitors historical data, and aids in future operational choice-making. The following is a collection of some of the literature about the applications of DT in the manufacturing stage.

2.3.2.1 Real-time Surveillance

Watching the various stages of production in a plant has been ongoing for some time. Nonetheless, the DT offers real-time monitoring that is distinct from and superior to other methods. To begin, DTs combine all necessary data with visual representations of 3D objects. Second, the DT combines data from the past, data from the current time, and data that is expected for the future to trace the past, evaluate the current, and forecast the future. DT for in-process actual analysis, various technologies, such as research, are being done into various technologies, including AR technology, three-dimensional in-nature visualization technology for tracking, and whole components of data perception tech. The high-fidelity model of the DT gives actual time information on the state of the actual item being observed, helping to understand the scenario and optimize it. The real-time analysis uses the DT.

2.3.2.2 Production Planning

Industrial control systems must consistently carry out pre-planned tasks and respond to unexpected events. By default, a manufacturing execution system (MES) based on static assumptions controls the production system. With a DT, it is possible to synchronize a system's real-world and digital representations, allowing for intelligent, real-time control from a more comprehensive viewpoint. With its high-precision model and real-time data, DTs are well-suited to the complexities of today's production systems. Most studies, however, were conducted purely theoretically or were only put into practice in a controlled laboratory setting. There is still a significant distance before a DT can be used for real-time intelligent control in a complicated production setting.

2.3.2.3 Prediction of Workpiece Performance

During the production process, the factory's internal and exterior environments may be disrupted by factors such as the deterioration of machinery and fluctuations in the raw materials. As a result, it is encouraging but challenging to forecast how the workpiece would function. The DT must provide sufficient data to properly predict the workpiece's performance before it is built, as well as measurable virtual structures production phenomena. Nonetheless, it might be challenging to determine the scientific laws governing some phenomena that occur throughout the manufacturing process, such as surface roughness and the assembly of complicated products. Computable abstraction of established physical rules, AI algorithms, and transdisciplinary models are all included in DTs. The accuracy of the predictions is improved by the models that are regularly updated.

2.3.2.4 Humans and Robots Working Together and Interacting

Both human and robotic motion planning has historically been developed using an open-loop approach. As a result, it is necessary to recalibrate human and robot movements whenever there is a change or disturbance in the assembly environment or the geometry of the item being assembled. Neither humans nor robots can see one another. By using a DT to run a closed-loop simulation, both the human and the robot may make real-time adjustments to their movement or behavior. Allocating tasks between humans and robots, optimizing workspace architecture, conducting ergonomic analyses of humans, and testing robot programs are all made easier with the help of a DT of the human-robot collaborative workplace.

2.3.2.5 Analyzing and Bettering a Process

The rising complexity of production makes conventional process planning challenging. The real processing conditions are not available to planners. Changing machining processes and variable production resources significantly affect product quality. It is essential to gather and map data in real time in the workplace to keep the DT up-to-date. DTs are created by integrating geometric, mechanical, and material models. The DT will then communicate with its manufacturing environment via plug-and-play mechanisms to evaluate and enhance the method and its parameters.

2.3.2.6 Asset Management

Better control, planning, and scheduling choices are made possible with the capacity to track and monitor assets. In addition to aiding in realizing production assets' business logic, the DT of such assets also permits their use in interactive and working together settings with other industrial property. DTs offer valuable data for various asset management processes, including initialization, reconfiguration, planning, commissioning, and status monitoring. Managing asset depreciation and reflecting the industrial assets' current state and performance in real time are two examples of the uses of DT in the literature.

2.3.2.7 Planned Manufacturing

Given production disturbances, advanced dynamic planning for production is needed. The virtual twin improves worldwide manufacturing timelines in reaction to system changes. In the virtual twin shop level (VTSL) paradigm, sensor, training, and business information system information enables production scheduling. After receiving the production schedule from the production schedules provider, the virtual shop floor thoroughly reviews it [33]. If there are fluctuations in the availability of resources, the production plan service will get recommendations for adjusting.

2.3.3 Applications in the Service Phase

Products throughout this stage of service are often distributed and independent from their original makers and suppliers. Their data is hard to maintain and retrieve, and it is impossible to create a closed-cyclic data flow. And although the existing digital model could be spot-on in depicting the product's design, it lacks any connection to a physical component that has been produced. Products from the same production run may exhibit varying functionality depending on the conditions in which they are utilized. At this stage, customers care most about the product's dependability and ease of use, while producers and distributors focus on the product's current operational condition, maintenance plans, and so on. It was proposed by Ali et al. that the DT technique may help predict and understand deterioration and abnormal occurrences [34]. A representation of the tail of each aircraft that is very accurate with the number was suggested by Majid et al. as part of a structural life prediction procedure for airplanes (a digital twin) [35]. At this phase, we determined the mission's expected flight path, and it would perform. It is thus possible to determine whether to deploy a given aircraft on a specific mission based on the possibility of the airframe satisfactorily surviving the operation's demands. According to NASA's plan [36], the DT will combine several best-physics models with onboard sensor suites. So, during the lifespan, decision makers will have a firm grasp on the physical processes associated with deterioration on the levels of the material, the structure, and the system. The DT technologies may reduce the likelihood of damage or deterioration by kicking off self-healing procedures or suggesting route adjustments to reduce loads.

2.3.3.1 Predictive Maintenance

No matter how a product is made or used, its safety and the risks associated with its usage are often established during the design process. Engineers may often advise costly regular maintenance based on safety considerations, even though this results in poor product usage and a high cost. A current time digital pair is a superior imitation of a virtual object. By combining several types of models (geometric, mechanical, material, electrical, and so on), DTs may precisely predict how a real item would react in each setting. From the early days of DT development to the current day, predictive maintenance has been the most widely studied and implemented use case. Several kinds of literature need to account for the impact of the design and manufacturing processes on product performance, even though many modern applications are centered on expensive machinery.

2.3.3.2 Finding and Fixing Problems

In most cases, these physical entities function correctly. Hence, the data needed to build a good model for defect detection needs to be increased. When it comes to physics-based defect diagnostics, it does an excellent job with known abnormalities but has yet to learn what to do with unexpected ones. Both are brought together in a DT, with the multi-physics model working with the product's data. Scientists simulate the motion of a physical product using deep transfer learning, dynamic Bayesian networks, and nonlinear dynamics. For lack of appropriate training data, Sasikumar et al. turned to simulation data [28]. Yet it is still unclear whether the variations in the distributions of the simulation data and the actual observed data match up.

2.3.3.3 State Monitoring

The DT can do more than compare collected data to expected values. It can also look at collected data from different points of view. Finding the cause of the failure is as simple as comparing the information obtained through the virtualization of the DT to the data that was obtained. The DT provides a model that is high-fidelity and accurate, and it is kept up to date over the whole of the product's existence. This means that a DT can show the present state of a real thing in a digital universe with less data. An information fusion algorithm was used by Mauro and Kana [37] and Saad et al. [38] to manage many physical items' semantic information. DT-state monitoring offers a unique identification for each product, allowing consumers to check on a product from everywhere.

2.3.3.4 Performance Prediction

Optimal designs at the outset may adapt poorly to changing operational circumstances over the service period. A DT is a similar copy of a virtual item that reflects how it was intended to be used before it was ever produced. Thanks to the entire lifespan data, the DT is fully aware of the physical twin's outcomes. In this scenario, a DT is useful for optimizing and forecasting how a product would perform. Nevertheless, the DT concept has yet to be used entirely by current research activities. The first thing to remember is that, just as with predictive maintenance, they fail to consider how designing and making the product could influence the outcome. Second, a digital twin should provide an ideal use plan considering the product's performance data, allowing the data's full potential to be realized.

2.3.3.5 Virtual Test

Due to its status as an electronic duplicate of the actual object, the DT may be used to simulate and verify critical processes in situations where their failure might result in catastrophic consequences. In this light, a DT is like a simulation but with greater fidelity to the actual world. Without worrying about how you are testing or exploring a system's features that would affect the real world, you may create a digital twin of that system.

2.3.4   Retiring Applications Phase

It is not uncommon for people to fail to recognize retirement as a distinct life stage. When a system or product is no longer used, most information on its operation is lost. In many cases, learning from the mistakes of the previous iteration would have prevented the difficulties that plagued the system or product in question. At the last stage of its existence, the DT may be kept cheaply and securely in cyberspace since it has all the data collected during the physical twin's life.

Our evaluation study identifies two publications discussing DTs in the retirement phase. During the entire life cycle (from design to recycling), Gehrmann and Gunnarsson created a groundbreaking technology based on digital twins for the improvement of WEEE to aid in activities related to production as well as refurbishment [39]. The development of product information models was guided by global standards from the very beginning through recycling. Akbarian et al. built an architecture for a DT-based refurbishment shop floor to address the unknowns in the remanufacturing process [40]. We also looked at the potential for a future paradigm shift in the way automobiles are remanufactured using DT technology.

2.4   IIoT-DT Combination with Blockchain-based Approach

2.4.1   Framework for Trust and Security Based on the Blockchain

Before the advent of intelligent IIoT, the manufacturing process was often portrayed as a mysterious "black box" with little to no transparency into its inner workings regarding data flow, functionality, and controls. This "black box" may be successfully divided into more compact "black boxes" for each piece of equipment, activity, and method. These once inaccessible "black boxes" are gradually opening because of developments in emerging technologies like enhanced sensors and DTs. Conversely, a business relationship may benefit all participants in the system. Although bargaining was possible in the decentralized form, building a central hub to oversee all nodes was a significant roadblock in the centralized approach. If there is not enough trust, the haphazardly managed industrial environment might swiftly fall apart due to even a little delay or mistake. Distrust hinders cooperative production's potential for global innovation, financial success, and operational efficiency. The need for more confidence in the system may be

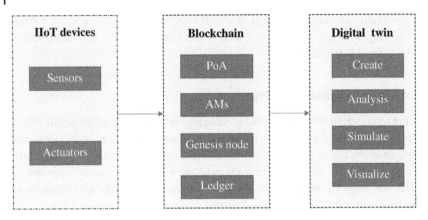

Figure 2.7 The infrastructure blockchain/digital twin integration for the IIoT network.

remedied by blockchain technology. Instead of agreeing on a single trusted party, participants distribute transaction data over a network of potentially dishonest nodes. Blockchain is a suitable environment for multi-party data exchange and the computations necessary for cross-organizational collaboration. The reliability of a public blockchain may be traced back to the interactions between its nodes. Members of the blockchain platform rely not on a central authority or an outside party, but rather than on the technology of distributed ledgers and the structure behind them to arbitrate transactions among producers working within an identical shared system. Figure 2.7 illustrates the suggested system.

2.4.2 IIoT-DT with Blockchain Layer Architecture

Specific manufacturing data may be moved to a publicly distributed blockchain network to reduce the high "trust cost" imposed on businesses. Organizations may increase their credibility by focusing on the safety of their data and the dependability of their creative employees. Blockchain can reduce trust costs in transactions. Several transactions occur between the firm and its client, agencies, suppliers, distributors, mobile carriers, and other manufacturers throughout the product's creation, assembly, and maintenance life cycle. Data security experts consider private and public keys necessary for asymmetric cryptography, the process of encryption as well as decryption of network traffic. Hash functions are another crucial tool built to fortify the safety of the information system. This ensures the security of the transactions and generates new blocks in the blockchain network. These safeguards will improve investor confidence in the network and ensure the confidentiality of all stock trades. Digital data representing untrusted variables like tasks, contacts, identities, transactions, assets, and contracts have all found a home in cryptographic representations. It is coupled with the blockchain to meet the requirements for confidentiality and identity verification. The freshly verified block will be added to the blockchain. All nodes in the network will eventually store it when it has been broadcast, and everyone will agree on the most current blockchain. If a blockchain disagreement is just partial, the shorter chain will be disregarded. The blockchain has uncovered trustworthy data forms that may facilitate business collaboration. Unreliable parties may be able to work together reliably via the consensus mechanism. Predetermined external events, such as job completion, may activate a public ledger. Cooperation between players is possible with well-stated rules.

For a greater understanding of optimization research, we have provided a summary of the general optimal framework for the study of incorporating blockchain technology into IIoT, as shown in Figure 2.8. This is based on the layered technological structure given earlier in this paragraph. The research efforts that go into optimizing the network layers, for the most part, focused on making the structure of a network better.

The research into optimizing the data layer includes making verification more efficient and reducing storage pressure. The research that is being done to optimize the consensus layer is primarily focused on boosting the performance of the blockchain as well as increasing its resilience to rogue nodes. When it comes

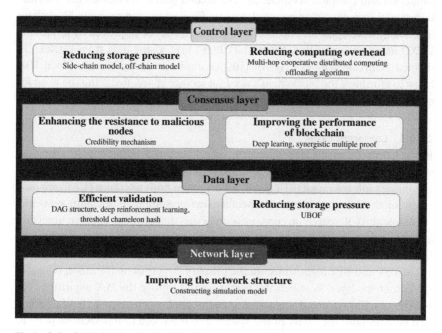

Figure 2.8 Digital twin with IIoT blockchain research framework that optimizes the general research process.

to the research that goes into optimizing the control layer, the primary research paths focus on lowering the amount of storage pressure and processing overhead.

2.4.2.1 Network Layer

In IIoT, network layer optimization research focuses on building applicable models and enhancing network structure. Papadis et al. studied networking structure. They created a blockchain-based IIoT distributed system model to explore block delay time during blockchain development and computing power's influence on blockchain performance. When public blockchains expand about the IoT installations that include many geographically dispersed nodes, this model can adequately anticipate system properties and efficiency and address block transfer and processing latency issues.

2.4.2.2 Data Layer

Data layer technologies may be separated into two broad groups: association certification structure and information model. There are two primary avenues for development based on the two distinct types of crucial technologies. One is to alter the data layer certification architecture to speed up blockchain technology validation and complete it efficiently. The second goal is to enhance the information model to reduce storage requirements for devices and promote blockchain technology development. The information model may be improved by improving UTXO using an innovative, Xu et al. (UBOF) [41]. Using data from UTXO, Liu defined "Unrelated blocks" (UB). UBOF's ability to identify and offload UBs helps to minimize the blockchain's need for storage space. Reducing the need for expensive data storage in the IIoT, significantly improves blockchain's scalability.

2.4.2.3 Consensus Layer

Blockchain's consensus layer allows the consolidation of ledger data in a trustless setting and defends against a wide range of potential assaults. Proof of X (PoX), Byzantine fault tolerance (BFT), and confirmation fault tolerance (CFT) are only a few of the leading consensus mechanisms present in blockchain networks (fault tolerant). Research on consensus processes that seeks to improve upon and expand prior studies fits under the category. Most existing consensus mechanism literature focuses crash-on refining or expanding this consensus's categorization.

The PoX and BFT algorithms are the primary targets of optimization efforts at the consensus layer. Research the best ways to improve the PoX algorithm. For IoT devices, Vakaruk et al. suggested a credit-based proof-of-work (PoW) method that ensures both the safety of the system and the speed of transactions [42]. To lower overall energy consumption and boost throughput, the algorithm may adjust the problems with trustworthy nodes while increasing challenges posed by attack nodes based on their behavior. The algorithm allows blockchain to function on low-power IoT devices.

This study addressed the issue of blockchain's demanding resource needs and the often-poor performance of IIoT devices. Janani and Ramamoorthy improved the throughput of the BFT method by transforming the consensus issue of BFT into a joint optimization problem, which they subsequently addressed using a Dueling Deep Q-learning strategy [43]. Because of this study, SD-IIoT may now use controller consensus to its advantage, allowing for efficient and secure network operation.

2.4.2.4 Control Layer

It emphasizes putting together and executing models and digital contracts at the control layer. Researchers have shown that by creating outside of the chain or on the opposite chain processing models, they may improve the blockchain's scalability without altering its basic functionality. Experts in this field have found that by identifying the weaknesses of intelligent contracts, they may improve the controlling layer's security.

Due to device computing storage and power limits, IIoT research is focused on upgrading standard blockchain processing models to adapt to broader devices connected to the IoT. MDIoTSP is Tong et al.'s harding's protocol. The blockchains were first fragmented into a significant number of smaller shards by the protocol. Each is generally considered a micro-blockchain according to the many fields of study. Afterward, it completed the multi-domain IoT blockchain environment by combining cluster sub-block hashing digests. This protocol met the need for large-scale information storage and transfer in multi-domain IoT architectures [44].

2.5 The Risk of Digital Twin Security

It is possible to categorize security risks associated with DTs according to the target, beginning with the system and progressing through the communication and data storage risks. The digital twin must meet critical security requirements regarding its availability and accessibility(s). The major causes of severe dangers may be high operational demand and poor computing power in IoT devices. Suppose digital copy access is denied or the DT is rendered inaccessible for whatever reason. In that case, the manufacturing life cycle may experience interruptions, resulting in significant financial losses. The data must maintain its integrity to guarantee that it has not been tampered with and contains no errors. Risks to the product's integrity might result in biased assessment and decision-making, exposing the company to increased risk. Keeping data secure is crucial to ensuring safety. In the worst-case scenario, a breach of confidentiality might result in irreparable financial harm to the company. Data leaks, IP disputes, and the disclosure of trade secrets are only some potential outcomes of such risks. The dangers that are presented by the digital duplicate might be made much worse by several additional factors. For instance, DT modules that rely on the need for real-time data interchange are more likely to be susceptible to more assaults than modules that do not rely on such specifications. Components that do not rely on such requirements are less likely to be subject to assaults.

2.5.1 Physical Attacks

Since they are vulnerable to being damaged, destroyed, or stolen, IoT device physical security is crucial. Interactions between a twin system's digital and physical components need additional security measures beyond those typically used in a network or computer system. Defects in a life-or-death product, for instance, may have catastrophic consequences. Hence, safety may be the most essential need for security. Safety is the absence of risk to people, property, and the environment due to system failures.

2.5.2 Data Modification Attack

Data stored on physical items may be safeguarded in several ways, including with tamper-proof and tamper-resistant devices. Yet, the attacker may still potentially alter the data. Essential shifts in predicting and analysis may need updating the data. Attacks using "data poisoning" are one method of modifying data, but vulnerabilities in the underlying infrastructure, applications, and communication channels may also be exploited.

2.5.3 System Attack

The DT virtual layer software may be used on a computer. There is a risk that attacks on the DT's virtual layer may have unintended consequences for the underlying operating system. The operating system flaws allow an attacker to steal or take over the digital twin's code. Malicious software may be used for a broad range of assaults and threats. They may cause severe damage to DT-based intelligent manufacturing systems, including full-service disruptions. Several firms have recently reverted to older production methods, abandoning smart manufacturing after a series of ransom assaults. Also, industrial control and operating systems are vulnerable to WannaCry and other forms of malicious software.

2.5.4 Software Attack

As the DT is the blueprint of the actual model, gaining illegal access to the software or programming might have devastating effects. As a result, the attacker will have all the information they need to exploit the system's flaws. If that happens, the attacker may be able to get much farther, into the system's back end. To provide one concrete example, "smart manufacturing" is a competitive business arena. The digital replica includes exact replicas of the physical product's data, information, and models. Threats to intellectual property and commercial secrets may arise if hackers access pre-release versions of DT software. The DT or program code may be stolen using several different methods. A prime example is the capability of insiders to make copies of software. Even if they do not want to, insiders pose a significant risk since their actions may compromise security directly. Weak software security might be the result of sloppy coding. When software has inadequate or nonexistent security, malicious actors may employ reverse engineering techniques to steal the program's source code.

2.5.5 Data Communication Attacks

Digital twins rely on disparate parts that must be seamlessly included in the network's nodes. These parts communicate with one another via a variety of protocols and interfaces. It is no secret that many of the most popular forms of electronic communication are riddled with security flaws. However, there are several well-known security and privacy flaws in the IoT and cloud computing. For instance, major vulnerabilities in IoT data transfer protocols might result from constrained resources in IoT devices. Similar dangers may exist when data is transferred to or from the cloud. The risks to data transmission may be broken down into five categories.

2.5.6 Man in the Middle Attack

A Man in the Middle (MITM) attack allows a third party to eavesdrop on communication in progress between any two communicating nodes. The MITM attack may be used across various protocols and technologies. The hack involving the reconfiguration of wireless keys, which might reveal sensitive information, is one example of a security breach caused by flaws in the IPv6 standard. The MITM attack might be used against Bluetooth and ZigBee Protocol.

2.5.7 DoS/DDoS Attack

The risk of a denial-of-service (DoS) or distributed denial-of-service (DDoS) assault is one of the most severe threats to the accessibility and usability of the

DT. It can cause problems for the system if data is supplied in a greater volume or size than intended. The three most common methods of causing a denial of service are: (i) providing so much traffic that the network crashes, (ii) delivering insufficient data that the services are shut down, or (iii) leveraging a compromised part of the system to provide signals ultimately resulting in the absence of service. Jamming attacks, in which specific radio signals are sent repeatedly or randomly, might prevent data transfer. DoS may also be caused by attacks that manipulate data gathered by spectral sensing. Unprotected IoT gadgets are easy prey for DDoS attackers.

2.5.8 Eavesdropping Attack

The data sent from sensors to controllers through the network may be recorded. With a passive attack, an adversary may learn something about the data flow from sensors to controllers. This exploit immediately compromises the privacy of potentially sensitive information sent in this traffic. Further, dangerous assaults may be launched using the gathered intelligence.

2.5.9 Spoofing Attack

To carry out hostile and dishonest acts, the attacker may often disguise himself and alter his identity. There are many ways spoofing attacks that may be launched against supervisory control and data acquisition (SCADA) systems. The modification of the sensor network's wireless node is an additional application. This may be done for one of two reasons: either to acquire entry to the connection or to divert the data that already exists on the connection. In the event that spoofing occurs, it is possible that the electrical signals or data that are sent from the various sensors that are located everywhere around the control system may become distorted.

2.5.10 Replay Attack (RA)

A replay attack (RA) is a type of post-hacker that requires a prior pre-hacker. The hackers gather and retain data for future retransmission. Because of this, the retransmitted signal or data might potentially result in malicious actions on controllers or sensors. For instance, as an example of an advanced persistent threat, this replay assault targeted the SCADA systems of Iranian nuclear sites. This kind of assault is referred to as the Stuxnet worm. In the future, similar assaults may disrupt the operation of other vital infrastructures and industrial systems.

2.5.11 Data Storage Attacks

Cloud computing is used for most of the data storage activities carried out by the apps for DTs. The data that is stored is private and vulnerable to a variety of threats. When cloud computing is finally breached, data breaches are sure to follow, especially when it comes to keeping data in the public cloud, which puts all the power in the hands of the service provider firm. There are a lot of trust and privacy problems to consider. Yet, the information can be stored in federated file structures, and apps such as inter planetary file system (IPFS) can be utilized for data storage and management.

2.6 Countermeasure

Table 2.2 outlines the many defenses that may be taken against the dangers posed by DTs. First, actual device integrity has to be addressed while transmitting information. There is a possibility that the danger of insider attacks may be mitigated by putting information security concepts like "providing the least privileges possible" and "using access restriction." A variety of techniques may neutralize threats involving data tampering. Tamper-proof or tamper-resistant components may be needed. Hash operations, blockchain technology, the Internet File System, and so on, maintain data integrity. The software comprises all parts' operating systems and the electronic twin's virtual layer's intelligent processing. Probably, specific crucial attack routes will not be accessible if appropriate methods for hardening operating systems are followed.

Table 2.2 Countermeasure DT-IIoT.

Attacks	Counter measures
Physical attack	Physical security
Data modification attacks	Hardware that is both blockchain, IPFS, and tamper-proof
Software attacks	Software, privacy software creation life cycle, and testing for security
Data communication attacks	Network redundancy, cryptographic method, blockchain, firewall, and intrusion detection system
System threats	System hardening, firewall, access control, intrusion detection system, and anti-malware
Data storing threats	Cryptographic solution
Machine learning threats	Sanitization of data, improvement of the robustness of algorithms, protection of information, evaluation mechanisms, and privacy-protecting techniques

A safe software development life cycle, often called an SDLC, will defend against most software flaws and vulnerabilities. To stop anyone from duplicating software and doing reverse engineering on the systems being used, security checks need to be run, and robust methods need to be implemented. In addition, penetrating testing, regular upgrades, and proper patch administration are crucial in avoiding attacks. They will help reduce the risks associated with software-related attacks. Software-defined networks (SDNs) and other intelligent provides high quality is recommended to increase the resistance of telecommunications networks against assaults. To ensure the robustness of a network, it is essential to use many redundant network pathways and communication channels. Encryption methods may be used as part of strategies to reduce the risks associated with data communication. These methods can protect data transmissions from a variety of dangers.

In addition, blockchain solutions may be used to guarantee the integrity of the underlying framework, and digitally signed contracts can be implemented to automate specific processes. Using network firewalls, intrusion detection system (IDS), antivirus software, and antimalware software is a realistic strategy for addressing threats to computer networks and systems. The hazards posed by data may be broken down into two categories: those posed by ML and those posed by data storage. Implementation of robust cryptographic solutions is one way to reduce the risks associated with data storage. Yet, the effectiveness of the solutions should be carefully studied, and the system's performance should be considered. The data may be exchanged between blockchain systems and the parties linked with them. Zero-knowledge proof (ZKP) algorithmic processes and digital agreements are two methods that may be used to ensure the data's privacy. Homomorphic encryption and other such developing technologies need to be taken into consideration as well.

The stages of ML should be structured to provide enough protection against assaults. To defend against data poisoning attacks, data cleansing and the improvement of algorithm robustness should be implemented. It is essential to construct robust models that can defend against adversarial ML approaches. Against evasion, impersonation, and inversion attacks, security assessment methods and privacy-preserving procedures should be employed.

2.7 Conclusion

The need for increased digitalization in the industrial and construction industries, as well as in green infrastructure and innovative healthcare, will lead to the widespread use of DT. Since these systems rely significantly on IoT and need decision assistance, they will all benefit from the implementation of DTs, including mobility, power, and even military systems. Multiple kinds of DTs may emerge, some of which will communicate with the physical assets they are linked with, while others will communicate with one another. It is essential to keep the security of these systems under check since their abuse might result in complications and holdups in the decision-making cycle, which could lead to significant unfavorable effects. A primary concern is establishing DTs on disparate infrastructures while maintaining their integrity. These kinds of structures are made up of a wide variety of technologies, each of which is susceptible to a different kind of assault. Most potential dangers represent a significant threat to the system's integrity. Nevertheless, a variety of responses may be taken to these dangers, and a summary of such responses can be found in this paper. The term "metaverse," which refers to the practice of merging VR goggles and other similar technology with a digital counterpart, is gaining popularity. Moreover, research must be done on the safety of the metaverse. In the next phase of the study, we will develop a complete security architecture for the DT and conduct in-depth research into various threat defenses. The use of blockchain technology and automated contracts to secure the metaverse will also be investigated.

Acronyms

UBOF   unrelated block offloading filter
UTXO   unspent transaction output
WEEE   waste electrical and electronic equipment

References

1 Fuller, A., Fan, Z., Day, C., and Barlow, C. (2020). Digital twin: enabling technologies, challenges and open research. *IEEE Access* 8: 108952–108971.

2 Kuts, V., Modoni, G.E., Otto, T. et al. (2019). Synchronizing physical factory and its digital twin through an IIoT middleware: a case study. *Proceedings of the Estonian Academy of Sciences* 68 (4).

3 Alcaraz, C. and Lopez, J. (2022). Digital twin: a comprehensive survey of security threats. *IEEE Communications Surveys & Tutorials* 24 (3).

4 Mylonas, G., Kalogeras, A., Kalogeras, G. et al. (2021). Digital twins from smart manufacturing to smart cities: a survey. *IEEE Access* 9: 143222–143249.

5 Platenius-Mohr, M., Malakuti, S., Grüner, S., and Goldschmidt, T. (2019). Interoperable digital twins in IIoT systems by transformation of information models: a case study with asset administration shell. *Proceedings of the 9th International Conference on the Internet of Things.*

6 Zeb, S., Mahmood, A., Hassan, S.A. et al. (2022). Industrial digital twins at the nexus of nextG wireless networks and computational intelligence: a survey. *Journal of Network and Computer Applications* 103309.

7 Jamil, S. and Rahman, M.U. (2022). A comprehensive survey of digital twins and federated learning for industrial Internet of Things (IIoT), Internet of Vehicles (IoV) and Internet of Drones (IoD). *Applied System Innovation* 5 (3): 56.

8 Vuković, M., Mazzei, D., Chessa, S., and Fantoni, G. (2021). Digital Twins in Industrial IoT: a survey of the state of the art and of relevant standards. In: *2021 IEEE International Conference on Communications Workshops (ICC Workshops)*. IEEE.

9 Kamble, S.S., Gunasekaran, A., Parekh, H. et al. (2022). Digital twin for sustainable manufacturing supply chains: current trends, future perspectives, and an implementation framework. *Technological Forecasting and Social Change* 176: 121448.

10 Perno, M., Hvam, L., and Haug, A. (2022). Implementation of digital twins in the process industry: a systematic literature review of enablers and barriers. *Computers in Industry* 134: 103558.

11 Wanasinghe, T.R., Wanasinghe, T.R., Wroblewski, L. et al. (2020). Digital twin for the oil and gas industry: overview, research trends, opportunities, and challenges. *IEEE Access* 8: 104175–104197.

12 Leng, J., Wang, D., Shen, W. et al. (2021). Digital twins-based smart manufacturing system design in Industry 4.0: a review. *Journal of Manufacturing Systems* 60: 119–137.

13 Boyes, H. and Watson, T. (2022). Digital twins: an analysis framework and open issues. *Computers in Industry* 143: 103763.

14 Thelen, A., Zhang, X., Fink, O. et al. (2022). A comprehensive review of digital twin—part 1: modeling and twinning enabling technologies. *Structural and Multidisciplinary Optimization* 65 (12): 354.

15 Metallidou, C., Psannis, K.E., Vergados, D.D., and Dossis, M. (2022). Digital twin and industrial internet of things architecture to reduce carbon emissions. *2022 4th International Conference on Computer Communication and the Internet (ICCCI)*. IEEE.

16 Yin, Y., Zheng, P., Li, C., and Wang, L. (2023). A state-of-the-art survey on augmented reality-assisted digital twin for futuristic human-centric industry transformation. *Robotics and Computer-Integrated Manufacturing* 81: 102515.

17 Attaran, M. and Celik, B.G. (2023). Digital twin: benefits, use cases, challenges, and opportunities. *Decision Analytics Journal* 100165.

18 Karaarslan, E. and Babiker, M. (2021). Digital twin security threats and countermeasures: an introduction. *2021 International Conference on Information Security and Cryptology (ISCTURKEY)*. IEEE.

19 Song, Q., Lei, S., Sun, W., and Zhang, Y. (2021). Adaptive federated learning for digital twin driven industrial Internet of Things. *2021 IEEE Wireless Communications and Networking Conference (WCNC)*. IEEE.

20 Lu, Y., Huang, X., Zhang, K. et al. (2020). Communication-efficient federated learning for digital twin edge networks in industrial IoT. *IEEE Transactions on Industrial Informatics* 17 (8): 5709–5718.

21 Sun, W., Xu, N., Wang, L. et al. (2020). Dynamic digital twin and federated learning with incentives for air-ground networks. *IEEE Transactions on Network Science and Engineering* 9 (1): 321–333.

22 Sun, W., Lian, S., Zhang, H., and Zhang, Y. (2022). Lightweight digital twin and federated learning with distributed incentive in air-ground 6G networks. *IEEE Transactions on Network Science and Engineering*.

23 Dang, H.V., Tatipamula, M., and Nguyen, H.X. (2021). Cloud-based digital twinning for structural health monitoring using deep learning. *IEEE Transactions on Industrial Informatics* 18 (6): 3820–3830.

24 Zhao, Y., Li, L., Liu, Y. et al. (2022). Communication-efficient federated learning for digital twin systems of industrial internet of things. *IFAC-PapersOnLine* 55 (2): 433–438.

25 Deebak, B.D. and Al-Turjman, F. (2022). Digital-twin assisted: Fault diagnosis using deep transfer learning for machining tool condition. *International Journal of Intelligent Systems* 37 (12): 10289–10316.

26 Altun, C., Tavli, B., and Yanikomeroglu, H. (2019). Liberalization of digital twins of IoT- enabled home appliances via blockchains and absolute ownership rights. *IEEE Communications Magazine* 57 (12): 65–71.

27 Jia, P., Wang, X., and Shen, X. (2020). Digital-twin-enabled intelligent distributed clock synchronization in industrial IoT systems. *IEEE Internet of Things Journal* 8 (6): 4548–4559.

28 Sasikumar, A., Vairavasundaram, S., Kotecha, K. et al. (2023). Blockchain-based trust mechanism for digital twin empowered industrial Internet of Things. *Future Generation Computer Systems* 141: 16–27.

29 Chen, H., Jeremiah, S.R., Lee, C., and Park, J.H. (2023). A digital twin-based heuristic multi-cooperation scheduling framework for smart manufacturing in IIoT environment. *Applied Sciences* 13 (3): 1440.

30 Feng, H., Chen, D., Lv, H., and Lv, Z. (2023). Game theory in network security for digital twins in industry. *Digital Communications and Networks* .

31 Janani, K. and Ramamoorthy, S. (2022). Integrated smart IoT infrastructure management using window blockchain and whale LSTM approaches. *Advances in Information Communication Technology and Computing: Proceedings of the AICTC 2021*. Singapore: Springer Nature Singapore. pp. 99-119.

32 Wright, S.A. (2023). IoT blockchains for digital twins. In: *Role of 6G Wireless Networks in AI and Blockchain-Based Applications*, 57–79. IGI Global.

33 Wang, Y., Su, Z., Guo, S. et al. (2023). A survey on digital twins: architecture, enabling technologies, security and privacy, and future prospects. *arXiv* preprint arXiv:2301.13350 .

34 Ali, A., Almaiah, M.A., Hajjej, F. et al. (2022). An industrial IoT- based blockchain-enabled secure searchable encryption approach for healthcare systems using neural network. *Sensors* 22 (2): 572.

35 Majid, M., Habib, S., Javed, A.R. et al. (2022). Applications of wireless sensor networks and internet of things frameworks in the industry revolution 4.0: a systematic literature review. *Sensors* 22 (6): 2087.

36 Janani, K. and Ramamoorthy, S. (2022). Threat analysis model to control IoT network routing attacks through deep learning approach. *Connection Science* 34 (1): 2714–2754.

37 Mauro, F. and Kana, A.A. (2023). Digital twin for ship life-cycle: a critical systematic review. *Ocean Engineering* 269: 113479.

38 Saad, A., Faddel, S., Youssef, T., and Mohammed, O.A. (2020). On the implementation of IoT-based digital twin for networked microgrids resiliency against cyber attacks. *IEEE Transactions on Smart Grid* 11 (6): 5138–5150.

39 Gehrmann, C. and Gunnarsson, M. (2019). A digital twin based industrial automation and control system security architecture. *IEEE Transactions on Industrial Informatics* 16 (1): 669–680.

40 Akbarian, F., Fitzgerald, E., and Kihl, M. (2020). Intrusion detection in digital twins for industrial control systems. *2020 International Conference on Software, Telecommunications and Computer Networks (SoftCOM)*. IEEE.

41 Xu, Q., Ali, S., and Yue, T. (2021). Digital twin-based anomaly detection in cyber-physical systems. *2021 14th IEEE Conference on Software Testing, Verification and Validation (ICST)*. IEEE.

42 Vakaruk, S., Mozo, A., Pastor, A. et al. (2021). A digital twin network for security training in 5G industrial environments. *2021 IEEE 1st International Conference on Digital Twins and Parallel Intelligence (DTPI)*. IEEE.

43 Janani, K. and Ramamoorthy, S. (2022). A secure multicontroller SDN blockchain model for IoT infrastructure. *Cyber Security, Privacy and Networking: Proceedings of the ICSPN 2021*. Singapore: Springer Nature Singapore. pp. 321-338.

44 Janani, K. and Ramamoorthy, S. (2023). A security framework to enhance IoT device identity and data access through blockchain consensus model. *Cluster Computing* 1–24.

3

Decision Support System for Digital Twin-based Smart Manufacturing Systems and Design in Industry 5.0

B. Akoramurthy[1], B. Surendiran[1], K. Dhivya[1], Subrata Chowdhury[2], Ramya Govindaraj[3], and Gautam Srivastava[4]

[1]*Department of CSE, National Institute of Technology Puducherry, Karaikal, Tamil Nadu, India*
[2]*Department of CSE, Sreenivasa Institute of Technology and Management Studies, Chittoor, Andhra Pradesh, India*
[3]*School of Computer Science Engineering and Information Systems (SCORE), Vellore Institute of Technology, Vellore, Tamil Nadu, India*
[4]*Department of Computer Science and Math, Lebanese American University, Beirut, Lebanon*

3.1 Introduction

While Industry 4.0 is still growing in popularity, the time has come for Industry 5.0 to shine. Industry 5.0, often known as the "Fifth Industrial Revolution," enables businesses to cater to individual consumers in accordance with their tastes and expectations. Large-scale customization levels of Industry 4.0 enlist the help of bots to carry out menial tasks, whereas Industry 5.0 aspires to conduct mass personalization with the aid of AI. With increased autonomy for collaborating robots, Industry 5.0 is predicted to radically alter the manufacturing process. As part of the current wave of industrialization, often known as Industry 5.0, robots will replace humans in many factory occupations. It is meant to make the most of people's innate ingenuity and intelligence. Swift advances in factory product development techniques, as well as the digitization and sophistication of product developing systems, are crucial as the sector moves from large-scale manufacturing to customized manufacturing.

Manufacturing procedures have become more intricate, programmed, and feasible since the beginning of the industrial age when water and steam-driven devices were the norms [1]. This has allowed machines to be operated with greater ease, efficiency, and longevity. Industry 5.0 facilitates this transition from mass production to mass personalization [2]. This is in response to the growing demand for products with a human touch in today's more impersonal marketplace. Since

Digital Twins in Industrial Production and Smart Manufacturing:
An Understanding of Principles, Enhancers, and Obstacles, First Edition. Edited by Rajesh Kumar Dhanaraj, Balamurugan Balusamy, Prithi Samuel, Ali Kashif Bashir, and Seifedine Kadry.
© 2024 The Institute of Electrical and Electronics Engineers, Inc. Published 2024 by John Wiley & Sons, Inc.

mass customization is now possible thanks to Industry 5.0 and rapid technological advancements in manufacturing, a fully digital and intelligent production system is a necessity [3]. Before Industry 4.0 came around, mass customization was triggered; nevertheless, this was insufficient. As an example, type 1 diabetes is challenging to control since individuals vary greatly in their metabolic rates, body size, skin elasticity, behavior patterns, and other aspects of their lifestyle. By moving to "Industry 5.0," we can provide individuals with apps that learn their routines and behaviors, which will lead to a lower, more covert, and dependable gadget designed specifically for them to use in managing their diabetes. Patients with diabetes would benefit tremendously from the development of an Industry 5.0 approach [4].

The following three crucial obstacles must be overcome immediately for safety management approaches in human-centered production on the path to Industry 5.0.

Problem 1: Traditional approaches to safety management are difficult to modify to meet changing circumstances. It is possible that safety management procedures that work well in one workshop will not be available in another. For instance, deep learning–based algorithms for detecting risky behavior need to employ datasets in accordance with various industrial contexts. The third obstacle is to develop risk management strategies that are adaptable and can be used in a wide variety of manufacturing settings.

Problem 2: Understanding the myriad ways in which people, machines, and the natural world all work together. Complicated electromechanical product factories are often vast in scale, and feature several workstations, employee behavior uncertainty, complicated settings, and high-risk procedures, all of which can contribute to safety mishaps [4]. Accidents like this can be avoided with the use of intelligent manufacturing systems that accurately represent human behavior, machine status, environmental factors, and their interplay.

Problem 3: Unsafe state causation. Machines do not comprehend and cannot detect dangerous situations caused by intricate interactions between beings, machinery, and the milieu. Computers with their associated technologies cannot forecast manufacturing safety mishaps in real situations as human beings do. The subsequent challenge, then, is a way to program industrial systems with human-level intelligence, meaningful reasoning, and the ability to report potentially life-threatening system states in a machine-understandable fashion.

Table 3.1 shows that digital twin technology (DTT) can solve problem 1 [5]. The factory is shown in both real and virtual forms by digital twin workshop (DTW). It optimizes these two viewpoints. Administration, design, and supervision of the factory's manufacturing procedure are all driven by the combined data [6]. Machine vision has the potential to replace human vision, prevent human

Table 3.1 Industry 5.0 workshop safety management problems.

Problems/obstacles	Remedies/solutions	Features of remedies
Flexibility/compatibility	A hybrid of digital twin and digital datasets	Virtual datasets provide the benefits of easy acquisition and wide use
Insights	Digital twin (DT)	The DTW allows for the actualization of the merger of the real factory, the digital factory, and the service system
Machine intelligence	Deep learning, machine learning, Artificial Neural Network (ANN)	The reliability of machine vision in identifying potentially dangerous situations is greatly improved by modern, deep learning-based algorithms

error in judgment and omission, and provide other benefits, such as simple implementation, adaptability, and low cost, making it an attractive solution to the second problem.

Computer vision's accuracy in detecting various dangers may be greatly improved by employing sophisticated deep learning techniques [7]. However, machine vision-based safety management is limited in its capacity to reason and learn from experience in the same way that humans do. Knowledge may be represented using semantic web technologies, and correlated data can share the same meaning [8].

Problem 3 may have some answers that may be gleaned via the usage of virtual datasets. Virtual datasets provide several advantages over physical ones, including easier collecting, fine-tuning, and automatic labeling [9]. In addition, DTW offers a wide range of virtual datasets amassed in the factory. Consequently, this article puts forward an analytical method for handling a factory's risky states dependent on the DT of the Fifth Industrial Revolution that incorporates digital-original fused datasets with taxonomy to realize semantic automatic reasoning of the factory's risky states, utilizing tried and true techniques for managing risk, such as the use of computer intelligence and taxonomies [10].

3.2 State of the Art

3.2.1 Applying Ontology to Risk Management

Semantic web technology [11] is gaining popularity because it can be used to convey a reason about meaning in a variety of contexts, making it ideal for research and applications involving representations of information. Semantic web technology relies on ontology, a formal system for defining the lexis of ideas used in a given realm and the meaningful interactions between defined [13]. The role of ontologies in risk management has been the subject of research by several academics. One such example is the [14] knowledge-based control system that relies on logic for optimal performance, which uses ontologies to provide smarter approaches to issues like energy and security management in commercial buildings. [15] offer structure care ontologies that may relate to building information modeling (BIM) and give an intellectual chance to assist in harmless and improved building mission implementation. Methods for Managing Dangers in the Building Sector are provided by Xing et al. [16], who draw taxonomy and enhanced strategies for reasoning with cases. To better characterize the objective world, Wang et al. [17] applies a strategy for building an incident taxonomy archetypal to the field of coal mine security.

Very few studies look at the use of ontologies to model workshop safety management, and the ontology that works for building sites may not be acceptable for factories. Furthermore, manual instance collection is still required in the present ontology modeling study of workplace safety management [18].

3.2.2 Computer Vision for Risk Management

Risks in the workplace's people, machines, and surroundings are the primary focus of machine vision–based factory safety management. As an illustration, Wang et al. [19] proposed a machine vision–based quick detection technique of workshop personnel DT model to accommodate the quantity and distribution of workers necessary in high-risk production workshops for the aerospace industry. Dandois and Ellis [20] demonstrate that high spatial resolution 3D measurements of vegetation structure and spectral characteristics can be produced by applying open-source computer vision algorithms to ordinary digital photographs acquired using inexpensive hobbyist aerial platforms. The concept will first be explained in creating and utilizing a DT of your well for drilling and how it will directly influence how drilling/well engineers, managers, and supervisors plan, and prepare and so on [21]. DES may be identified and measured using one of several available questionnaires, or objective evaluations of parameters such as critical flicker-fusion frequency, blink rate and completeness, accommodative function and pupil characteristics may be used to provide indices of visual fatigue [22]. Coles-Brennan et al. [23] aim to provide an overview of the extensive literature on digital eye strain research with particular reference to the clinical management of symptoms. Hughes [24] discuss the approach of linking the simulation model to physical systems to achieve the creation of a DT factory. Lepikhin et al. [25] discuss the main methodological aspects of risk analysis of technical systems using DTs. The physical demands of military service place soldiers at risk of musculoskeletal injuries and are major concerns for military capability. Lloyd et al. [26] outline the development of new training technologies to prevent and manage these injuries. Other influential work includes Digital Twins-based Risk Management [27–29].

Based on these findings, it is clear that DT is committed to developing secure industrial facilities. Despite the potential of machine vision and ontology to increase human and machine environmental safety, relatively little research has been done in these areas.

3.3 The Suggested Design

The four main components of the suggested techniques are depicted in Figure 3.1. A real-world workshop a digital workshop, the risky circumstances of the workshop's ontology approach, and the acquisition of instances using a hybrid dataset of digital and physical data.

3.3.1 A Real-World Workshop

The manufacturing workshop serves as the study's object, and its dangerous statuses feed both the ontology model and the raw data. Dangerous circumstances may be broken down into numerous categories depending on the type of manufacturing site they are present at, such as dangerous conduct, lopsided attire, hazardous maneuvers, and an insecure milieu. As can be seen in Table 3.2, each dangerous condition is associated with a unique combination of content, probable manufacturing accidents, and safety control measures.

3.3.2 A Digital Workshop

A computer representation of a real workshop, or "Digital workshop," can be used to create an identical copy in terms of shape, behavior, and surroundings. In this article, the virtual workshop is primarily a geometric representation of production models, an environmental restoration of the production space, and a behavioral simulation of potentially dangerous conditions. As a result, as stated in Sections 3.5 and 3.6, the digital factory provides deep learning algorithms with exceptional synthetic data.

3.3.3 Risky Circumstances of the Workshop's Ontology Approach

As shown in Table 3.2, the taxonomy design is a meaningful representation of the ideas, attributes, connections, adages, rules of thought, and added information

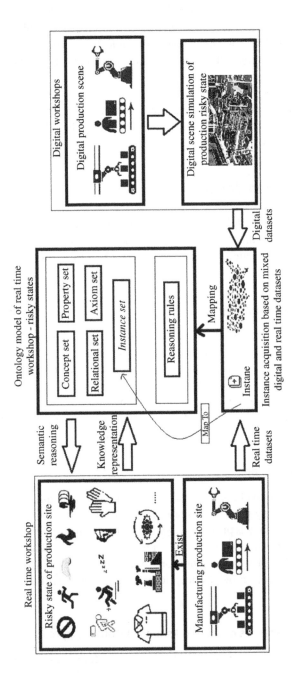

Figure 3.1 Exhibits the proposed simulation model which encompasses four components such as 1. A real-world workshop, 2. A digital workshop, 3. the Risky circumstances of the workshop's ontology approach, and 4. the acquisition of instances using a hybrid dataset of digital and physical data.

Table 3.2 Categorization and detailed information about dangerous workshop conditions.

Tier	Name	Description	Unforeseen production mishaps	Methods for managing and regulating risk
Risky atmosphere (RA)	Blockage by strange objects, the flow of a dangerous liquid, fire accident	Unauthorized people and equipment incursions are prohibited. The manufacturing workshop had a leak of a potentially dangerous liquid. Factory fire	Accidental run-ins with factory workers or machinery, toxins, flames, explosions, etc., and disasters like an explosion or fire	Informing the workshop manager to come to check things out. Following protocol, we must evacuate the building in an orderly fashion
Risky operations (RO)	Lack of backup. Getting close to the edge of danger	Only one worker at the workshop regularly uses the floor as a climbing surface	A really large drop. Accidental run-ins with machinery used in production, etc.	Calling in the factory supervisor to check things out and issue a public alert
		Workers in a manufacturing facility risk injury by approaching an operating arm		Send out alerts, reduce equipment speeds, or shut it down completely based on how dangerous the situation is
Risky conduct (RC)	Accident at the factory, in the middle of the garage, rushing inside the garage, and conversations that go on for hours over the phone	The workshop employees had trips and falls. The personnel of the workplace are known for their high-speed running	Concussion, fall, and manufacturing unit crash accident, injury from production machinery	Notifying the supervisor, who will then go to the scene and notify the medical personnel. Send forth a little nudge

regarding various dangerous situations in a workshop. We can automatically infer persons engaged, potential manufacturing mishaps, and appropriate safety control measures, as well as the existing dangerous condition.

3.3.4 Acquiring Instances Using a Hybrid Dataset of Digital and Physical Data

Using the digital-real mixed dataset, this section trains a deep learning model that detects instances in dangerous situations, which are then used as ontology instances to propel semantic reasoning.

3.4 Inference Engine Framework

In this part, we lay the groundwork for semantic deduction by establishing the Workshop Risky Situation Ontology (OWRS). Due to a large number of symbols, this article makes use of italics to denote the idea, an open subset ⊚ that explains the connection or attribute of the notion, a white concave-sided diamond to represent the group, ◇, and a white concave-sided diamond with a rightward sideline to specify the items in a group.

To paraphrase (3.1), "OWRS is a group of notion, Association groups, Attributes groups, postulate groups, deductive rules groups, and events groups about the risky state of a Factory." This includes concepts like the casting procedure, fabricating, kinetic processing, welding process, and assembly.

$$O_{WRS} = \left\{ DEP_{WRS} + N_{WRS} + N^A{}_{WRS} + N^A{}_{WRS} \right\} \tag{3.1}$$

In OWRS, where WRS serves as the central idea, NWRS stands for the set of related concepts. A WUS can be classified as either risky conduct (RC), risky operations (RO), or a risky atmosphere (RA), as shown in Table 3.2. Workshop workers (WW), workstations (W), manufacturing machinery (MM), probable workplace incidents (PWA), and associated treatments (T) are all included in NWRS in addition to the fundamental kinds. The NWRS definition is (3.2).

$$N_{WRS} = \{WRS, WW, W, MM, T, PWA\} \tag{3.2}$$

With the use of OWL2's object attributes, NAWRS delivers OWRS's "associated" logical relationship set of notion instances. A portion of $N^A{}_{WRS}$ is based on fundamental OWL2 relations like the "subClassOf" inheritance relation. The other is made to order to account for the hazardous conditions at the factory where it is made. The "cause peril" connection, for instance, can be utilized to identify the WRS that may have triggered a manufacturing mishap.

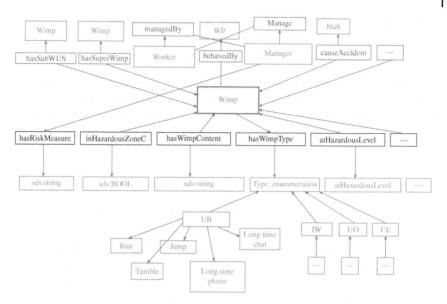

Figure 3.2 O_{WRS} and its associations.

The "behaved" connection can stand in for the workers in a given WRS. The "hasRiskTreatment" connection reveals the processing strategy that goes along with a WUS. The third clause gives the official definition of N^A_{WRS}.

$$N^A_{WRS} = \{ hasSubWRS, hasSuperWRS, behavedBy,$$
$$causePeril, hasRiskTreatment \dots \} \tag{3.3}$$

N^A_{WRS} represents in OWL2 [30] the set of connections between concept instances and data kinds (termed "properties") in OWRS. Equation (3.4) provides a formal description of NAWRS, and the "hasWRSType" attribute links a WRS to its type; for example, "uncertain objects invasion" is a member itself of the group defined.

The "hasWRSContent" feature connects a WRS with its explanation, providing context for the WRS's language of nature. When referring to the foundry, the IW of "no safety helmets" might mean that employees are not wearing the mandatory yellow helmets for protection. Furthermore, "atHazardousLevel" may be used to describe the danger level of each WRS as an integer that is positive (Figure 3.2).

$$N^A_{WRS} = \{ hasWRSContent, atHazardousLevel, hasWRSType \} \tag{3.4}$$

3.5 A Digital-Original Dataset Amalgamation

3.5.1 Recreation of Digital Workshop for Risky States

To acquire premium digital datasets and to train highly rated deep learning models, a high-fidelity digital factory is needed. A risky digital workshop must be constructed from the components', behaviors', and rules' points of view in accordance with Wu et al. suggested the digital factory's 5D twin model. The graphical representations of people, machines, and materials, as well as the modeling of the manufacturing setting, including the light setting, are all part of the simulated workshop's dangerous condition from an element's perspective. When a simulated production environment takes into account a hazardous situation, the resulting dynamic simulation model may be seen from the perspective of the state's behavior, which depicts the state's sequence, concurrency, and linking. Based on this understanding of safe production practices, a rule-based digital factory may develop a taxonomy design that complies with the norms for risk management governing the entire manufacturing process. The following sections provide elaborate discussions on all the aforementioned topics.

3.5.2 Digital Workshop Lighting and Texture Modeling

The model of the production equipment is created using parametric modeling, which not only guarantees geometry and assemblage correctness along with it but also allows mobility and communication between humans and machines. Furthermore, it is critical to represent a virtual person and provide the relevant behavior, as the factory crew is the main identification item of the dangerous condition. More importantly, the effect of object detection and the accuracy of virtual datasets are both affected by the similarity between the virtual and real lighting environments. So, in this part, we will talk about how to model a virtual workshop with high fidelity, paying special attention to details like lighting and textures (see Figure 3.3).

The classic lighting model is often computed for available light, soft light, and target light, and is typically broken down into the Lambertian reflectance, the Specular reflection model, and the modified Phong reflection model. Although the Lambertian reflectance model can demonstrate certain surface roughness phenomena, it is not well suited for metals due to its lack of target light control. The modified Phong reflection model improves upon the Phong model that utilizes a separate collection of vectors, which is always smaller than 90° in order to do calculations. However, modified Phong reflection is a physical-experience-based lighting model that does not fulfill the first law of thermodynamics and has negative effects on character model detection in low-light settings.

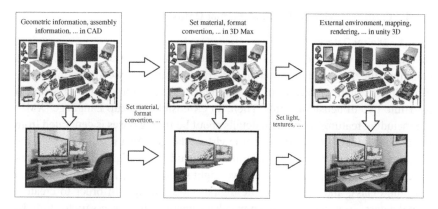

Figure 3.3 Simulated workshops with strong-fidelity lighting and texturing.

Therefore, the physically based shading (PBR/S) method is utilized, and the reflection formula is written as in the below-mentioned equation to achieve a more lifelike appearance. The rendering equation describes the global lighting of a scene. This equation is modified to be utilized in PBR; this form, known as the reflectance equation, is used for this purpose.

The basic form of the reflectance equation for dark surfaces looks like this:

$$L_0(p, \omega_0) = L_e(p, \omega_0) + L_r(p, \omega_0) \tag{3.5}$$

This formula explains the quantity of light radiating from a surface point in a given direction. The radiance of a point is the quantity of light emitted from that location.

3.5.3 Analyzing Risky Conditions in a Workshop

When the digital workplace model has been built, the final step in the dynamic rendering of the risky state is to construct the required behavior model in accordance with the real dangerous situation of the real-world factory. In this study, the actual hardware driver is examined to determine the best gesture commands to utilize while creating a strong-fidelity digital workshop simulation. When using Unity3D, the model's motion may be governed by a wide variety of commands. Scripts must be written to teach models how to move in each frame or second. Once the code is loaded into the target object, the script is executed at the appropriate time. There are many different types of controlling elements in Unity 3D; we are ignoring to mention the names here since almost every controller is used up for the progression. The motion of the workshop staff model, for instance, is managed by means of the animator controller components.

3.5.4 Creating a Hybrid Original-Virtual Dataset

3.5.4.1 Digital Dataset

In Figure 3.4, we see the steps required to produce a virtual dataset. Before obtaining the type and position of the dynamic model, we first convert the three-dimensional renderings of the simulated environment created in Segment 3.5.1 into two-dimensional stills. Then, an eXtensible Markup Language (XML) file including features along with details about bounding boxes is generated from the digitized data of each image in the virtual dataset. As a result, the XML file contains a wealth of data for model training, including the digital dataset picture's upper left side (i.e., a_{min}, b_{min}) and lower right side (i.e., a_{max}, b_{max}), the intended attribute's class, and the size of the digital dataset image. In order to create a digital dataset from the labeling knowledge generated, it is necessary to store the picture with the same name as the top-level element comments stored file. In the simulated workshop, prepping the digital datasets can improve the resilience and variety of the supervised ML used in the original factory. As an illustration, the script may be used to manage the digital camera's location so that it is identical to that in the actual workshop, and the brightness of the digital workshop can be modified by turning up or down the source of illumination.

3.5.4.2 Genuine Dataset

Because of the dynamic nature of the workshop's human-machine-environment states, as well as the fluidity of the work environment and production circumstances, the model may tolerate some variance from the real-life production scene in regard to visualization, precision, and so on. Consequently, the overfitting constraints, the effect on generalization capabilities of the preliminary drill system on the test group, and the poor discovery rate all derive from using a simulated dataset for conditioning the network.

Since this might lead to overfitting, it is best to combine a digital dataset with a real one to create a simulated-real mixed dataset. To get a true representation of the dataset, we are utilizing the live feed from the workshop security camera. The captured picture dataset is then labeled, and the image's target details are highlighted in accordance with the various danger levels.

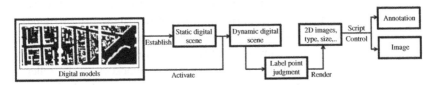

Figure 3.4 Infrastructure for the generation of simulated datasets.

3.5.4.3   Assigning an Ontology Term to a Detected Object

The deep learning network is trained using the blended dataset of synthetic and real-world data that was constructed in Segment 3.5.2. Factory employees, unsafe situations kinds, worker locations, manufacturing machines, and other events linked with a risky workshop setting may all be identified using a skilled network. In order to put into action the semantic reasoning rules, it is necessary to first map the instances discovered by the deep learning system to the situation I_{WRS} belonging to OWRS created in Segment 3.4 (inference engine Framework). As a means of exchanging information, XML has gained widespread support. As a result, we use XML to export the discovered instance data from the hybrid dataset. Many different mapping techniques have been offered. Some of these efforts focused primarily on a broad linkage between XML and RDF.

Some try to translate document structure to semantic web language (SWL) deprived of taking instances of XML data document into account. To do this, the study proposes a system that allows for the mapping of XML with created XML schema to the OWRS designed with SWL instances containing all the necessary data.

3.6   Discussion and Demo (DD)

3.6.1   Ontology Construction Using eNanoMapper Slimmer

As can be seen in Figure 3.5, the eNanoMapper slimmer tool is used to generate an OWRS, complete with class, object, and data type attributes, that corresponds to the OWRS of the risky state of the factory-making spot established in Segment 4. Further, the SWRL reasoning rules described in OWRS are crafted with the use of the rule editor in eNanoMapper Slimmer. The axioms for classes, properties, and relations are imported into eNanoMapper Slimmer from OWRS.

3.6.2   Construction of a Photorealistic Simulated Environment for DTW

Figure 3.6 depicts a demonstration of the suggested method. Solidworks, 3DMax, and Unity3D are used in tandem with the designing and virtual reality approach for creating the same as the original digital scene of the factory at 1:1. Workshop personnel are imported as dynamic models into Unity3D, while static models include manufacturing automatons, resources, components, and illumination paraphernalia. When it comes to original illumination, Unity 3D enhances the original illumination model based on Bi-directional Reflectance Distribution Function (BDRF) which boosts the intensity of the lighting. When it comes to indirect lighting, Unity 3D has your back thanks to its support for shadows, ambient lighting, and even multiple light sources.

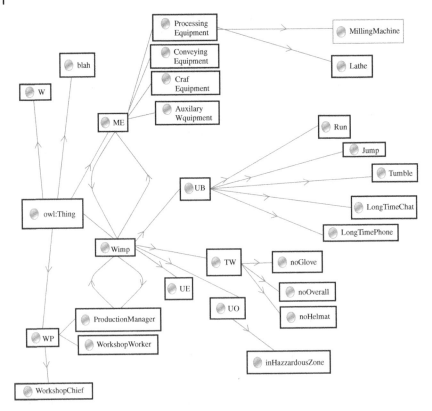

Figure 3.5 Creation of O_{WRS} using eNanoMapper.

3.6.3 Generation of Hybrid Synthetic-Real Datasets

Figure 3.7 depicts the steps taken to create the mixed-reality virtual dataset. The following are the individual stages of constructing a virtual dataset:

1. Using the simulated laboratory setup described in Section 3.5.2, a virtual dataset may be retrieved. To move and take pictures with virtual cameras in Unity 3D, developers use C# scripts to activate static settings. As a result, the virtual scene's 3D model is converted into a 2D picture, and data about the image's dimensions, the model's kind, and its position are recorded. Next, an XML file, complete with metadata on the dataset's boundaries and the categories into which they fall, is generated digitally.
2. Checking the integrity of digital data collections. The virtual dataset is culled from low-quality pictures and annotation files using the no-reference image quality assessment (NR-IQA) described in the previous section aims to make an estimate of how well a picture will look to human viewers without having access to any perfect examples.

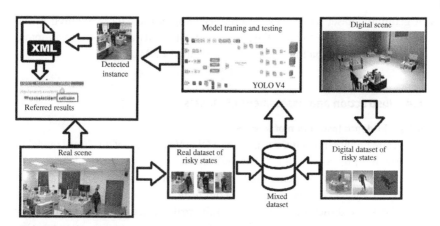

Figure 3.6 Computer-generated setting and presentation method.

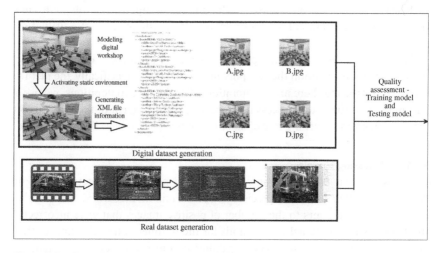

Figure 3.7 Digital-original mixed dataset production.

The following are the distinct procedures involved in the actual development of a dataset:

- Datasets obtained from live camera footage are authentic. Adobe Premiere Pro is used to edit the footage of dangerous situations such that it is cut together into segments that correspond to the categories in Table 3.2. Real datasets may be easily generated by segmenting recorded videos according to their type.
- Gathering photos from an actual dataset. At 5-second intervals, the actual dataset video feed frames the real workshop scene. Section 3.5.2 details the assessment technique used to determine which photographs should be discarded due to their low quality.

- A genuine dataset with labels. Manual labeling of the data is required in the actual world, unlike the simulated one. Labeling software is used to annotate the actual dataset's picture in order to generate the XML file.

3.6.4 Instruction and Assessment of Models

3.6.4.1 In-depth Instruction of Models

To train and analyze the data, we used the YOLO V4 [33] target discovery model in the PyTorch framework because of its high real-time performance and the need to recognize tiny targets. We generated a range of various amounts of mixed datasets to see how they affected the outcomes of our tests. The model is pretrained on the mixed dataset and then put to the test on the real dataset. Over the course of 10 iterations of training, 500 original photos are used to represent a constant training set, and the digital dataset is expanded by k times ($k = 1, 2, 3..$). Simultaneously, the fused dataset (a new one) is used as the training group, whereas all photos not found in the actual dataset library's preset training group make up the evaluation group.

The values that follow are the parameters for the synthetic dataset's training that take into consideration the components' varied sizes. Specifically, the picture size is 416×416, and there are 100 iterations. For the first 50 iterations, we use a learning rate of 0.001 and process four photos at a time. Each of the previous 50 iterations has had a learning rate of 0.0001, and two photos have been processed.

3.6.4.2 Assessing Models

Model assessment has been done based on the performance metrics used in data analysis such as recall, precision, and so on. In this work, we used "α error" term which refers to the number of positive images that were incorrectly prophesied, whereas "sensitivity" stands for the number of negative images that were incorrectly anticipated. The reliability of the overall detection findings is determined by two sets of numbers in which the first one is True positive rate serves as the X-axis, while the precision serves as the Y-axis in this case. It is possible to calculate the precision-recall (PR) curve, which shows the link between the two metrics. The PR curve allows one to determine the weighted mean of the precision scores by summing the points along the curve. If the PR region is large, then the healthier is the trained model's detection capacity. Using the false positive per image (FPPI), we took a representative sample of nine points over the assessment array; also, the overall failure rate was calculated as the mean of these nine. For curves that flatten down before reaching the specified FPPI rate, the minimum miss rate is determined. To name a few: Each image has an average false positive detection rate or FPPI. It presumes that N pictures exist and that FP false positives exist in the results.

The F1-Score, often called the Balanced F-Score, is the mathematical arithmetic mean (Σ) of the quantity measures and how accurate the values are. The number can take on the range of -1 to $+1$, with $+1$ indicating the model's optimal performance and -1 indicating its poorest performance. As the formula for determining the F1-Score is universally known, we neglect to mention the paper.

3.6.4.3 Verdicts from Model Training

Test outcomes are obtained in accordance with the training model and evaluation indices after putting the model that was trained on a new dataset to see how it performs. The greatest F1-Score is achieved at a 3:1 virtual-to-real picture ratio, and then gradually drops as more virtual data is added (see Figure 3.8). When the ratio of virtual to actual photos is 2 to 1, the workshop staff's logarithmic average missed detection rate is at its best, and it rises gradually as the number of virtual data points grows (Table 3.3).

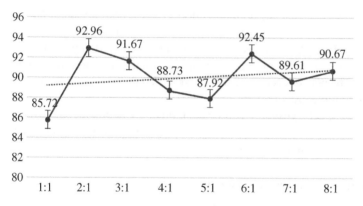

Figure 3.8 Evaluations of the model-in-training (MIT).

Table 3.3 Metrics and evaluations of the model-in-training (MIT).

Digital:original	Precision	Recall	F1−score	OD metric
400:400	85.72	86.94	85.78	84.78
800:400	92.96	93.65	92.45	91.89
1200:400	91.67	95.45	94.78	93.67
1600:400	88.73	92.78	91.43	89.7
2000:400	87.92	90.56	89.56	88.67
2400:400	92.45	95.34	94.78	93.56
2800:400	89.61	90.56	89.37	88.76
3200:400	90.67	91.45	90.45	89.67

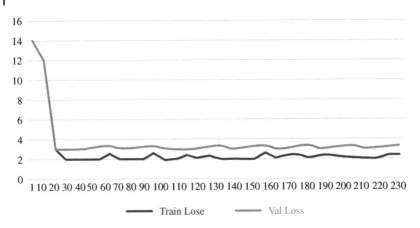

Figure 3.9 The loss curve.

The model tests show that the optimum effect is achieved at a ratio of 3:1 for the simulated picture to the original picture and the object-detecting metric is maximized at a ratio of 2:1. It is possible that the model's detection performance in real-world settings will deteriorate if only the percentage of simulated datasets keeps growing. The network is trained using 1200 different datasets. In Figure 3.9 we can see the training lost with time.

As a result, we select a virtual-to-real image ratio of 3:1 to provide the highest feasible percentage of digital datasets with minimal impact on the simulation's actual capacity to detect abnormalities.

3.6.5 Insights into Semantics Reasoning

Detection of the real-world workshop makes use of the training model, as shown in Figure 3.10. Every 1.5 seconds approximately (it would take 300 frames, or 10 seconds, to depict an event), the factory's state is sent to an XML file, where it may be analyzed for signs of dangerous behavior, improper attire, and so on. An unsafe state is outputted, for instance, when a hazard zone is detected, and the level of the hazard determines which hazardous state is communicated. Once the danger level reaches 3, as depicted by the yellow region, the XML file is generated mechanically.

A mapping from the discovered hazardous state instances in XML format is exported and imported into an OWRS ontology instance. Multiple linked imports are a significant challenge for an intricate taxonomy like eNanoMapper because they introduce redundant material. If a worker in a workplace is running, as seen in Figure 3.10, it is assumed that they are engaging in risky behaviors and have a "running" type WRS, which increases the potential for a "collision" to occur between the worker and the robot.

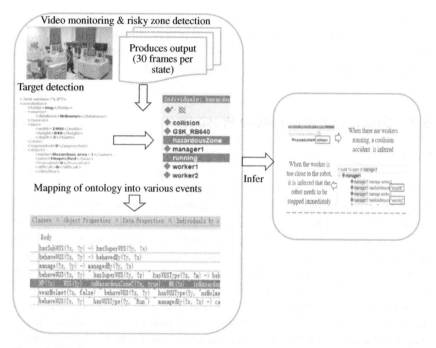

Figure 3.10 Ontology inference and risky states detection.

3.6.6 Discussions

We embrace a combination of digital-original blended datasets and taxonomy to identify the factory's risky state while providing information regarding the automatic logic of safeness and its assessment, all with the goal of bettering the empathetic human experience that has traditionally underpinned the administration of factory sites during production. We construct OWRS in eNanoMapper using the DT architecture to accurately and explicitly record the workshop's precarious condition and connectivity. We are able to swiftly and accurately identify a variety of harmful scenarios in the workshop by using the YOLO target detection network in place of human detection and by linking identified cases to the taxonomy for meaningful interpretation. At last, two potentially hazardous scenarios—"running" and "entering the danger zone"—are shown to illustrate the production safety mishaps that may occur and the operation parameters that factory management should be considered for the risk-free environment.

Additionally, the original dataset is not used, but rather the fused dataset is used for both schooling and examining the deep learning model. Our virtual workshop allows us to rapidly and automatically configure the potentially dangerous work environment and record data about it. In addition, we showed that the best impact occurred when the virtual dataset to actual dataset ratio was 3:1 by training and testing the deep learning model. Overfitting has been eliminated, and the reliability of the test findings has been assured because at present the virtual dataset accounts for the greatest possible percentage of the total.

3.7 Conclusion and Future Prospects

We have suggested DT-based risk-free management systems for industrial processes as an improvement over more subjective forms of safety management.

Insight, device comprehensibility, and agility are the three major obstacles to information-driven, person-machine environmental safety convergence necessary for Industry 5.0. This research proposes a DT-based technique for reasoning about workshops' dangerous conditions and builds a semantic cognitive context by using digital-original fused datasets and also taxonomies to address these difficulties. With its informal formulation, the WRS ontology discussed in this study is able to tackle the problem of machine comprehensibility. To supplement the original dataset used for schooling and examining the system for identifying targets, we model the hi-fi digital scene of DT, which can adaptably replicate different workshop hazardous conditions and record the digital information set. Importantly, we substitute the original dataset with the digital-original fused dataset and try several configurations to see what works best. There is, nevertheless, room for development in the approach. The O_{WRS} is incomplete; further work is to construct a more exhaustive model of taxonomy using higher fine-grained principles of logic that may be used in all possible workshop production locations.

O_{WRS} instances can be detected automatically, but mapping them to the appropriate ontology instance is done by hand, which adds extra effort and slows down real-time performance. In this research, we use YOLO V4, a deep learning network that, without further context, can only recognize a single scenario of employees and equipment. This opens the door to the employment of multi-target detection and instance segmentation techniques. When eNanoMapper infers possible safety incidents and the accompanying treatment actions, further work has to be done on a response from the DT system. We plan to investigate these restrictions more in the future.

References

1 Zhang, H., Qi, Q., and Tao, F. (2022). A multi-scale modeling method for digital twin shop-floor. *Journal of Manufacturing Systems* 62: 417–428. https://doi.org/10.1016/j.jmsy.2021.12.011.

2 Demir, K.A., Döven, G., and Sezen, B. (2019). Industry 5.0 and human-robot co-working. *Procedia Computer Science* 158: 688–695. ISSN 1877-0509, https://doi.org/10.1016/j.procs.2019.09.104.

3 Sachsenmeier, P. (2016). Industry 5.0—the relevance and implications of bionics and synthetic biology. *Engineering* 2: 225–229.

4 Adel, A. (2022). Future of industry 5.0 in society: human-centric solutions, challenges and prospective research areas. *Journal of Cloud Computing* 11: 40. https://doi.org/10.1186/s13677-022-00314-5.

5 Ge, S., Zhang, F., Wang, S., and Wang, Z. (2020). Digital twin for smart coal MiningWorkface: technological frame and construction. *Journal of China Coal Society* 45 (6): 1925–1936. Available at: https://doi.org/10.13225/j.cnki.jccs.ZN20.0327.

6 Li, H., Liu, G., Wen, X.Y. et al. (2021). Industrial safety control system and key technologies of digital twin system oriented to human machine interaction. *Computer Integrated Manufacturing Systems* 27 (2): 374–389. https://doi.org/10.13196/j. cims.2021.02.006.

7 Angjeliu, G., Coronelli, D., and Cardani, G. (2020). Development of the simulation model for Digital Twin applications in historical masonry buildings: The integration between numerical and experimental reality. *Computers and Structures* 238: 106282.

8 Ruzsa, C. (2021). Digital twin technology - external data resources in creating the model and classification of different digital twin types in manufacturing. *Procedia Manufacturing* 54: 209–215. ISSN 2351-9789, https://doi.org/10.1016/j.promfg.2021.07.032.

9 Grieves, M. and Vickers, J. (2017). Digital twin: mitigating unpredictable undesirable emergent behavior in complex systems. In: *Transdisciplinary Perspectives on Complex Systems: New Findings and Approaches*, 85–113. Cham, Switzerland: Springer International https://doi.org/10.1007/978-3-319-38756-7_4.

10 Xiang, L., Jinsong, D., Xiaolong, W. et al. (2019). Research on digital twin technology for production line design and simulation marine science and engineering. 9 (3): 3–64. https://doi.org/10.3390/jmse9030338.

11 Lan, G., Liu, T., Wang, X. et al. (2022). A semantic web technology index. *Scientific Reports* 12: 3672. https://doi.org/10.1038/s41598-022-07615-4.

12 Wanasinghe, T.R., Wroblewski, L., Petersen, B.K. et al. (2020). Digital twin for the oil and gas industry: Overview research trends opportunities and challenges. *IEEE Access* 8: 104175–104197.

13 Jiang, X., Wang, S., Wang, J. et al. (2020). A decision method for construction safety risk management based on ontology and improved CBR: example of a subway project. *International Journal of Environmental Research and Public Health* 17 (11): 3928. https://doi.org/10.3390/ijerph17113928.

14 Zolhavarieh, S., Parry, D., and Bai, Q. (2017). Issues associated with the use of semantic web technology in knowledge acquisition for clinical decision support systems: systematic review of the literature. *JMIR Medical Informatics* 5: e18.

15 Ding, L.Y., Zhong, B.T., Wu, S., and Luo, H.B. (2016). Construction risk knowledge management in BIM using ontology and semantic web technology. *Safety Science* 87: 202–213. ISSN 0925-7535 https://doi.org/10.1016/j.ssci.2016 .04.008.

16 Xing, X., Zhong, B., Luo, H. et al. (2019). Ontology for safety risk identification in metro construction. *Computers in Industry* 109: 14–30.

17 Wang, J., Huang, Y., Zhai, W. et al. (2023). Research on coal mine safety management based on digital twin. *Heliyon* 9 (3): e13608. ISSN 2405-8440, https://doi.org/10.1016/j.heliyon.2023.e13608.

18 Su, Y., Yang, S., Liu, K. et al. (2019). Developing a case-based reasoning model for safety accident pre-control and decision making in the construction industry. *International Journal of Environmental Research and Public Health* 16: 1511.

19 Wang, S., Rodgers, C., Zhai, G. et al. (2022). A graphics-based digital twin framework for computer vision-based post-earthquake structural inspection and evaluation using unmanned aerial vehicles. *Journal of Infrastructure Intelligence and Resilience* 1 (1): 100003. ISSN 2772-9915, https://doi.org/10 .1016/j.iintel.2022.100003.

20 Dandois, J.P. and Ellis, E.C. (2010). Remote sensing of vegetation structure using computer vision. *Remote Sensing*. (IF: 5).

21 Cunha-Vaz, J., Bernardes, R., Santos, T. et al. (2012). Computer-Aided Detection of Diabetic Retinopathy Progression. (IF: 3).

22 Nadhan, D., Mayani, M.G., and Rommetveit, R. (2018). Drilling with Digital Twins, Day 1 Mon, August 27, 2018.

23 Sheppard, A.L. and Wolffsohn, J.S. (2018). Digital eye strain: prevalence, measurement and amelioration. *BMJ Open Ophthalmology*. (IF: 5).

24 Coles-Brennan, C., Sulley, A., and Young, G. (2018). Management of Digital Eye Strain. *Clinical & Experimental Optometry* (IF: 4).

25 Ivanov, D., Dolgui, A., Das, A., and Sokolov, B. (2019). Digital supply chain twins: managing the ripple effect, resilience, and disruption risks by data-driven optimization, simulation, and visibility. In: *Handbook of Ripple Effects In The Supply Chain*. (IF: 3).

26 Hughes, R. (2019). *Virtual Simulation Model of the New Boeing Sheffield Facility*. Springer Series In Advanced Manufacturing.

27 Lepikhin, A.M., Makhutov, N.A., Shokin, Y.I., and Yurchenko, A.V. (2020). Analysis of risk concept for technical systems using digital twins. *Computing Technologies* 25 (4): 99–113.

28 Agha, H., Geng, Y., Ma, X. et al. (2022). Unclonable human-invisible machine vision markers leveraging the omnidirectional chiral Bragg diffraction of cholesteric spherical reflectors. *Light, Science & Applications* 11: 309.

29 Lloyd, D.G., Saxby, D.J., Pizzolato, C. et al. (2023). Maintaining soldier musculoskeletal health using personalised digital humans, wearables and/or computer vision. *Journal of Science and Medicine in Sport* 4: 123.

30 Wu, C., Zhou, Y., Pessôa, M.V.P. et al. (2021). Conceptual digital twin modeling based on an integrated five-dimensional framework and TRIZ function model. *Journal of Manufacturing Systems* 58: 79–93. ISSN 0278-6125, https://doi.org/10.1016/j.jmsy.2020.07.006.

31 Bilberg, A. and Malik, A.A. (2019). Digital twin driven human–robot collaborative assembly. *CIRP Annals* 68: 499–502.

32 Barricelli, B.R., Casiraghi, E., and Fogli, D. (2019). A survey on digital twin: definitions characteristics applications and design implications. *IEEE Access* 7: 167653–167671.

33 Tao, F., Sui, F., Liu, A. et al. (2019). Digital twin-driven product design framework. *International Journal of Production Research* 57 (12): 3935–3953.

34 Jeon, S.M. and Schuesslbauer, S. (2020). Digital twin application for production optimization. *Proceedings of the IEEE International Conference on Industrial Engineering and Engineering Management (IEEM)*. pp. 542–545.

35 Mandolla, C., Petruzzelli, A.M., Percoco, G., and Urbinati, A. (2019). Building a digital twin for additive manufacturing through the exploitation of blockchain: a case analysis of the aircraft industry. *Computers in Industry* 109: 134–152.

36 Liu, Z., Meyendorf, N. and Mrad, N. (2018). The role of data fusion in predictive maintenance using digital twin. *Proceedings of the Annual Review of Progress in Quantitative Nondestructive Evaluation*.

37 Bianconi, C., Bonci, A., Monteriù, A. et al. (2020). System thinking approach for digital twin analysis. *Proceedings of the IEEE ICE/ITMC* 1–7.

38 Ding, Y. and Foo, S. (2002). Ontology research and development. Part 1-A. *Review of Ontology and Gerantology* 28 (2): 123–136. https://doi.org/10.1177/016555150202800204.

39 Zhang, J., Li, L., Lin, G. et al. (2020). Cyber resilience in healthcare digital twin on lung cancer. *IEEE Access* 8: 201900–201913.

40 Minerva, R., Lee, G.M., and Crespi, N. (2020). Digital twin in the IoT context: a survey on technical features scenarios and architectural models. *Proceedings of the IEEE* 108 (10): 1785–1824.

41 Lu, Y., Zhang, H., Chand, S. et al. (2022). Outlook on human-centric manufacturing towards Industry 5.0. *Journal of Manufacturing Systems* 62: 612–627. https://doi.org/10.1016/j.jmsy.2022.02.001.

42 Tao, F., Zhang, H., Liu, A., and Nee, A.Y.C. (2018). Digital twin in industry: state-of-the-art. *IEEE Transactions on Industrial Informatics* 15 (4): 2405–2415. https://doi.org/10.1109/TII.2018.2873186.

43 Vrabič, R., Erkoyuncu, J.A., Butala, P., and Roy, R. (2018). Digital twins: understanding the added value of integrated models for through-life engineering services. *Procedia Manufacturing* 16: 139–146.

44 Maskooni, E.K., Naghibi, S.A., Hashemi, H., and Berndtsson, R. (2020). Application of advanced machine learning algorithms to assess groundwater potential using remote sensing-derived data. *Remote Sensing* 12 (17): 1–25.

45 Wu, H., Zhong, B., Li, H. et al. (2021). Combining computer vision with semantic reasoning for on-site safety management in construction. *Journal of Building Engineering* 42: 103036. https://doi.org/10.1016/j.jobe.2021.103036.

46 Pires, F., Cachada, A., Barbosa, J., et al. (2019). Digital twin in industry 4.0: Technologies applications and challenges. *Proceedings of the IEEE 17th International Conference on Industrial Informatics (INDIN)*, vol. 1. pp. 721–726.

47 He, B., Cao, X., and Hua, Y. (2021). Data fusion-based sustainable Digital Twin system of intelligent detection robotics. *Journal of Cleaner Production* 280: 124181.

48 Mohammadi, A.F., Rezayati, M., Venn, H.W., and Karimpour, H. (2020). A mixed-perception approach for safe human–robot collaboration in industrial automation. *Sensors* 20 (21): 6347. https://doi.org/10.3390/s20216347.

49 Schroeder, G.N., Steinmetz, C., Rodrigues, R.N. et al. (2021). A methodology for digital twin modeling and deployment for industry 4.0. *Proceedings of the IEEE* 109 (4): 556–567.

50 Bohring, H. and Auer, S. (2015). Mapping XML to OWL ontologies. Marktplatz Internet: Von e-Learning bis e-Payment. 13 Leipz Inform-Tage (LIT 2005). pp. 147–156.

51 Brosinsky, C., Westermann, D., and Krebs, R. (2018). Recent and prospe ctive developments in power system control centers: adapting the digital twin technology for application in power system control centers. *Proceedings of the IEEE International Energy Conference (ENERGYCON)*. pp. 1–6.

52 Moyne, J., Qamsane, Y., Balta, E.C. et al. (2020). A requirements driven digital twin framework: Specification and opportunities. *IEEE Access* 8: 107781–107801.

53 Xia, M., Shao, H., Williams, D. et al. (2021). Intelligent fault diagnosis of machinery using digital twin-assisted deep transfer learning. *Reliability Engineering and System Safety* 215.

54 Aheleroff, S., Xu, X., Zhong, R.Y., and Lu, Y. (2021). Digital twin as a service (DTaaS) in industry 4.0: an architecture reference model. *Advanced Engineering Informatics* 47.

55 Bochkovskiy, A., Wang, C.-Y., and Liao, H.-Y.M. (2020). YOLOv4: Optimal Speed and Accuracy of Object Detection. https://doi.org/10.48550/arXiv.2004 .10934.

56 Umeda, Y., Ota, J., Kojima, F. et al. (2019). Development of an education program for digital manufacturing system engineers based on 'digital triplet' concept. *Procedia Manufacturing* 31: 363–369.

57 Kim, R.-W., Kim, J.-G., Lee, I.-B. et al. (2021). Development of three-dimensional visualization technology of the aerodynamic environment in a greenhouse using CFD and VR technology. Part 2: Development of an educational VR simulator. *Biosystems Engineering* 207: 12–32.

4

Industrial Internet of Things: Enhancement of Industries with Hyperautomation for Smart Manufacturing Machines

G.S. Gopika[1], M. Sree Krishna[1], R.S. Rajasree[2], S. Gnanavel[3], and Achyut Shankar[4]

[1]*Department of CSE, Sathyabama Institute of Science and Technology, Chennai, India*
[2]*Department of AI & ML, New Horizon College of Engineering, Bangalore, India*
[3]*Department of Computing Technologies, SRM Institute of Science and Technology, Chennai, India*
[4]*University of Warwick, Coventry, UK*

4.1 Introduction

Industries all over the world are embracing the Internet of Things (IoT) more and more to identify innovative networks for creation and services that are enabling new forms of marketing and generating new sources of income [1]. The temporal variety is quickly affecting the approach paradigm through how industries structure commercial activities and reach possible customers. Although the IoT interconnection supports enhanced business operations, for enterprises to appreciate the full potential of activating IoT, industrial Internet of Things (IIoT), and hyperautomation technologies must be linked.

A technology paradigm considered as the IoT provides a world connection whereas machines and items can coordinate and correspond with each other. All industrial domains, consisting of smart homes, smart cities, agriculture, cars, health care, industrial output, and transportation are affected by the IoT, which is prepared to be operated as technology advancement.

In this condition, companies are taken over to reconsider their development methods, which have the potential to lead to manufacturing systems improvements on an unprecedented scale. The concept of Industry 4.0, which underscores the idea of dependable digitization and the interconnection of all industrious units, combines the strengths of conventional industry with the Internet [8–10]. Several IoT techniques are already incorporated in the client automated products like smart homes, connected autos, and smart wearables. Yet, it is projected

Digital Twins in Industrial Production and Smart Manufacturing:
An Understanding of Principles, Enhancers, and Obstacles, First Edition. Edited by Rajesh Kumar Dhanaraj, Balamurugan Balusamy, Prithi Samuel, Ali Kashif Bashir, and Seifedine Kadry.

that IIoT, also known as IoT for Business, will have the power to alter multiple industries, notably manufacturing, petroleum and natural gas, farming, and mining [16, 21].

Hyperautomation is a digitalization trend that will increase the value of automation. The conceptual workers—i.e., those who are educated and work to extend basic (such as data input) and complicated (such as report preparation) tasks—are being replaced and assisted by this automation. Hyperautomation actually cannot completely replace a worker because artificial intelligence (AI) is not that effective at doing it. Nonetheless, it can enable the employees and work with them to achieve a higher degree of efficiency and fewer mistakes [2–4]. As previously said, hyperautomation is a concept that integrates a variety of technologies with the aim of simulating human operations clearly shown in Figure 4.1.

As automation technologies advance, hyperautomation is becoming more prevalent. Businesses are transforming their operations to create intelligent, human-centered workplaces. For organizations dependent on software and automation tools to maintain their competitive edge, this transformation has

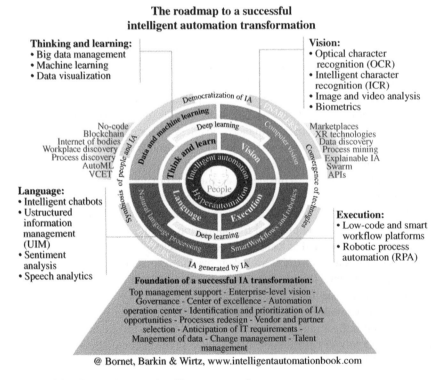

Figure 4.1 The road map to intelligent automation.

ushered in a new era [5–7]. Businesses may move past technology's special benefits to complete digital mobility and scale flexibility by integrating all forms of automation closely. Businesses can accurately monitor using critical analysis predicated on automated systems, saving time and money every week, month, and year. Since automation has taken the lead in most sectors, many companies can leverage this new technology.

An effect of technology is that it performs tasks in a different manner than automated robotic processes. This technological advancement is still in its infancy. Hyper-ability automation to involve people inside the procedure is an important feature. When people and technology work together, workers may begin programming automation tools and other software, eventually progressing to the stage of intelligence judgment through machine learning (ML).

An informed, flexible, and quick-thinking workforce might be created by hyperautomation. Accurate judgments can be made based on information and insights [20]. Model recognition relies on the least amount of human input to determine the next step and enhance efficiency. With training data, the system's algorithm is trained before creating a working model. One essential element of hyperautomation is the ability to create a confluence of various forms of comput-erization, which naturally go together with one another at the maximum level. This comprises sophisticated optical character recognition, natural language processing, and ML, which aids the software in recognizing data correctly.

4.2 Related Work on Hyperautomation

The current technology companies can be quickly covered with hyperautoma-tion systems. The preeminent way to implement computerization in a particular situation is using traditional business automation strategies. These automata were created for a particular program element. As firms are becoming more hyperauto-mated, there are several methods to enhance internal operations. Being a relatively new concept, firms are still figuring out how to implement hyperautomation. While some companies perceive better results with a more federal or dispersed approach to managing huge projects, others may experience the opposite. There are several additional issues with privacy and security that could arise.

4.2.1 Primary Requisites of Hyperautomation

The important goal of the industry method, which is to design and improve end-to-end algorithms that provide creative new business models, can be uniquely identified by hyperautomations. Industries are aware of the method of digital technologies in their present work environment. The suggestion must be always developing and evolving the product market is going to be a big problem. It is exhausting to decide which work or products should be attainable to the customers. The goal of integrators and collections are anticipated to avoid the duplication across their items due to the huge market and improve the client thoughts of potential givers. The healthcare industry may benefit from automated things because it provides more and better patient outcomes, more complex findings, and more accurate data.

4.2.2 Technological Perspective of Hyperautomation

The major four perspectives are used in the hyperautomation domain and they are clearly depicted in the below-mentioned Figure 4.2.

4.2.2.1 Robotic Process Automation (RPA)

Robotic process automation (RPA) is really a tool that will follow your instructions like an extremely obedient rule-follower. It allows users to employ computer "bots" that can understand, copy, and then do business tasks. Similar to people, RPA software robots may interact with just about any application interface. However, RPA bots may work continuously, far more quickly, and with 100% accuracy. Bots may perform calculations, access and transmit files, analyze emails, log into programs, connect to APIs, and extract complex data in addition to simply copying and pasting [13–15]. Automating processes with robots can therefore greatly benefit any firm by hyperautomating documentation and process management.

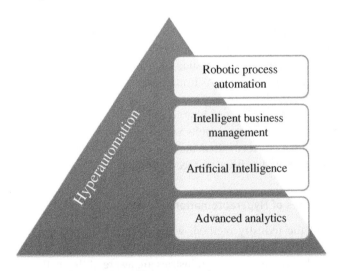

Figure 4.2 Perspectives of hyperautomation.

In order to provide loans, banks need to incorporate a lot of the documentation work. Banks used to keep a lot of the paperwork that is to be used for the consumer or company financing. RPA supports the banks in analyzing the loan documentation and suggesting good decisions. Additionally, this permits banks from the preceding difficulties like human mistakes and assists in lowering the cost associated with hiring staff. The advantages of this hyperautomation of industries are better decisions about investment patterns, cash inflow, and so on.

4.2.2.2 Natural Language Processing (NLP)

Natural language processing (NLP) does have the capacity to extract and arrange information from written materials, such as determining the company from which a payment originated and the invoice's subject, and then mechanically inserting this data into the accounting system. Sentences are torn down into their component pieces using text analytics techniques including sentiment analysis, text categorization, and named entity recognition. These elements are then decrypted using NLP to examine the text and then further classify the message being conveyed. For sophisticated document comprehension and analysis, businesses can benefit from NLP.

The dominant technology in this division is NLP [13]. There is a capability to comprehend the content of the topics by reading a text or listening to an audio file. As a result, the machine employs a variety of inputs, including a microphone, the text of such an email or telephone call, as well as other relevant sources. With the appropriate attributes, it can identify the language being used at the time. It can comprehend informal as well as more prevalent forms of writing and speaking.

The capability of this technology to comprehend sentiment in a circumstance is another intriguing characteristic. In order to extract the sentiment, it blends many words in a specific context. A person can use specific words to describe a situation or communicate their sentiments. These words convey a person's desire to have the listener understand their mental state. It may be emotionally charged, making it relevant to human feelings like joy, rage, rudeness, and politeness, among others. It could also be pertinent to the subject, whether it be favorable, negative, or neutral.

4.2.2.3 Artificial Intelligence and Machine Learning

AI is necessary to achieve automation levels that are inaccessible to humans. The rapid grow with AI's aid by expanding automation's capabilities and uses. The results are more accurate, rapid, and efficient. The fast-moving document processing software is ideal for handling the constantly changing ML algorithms. Businesses can now be used by their data in new ways thanks to ML. Instead of focusing on a variety of diverse data points, they can find valuable information.

ML analyzes the data using a variety of algorithms, and by looking at the numbers it can identify or deduce the significance of the information based on prior knowledge. The machine should initially be given samples of data with variables and results or outputs in order to obtain this result. The machine then requires test data so that it may estimate results, compare them to actual results, and report on the algorithm's accuracy when applied with the specified data model. It is feasible to train the computer using various algorithms and determine which has the higher precision using the same model.

4.2.2.4 AI and ML in IIOT

The names "AI" and "IoT" have been created, and they are frequently used in the media. As a result, people have higher expectations for these technologies and more fantastical ideas in their heads. Industrialists anticipate that AI will provide more answers to issues and needs that are currently existent with the interconnection of components through the use of IoT techniques, in addition to transfer learning and ML techniques.

The fundamental AI components of ML and learning techniques assist in managing the data obtained from IIoT by identifying, categorizing, and drawing conclusions. Artificial neural network (ANN) use in IoT and business analysis is essential for successful and effective decision-making, especially in the area of real-time data analysis and streams in relation to edge computer networks.

Including essential technologies, equipment, software, and applications improve dependability, outcome, and customer happiness in the IIoT, significantly thanks to ML and ML algorithms. Major issues with IIoT include security, interoperability, real-time reaction, and future readiness. Contextual analysis should be used in conjunction with ML algorithms to address these issues and improve IIoT.

One of the key elements of the conceptual framework in IIoT is ML, which aids in the timely evaluation and diagnosis of network and IoT device problems. The creation of estimation techniques, anomaly detection, and suggestion models are all ML capabilities. This can aid in spotting system dangers in the future and hidden flags. Also, in order for the system to generate predictions and suggestions with high accuracy, it is crucial to teach it to assess the context.

4.2.2.5 Low-Code Platforms

Technologies for hyperautomation would be applicable on a vast scale. Applications are the kind that boost the availability of hyperautomation methodologies by reducing the struggles that stand in the way of a concept and its implementation. For instance, developing a self-service schedule maker

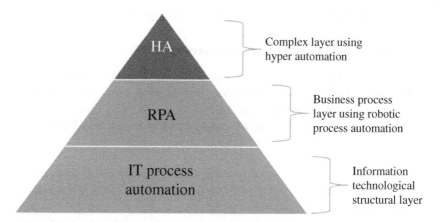

Figure 4.3 Depicts the structure of hyperautomation in industries.

once required a large financial investment, a team of programmers, and a protracted development cycle. It is now possible to construct scenarios in order such as this or a workflow with several steps, such as payment and account setup.

The goal of low-code and it's concept to provide business users as much freedom as possible to create their own applications without having any coding or IT background. Low-code applications are made with very little code; for instance, users only need to declare and then use variables. Instead, the entire development process is carried out using visual instructions and drag-and-drop of click features in a no-code platform.

Everyone who can use a PC, a website, or a web service may utilize these development platforms because they have an intuitive user interface. They are typically cloud-based, and business users can select a ready-to-use command or action before dragging and dropping it in the proper order to carry out a certain task [19]. For everyone's benefit, the merchants provide a tone of activities and even whole projects that may be used right away.

The fundamental idea behind this subject is that anyone may design automation because the best guy to perform tedious activities on a regular basis is also the person who knows them best. By doing this, it is possible for any business user to have a robot that assists them in their work.

Some automation projects could start off as straightforward tasks, such as adding a new task to the user's to-do list whenever they get an email with a specific subject. How the projects are carried out is shown in Figure 4.3. Yet, they may be scaled effectively in ever more complicated projects with numerous interconnected activities that run to support the business customer in a specific process.

4.3 Standardization of Industrial IoT and its Initiatives

4.3.1 Industrial Internet Reference Architecture (IIRA)

The four views of business, usage, functionality, and implementation are used to [17, 18] describe and classify typical architectural issues as depicted in Figure 4.4. According to IIRA, the functional area is the most crucial aspect to take into account for an industrial IoT system which is divided into five layers as follows:

- Work planning, forecasting, enterprise resource estimation (ERE), and product lifecycle management are just a few of the features offered by the business layer, which enables full-cycle industrial operations.
- It contains the definite purpose reason, such as interfaces for relevance programming, user interfaces, and so on, to offer the functionalities necessary to enable the implementation of a task or goal.
- The information layer provides functionality for gathering and storing data as well as semantics, analyzing data, and providing and deploying data. This layer offers the features necessary for assets to function properly over their entire life cycle. Additionally, it is in charge of updating, diagnosing, and deploying assets.
- Control Layer functions enables the management of industrial resources like sensors, actuators, and connectivity.

4.3.2 RAMI 4.0

RAMI 4.0 is a 3D condensation of Industry 4.0's most critical elements that make sure all participants have a shared understanding and point of view [19, 37]. The three axes stand in for all of Industry 4.0's key components. They enable the classification of an object in the model, such as a machine. As a result, RAMI 4.0 can be used to describe and implement extremely flexible Industry 4.0 concepts. A gradual transition from the current world to the one of Industry 4.0 is possible thanks to the standard architecture model.

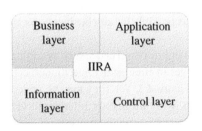

Figure 4.4 Depicts the layers of IIRA.

4.3.3   One M2M

The oneM2M world standards initiative is responsible for creating the technical requirements for a standard M2M server side. Your business is able to take part in oneM2M thanks to Alliance for Telecommunications Industry Solutions (ATIS) participation [11]. OneM2M seeks to define a vertical service layer that will enable communication and connection with various IoT systems. The market is divided since different suppliers offer IoT systems that cannot communicate with one another or construct vertical pipelines.

4.3.4   Framework Using Arrowhead

The definition for the client applications transporting information essential for the activity being automated is incorporated into the Arrowhead Framework together with standard solutions for the fundamental functionality in the areas of information infrastructures, systems management, and information assurance [12].

Integrated system creation, implementation, and automation based on the Service Oriented Architecture (SOA) idea are made easier by the Arrowhead Framework in a suitable way. Keeping a record, authorization, and orchestration systems are the three essential core systems required by the Arrowhead Framework to create SOA-based local clouds.

4.4   Customized Hyperautomation Workflows Approach

The client as well as the employee, who are approximately roughly proportional to that of the company, achieve greatly from the hyperautomation. The development of tools to achieve business and IT necessities to attain, save, and regularize the process can be incorporated to other domains for their improvement and to organize them. Until the whole development process is automated teams may get automated to check the enablement of decision-modeling software. Developed process automation, human data collection, and automation, everything is developed by AI (Figure 4.5). An important toolkit for efficient automation is to focus on every stage of the achievement to get better and quicker results. Several novel methods can be collaborated with new automated technology to help businesses create as many end-to-end companies as possible [35, 36].

The concept of a virtual tool can automate and reduce human mistakes that happen and preserve the time and their money. It is used to perform the management actions and enhances the consumer experience. Automation can be appropriately compared with the robotic arms that enable the tasks more quickly and

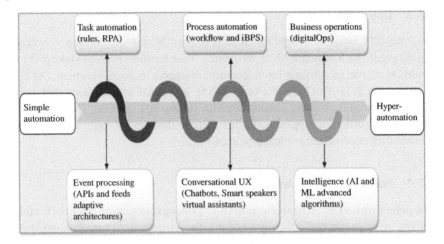

Figure 4.5 Graphical representation of workflow in hyperautomation.

accurately. The concept of the robotic automated brain is obtained as the nature of hyperautomation for achieving the best actions. This includes the variety of automated AI technologies [34]. When it is used in integrating with software products, the above technologies efficiently expand scalability of the automation. The online banking and banking automated at any time, to keep the regular practice for procedures and automation, are achieved by a difference of backend efficiencies given the hyperautomation. Sophisticated data can be analyzed and handled by the banking and financial sectors.

4.4.1 Use Case Scenarios of Automation

4.4.1.1 Health Care

Effects of technology can benefit the healthcare industry by enhancing treatment satisfaction, boosting revenue, and producing more precise data. Hyperautomation is utilized to manage client communications, billing cycles, and collection. Additionally, it can deal with managing patient records, gathering information, and producing meaningful output for more precise treatment plans. Hyperautomation is frequently implemented to ensure regulatory compliance, which is essential for any healthcare organization's survival and success [27]. Additionally, it can be used to schedule personnel and other resources as well as control medication inventories and procurement. The concept of virtual has countless applications in the healthcare sector, and its advantages can benefit the company, its partners, and its clients.

4.4.1.2 Supply Chain

Low personnel numbers have caused process delays as a result of the pandemic, which has significantly hampered the capacity to get materials on schedule and has at worst created logistical challenges [22]. Effects of technology may speed up, be more effective at, and be more accurate in repeated processes by eliminating the need for user intervention.

4.4.1.3 Banking and Finance

The financial and financial services sectors are under continuous pressure to lower costs, boost productivity, and offer a more accessible and individualized customer satisfaction. Nowadays, automation can be the important equipment of every industry [23, 24]. Work protocols have used a procedure to become more achievable. From programming services through test AI-based solutions, the use of toolkits are used.

It may focus on other jobs that are more important than repetitive ones, like data entry or performed on the basis, by using automation to save time, money, or effort.

Hyperfunction automated technology is to enhance how fast the banks can do the transactions, manage their profits, and provide the customers sales and operations. In addition to that, it guides them to achieve their goal by improving the accuracy and operations of the processes they are doing with.

4.4.1.4 Retail

There are numerous prospects for automation in the retail sector. Order processing is heavily influenced by e-commerce, as more individuals than ever use customer loyalty and place their orders online [25, 26]. Hyperautomation supported by AI can speed up front-end operations including targeted marketing via social media ad placement and focused email marketing, online loyalty recognition, face detection when a customer approaches a shop, and more. Back-end retail activities like purchasing, accounting, supplier relationship management, inventories, and shipping can be made more accurate and efficient while also being less expensive thanks to hyperautomation.

4.4.1.5 Lending Operations in Hyperautomation

Initiatives to cut costs and enhance client experiences have both contributed to the shift toward robotics in lending operations. Inefficient paper-based procedures have historically been used extensively in both the consumer lending process and the commercial loan process [27]. Technologies like RPA can extract relevant pertinent loan data from large amounts of data, carrying out work that may take a human worker the majority of their week to do.

4.5   Advantages of Hyperautomation for Security

4.5.1 Human Error Deduction

Effects of technology lessen the need for human personnel in data-intensive daily processes. This focus on automation helps avoid errors that even the most experienced and well-intentioned workers can make [28]. IT professionals may feel more comfortable in the ability of their technical protections to prevent breaches with less opportunity for error.

4.5.2 Improvement in Response Time

The rapidity of hyperautomation offers a safety benefit. A human cannot identify something unusual as quickly as an automated process. Teams may then react to possible security risks more quickly, increasing the likelihood that they will be capable of helping stop a breach.

4.5.3 Increased Visibility

The effects of technology enhance teams' understanding of their digital infrastructure. One of the seven essential elements of hyperautomation is process modeling and mining since firms must comprehend their processes before automating them. They will achieve even greater transparency as a result of automation's consolidation and quick reporting. It is simpler to identify and fix issues thanks to this visibility [29]. Additionally, it enables teams to understand how moving up or down could change access controls, create new dependencies, or have other effects on their IT environments. They can then put in place the necessary security controls and measures to maintain their security in the face of fluctuating conditions.

4.6   Future Work Prospects and Barriers

To offer various research recommendations in the area of IIoT, this section examines the open difficulties. There are numerous problems that the IoT's enabling technologies, including such sensor networks like wireless, cloud technology, big data analytics, interoperability, and integrated devices, are constantly working to solve in both academia and business.

4.6.1 Connectivity and Interoperability

IoT interoperability provides the ability of various IoT implementation components to efficiently interact to achieve the efficiency, share the huge amount of data, and work in concert to achieve a universal goal. Industries need to be capable to comprehend and send data transversely across all interfaces from gadgets to the Internet. In addition, the consumer and home IoT markets experience a delay due to a lack of automated, universal compatibility, but also municipal and commercial IoT installations [30]. For smart city projects to fully benefit from technology, hundreds or thousands of sensors across several categories may need to collaborate. Several communication protocols may be required for industrial IoT deployments with various device needs and physical constraints.

4.6.2 Scalability

Due to problems with IoT technology, the capacity of a protocol to handle an enlarging amount of work by incorporating more resources continues to be problematic for many developers. Furthermore, if not addressed quickly enough, these vulnerabilities may develop into issues that put increased maintenance requirements and latency concerns at risk [31].

4.6.3 Fault Tolerance

The main goal of high availability in the IoT is to easily adapt to constantly changing conditions and create reliable redundancy. IoT will be more vulnerable to assaults because billions of devices now are producing and consuming services [32]. Very limited devices will be most vulnerable to attacks, and malevolent systems may attempt to gain direct or indirect control over other devices. As a result, fault tolerance may provide a problem for scalability.

4.6.4 Flexibility

Due to the number of current IoT applications, it has been challenging to provide services that meet each application's needs. On-the-go apps that are continuously improved, customized, value-added, and autonomous are often needed by IIoT users. Using a flexible framework that will change in the future is the finest technique to properly create an IIoT solution that expands and aids the business in integrating various applications [5]. The worksites are constantly changing and present a variety of dangers in the framework of the mining industry. The IoT gadgets and networks should be adaptable to alter as necessary, for instance, there is an alarm for some mishaps in the mine.

4.6.5 Security and Safety

Many security issues are exacerbated by the sophistication of IIoT methods and the variety of networking infrastructures. To prevent malicious code from

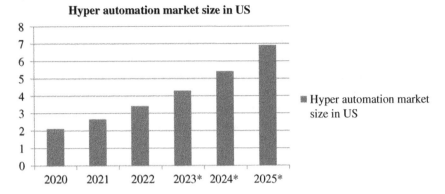

Figure 4.6 Prediction of the hyperautomation technologies in the US.

being injected to carry out risky operations like managing the blast system, cars, drones, and ventilators in the mines, the control mechanisms, IoT devices, and applications in the mining sites must be secure. The IIoT requires underside security, thus the IIoT software must adopt a secure boot process, firewalling, user access rules, device identification, and the ability to run safety patches and updates [33]. Furthermore, security confirmation and validation provide confidence in the effectiveness of safety checks created to close identified security gaps. It is already utilized in many foreign countries including in the United States. Figure 4.6 shows the statistics of the hyperautomation in the United States.

Future perspectives of a technology like hyperautomation will have a practical effect on companies in every sector. When there are no redundant or low-value jobs, people may provide those that have a greater optional value for the company [6, 7]. The integration of computerization and manual work can improve the client experience for business, lower developmental rate, and raise the profit in any case.

4.7 Conclusion

IoT is thought to be a key factor in streamlining and automating the demand and industry development processes. Everything will be intelligently connected through the Internet within a given sector. Any business will strive to offer a significant value proposition through competencies. Management of supply chains, quality assurance, predictive maintenance, and electricity consumption are just a few of the company operations that are greatly aided by AI and ML. These technologies increase accountability, hasten decision-making, and offer a way to make precise predictions. Additionally, ML and AI help to speed up reaction times by removing anticipated failures.

Automation was integrated with novel technologies like automated AI to solve difficult problems and organize the processes. Causes of technology may find the individuals by entering them to optimize the work alongside technology in an integrated collaborative environment. It enhances and uses tools to detect very huge amounts of information and apply the deep insights to its business as efficient powerful stakeholders and policymakers; hyperautomation will transform the enterprises by enhancing corporate procedures by reducing reluctant tasks and automating human created ones. It makes the industries to operate consistently, wisely, and quickly. Prices are reduced as an outcome and clients generally have an optimized overall experience. Hyperautomation which provides more risks for any new company or business process methodology. The technology automates new different tools that can be mechanized and optimizes complex problems rapidly. The article also included a list of important open research questions and future research objectives.

References

1 Akshatha, N., Harishree, K., Rai, J. et al. (2019). Tactile internet: next generation IoT. *Proceedings of the 3rd International Conference on Inventive Systems and Control*, ICISC 2019, no. ICISC. pp. 22–26.

2 Bornet, P., Barkin, I., Wirtz, J. et al. (2021). *Intelligent Automation – Welcome to the World of Hyperautomation*, 1e. World Scientific Publishing Pvt Ltd.

3 Büchi, G., Cugno, M., Castagnoli, R. et al. (2020). Smart factory performance and industry 4.0. *Technological Forecasting and Social Change* 150: https://doi .org/10.1016/j.techfore.2019.119790.

4 Choi, S., Kim, B.H., Noh, D. et al. (2015). A diagnosis and evaluation method for strategic planning and systematic design of a virtual factory in smart manufacturing systems. *International Journal of Precision Engineering and Manufacturing* 16 (6): 1107–1115. https://doi.org/10.1007/s12541-015-0143-9.

5 Davis, J., Edgar, T., Porter, J. et al. (2012). Smart manufacturing, manufacturing intelligence and demand-dynamic performance. *Computers & Chemical Engineering* 47: 145–156. https://doi.org/10.1016/j.compchemeng.2012.06.037.

6 De Moura, R.L., Ceotto, L.L.F., Gonzalez, A. et al. (2017). Industrial IoT and advanced analytics framework: an approach for the mining industry. *Proceedings of the IEEE International Conference on Computational Science and Computational Intelligence (CSCI)*, Las Vegas, NV, USA (14–16 December). pp. 1308–1314. https://doi.org/10.1109/CSCI.2017.228.

7 Delsing, J. (2017). Local cloud internet of things automation: technology and business model features of distributed internet of things automation solutions.

IEEE Industrial Electronics Magazine 11 (4): 8–21. https://doi.org/10.1109/MIE .2017.2759342.

8 Evans, D. (2011). The internet of things: how the next evolution of the internet is changing everything. *CISCO White Paper* 1: 1–11.

9 Ghosh, A., Chakraborty, D., Law, A. et al. (2018). Artificial intelligence in Internet of things. *CAAI Transactions on Intelligence Technology* 3 (4): 208–218. https://doi.org/10.1049/trit.2018.1008.

10 Gokhale, P., Bhat, O., and Bhat, S. (2018). Introduction to IoT. *International Advanced Research Journal Science and English Technology* 1: 1–24.

11 Husain, S., Kunz, A., Song, J. et al. (2014). Interworking architecture between oneM2M service layer and underlying networks. *Globecom Workshops (GC Wkshps)*. New York, NY, USA: IEEE. pp. 636–642. https://doi.org/10.1109/ GLOCOMW.2014.7063504.

12 Javaid, M., Haleem, A., Singh, R.P. et al. (2021). Significance of sensors for industry 4.0: roles, capabilities, and applications. *Sensors International* 17: 100110.

13 Javaid, M., Haleem, A., Singh, R.P. et al. (2021). Significance of quality 4.0 towards comprehensive enhancement in manufacturing sector. *Sensors International* 2 (4): 100109.

14 Koubâa, A. (ed.) (2020). *Robot Operating System (ROS): The Complete Reference Volume 5*, 895. Springer Nature.

15 Lee, I. and Lee, K. (2015). The Internet of Things (IoT): applications, investments, and challenges for enterprises. *Business Horizons* 58 (4): 431–440. https://doi.org/10.1016/j.bushor.2015.03.008.

16 Lesi, V., Lesi, V., Jakovljevic, Z., and Pajic, M. (2019). Reliable industrial IoT-based distributed automation. *Proceedings of the International Conference on Internet of Things Design and Implementation*, Montreal, QC, Canada (15–18 April). pp. 94–105. https://doi.org/10.1145/3302505.3310072.

17 Lin, S.W., Miller, B., Durand, J. et al. (2015). The Industrial Internet of Things Volume G1: Reference Architecture Version 1.9 June 19, 2019.

18 Liu, C. and Jiang, P. (2016). A cyber-physical system architecture in shop floor for intelligent manufacturing. *Procedia CIRP* 56: 372–377. https://doi.org/10 .1016/j.procir.2016.10.059.

19 Marzband, M., Parhizi, N., Savaghebi, M., and Guerrero, J.M. (2016). Distributed smart decision-making for a multi microgrid system based on a hierarchical interactive architecture. *IEEE Transactions on Energy Conversion* 31 (2): 637–648. https://doi.org/10.1109/TEC.2015.2505358.

20 McIver, D., Lengnick-Hall, M.L., Lengnick-Hall, C.A. et al. (2018). A strategic approach to workforce analytics: integrating science and agility. *Business Horizons* 61 (3): 397–407. https://doi.org/10.1016/j.bushor.2018.01.005.

21 McNinch, M., Parks, D., Jacksha, R., and Miller, A. (2019). Leveraging IIoT to improve machine safety in the mining industry. *Mining, Metallurgy & Exploration* 36 (4): 675–681. https://doi.org/10.1007/s42461-019-0067-5.

22 Moghaddam, M., Cadavid, M.N., Kenley, C.R., and Deshmukh, A. (2018). Reference architectures for smart manufacturing: a critical review. *Journal of Manufacturing Systems* 49: 215–225. https://doi.org/10.1016/j.jmsy.2018.10.006.

23 Panetta, K. (2019). Gartner Top 10 Strategic Technology Trends for 2020. Available at: www.gartner.com.

24 Richardson, S. (2020). Affective computing in the modern workplace. *Business Information Review* 37 (2): 78–85. https://doi.org/10.1177/0266382120930866.

25 Rimol, M. (2021). Gartner Forecasts Worldwide Hyper Automation-Enabling Software Market to Reach Nearly $600 Billion by 2022. Available at: www.gartner.com.

26 Roy, C., Misra, S., Pal, S. et al. (2020). Blockchain-enabled safety-as-a-service for industrial IoT applications. *IEEE Internet of Things Magazine* 3 (2): 19–23. https://doi.org/10.1109/IOTM.0001.1900080.

27 Rymaszewska, A., Helo, P., and Gunasekaran, A. (2017). IoT powered servitization of manufacturing–an exploratory case study. *International Journal of Production Economics* 192: 92–105. https://doi.org/10.1016/j.ijpe.2017.02.016.

28 Rzepka, A., Borowiecki, R., Miskiewicz, R. et al. (2021). Changes in management during transformation of power industry. *European Research Studies Journal* XXIV (2): 1149–1162. https://doi.org/10.35808/ersj/2179.

29 Schmiedbauer, O. and Biedermann, H. (2020). Validation of a lean smart maintenance maturity model. *Tehnicki Glasnik* 14 (3): 296–302.

30 Siderska, J. (2020). Robotic process automation—a driver of digital transformation? *Engineering Management in Production and Services* 12 (2): 21–31. https://doi.org/10.2478/emj-2020-0009.

31 Srivastava, A., Kumar, A., and Damle, M. (2020). Hyperautomation in transforming underwriting operation in the life insurance industry. *PalArch's Journal of Archaeology of Egypt/Egyptology* 17 (6): 4928–4944.

32 Swetina, J., Lu, G., Jacobs, P. et al. (2014). Toward a standardized common M2M service layer platform: introduction to oneM2M. *IEEE Wireless Communications* 21 (3): 20–26. https://doi.org/10.1109/MWC.2014.6845045.

33 Xu, X. (2017). Machine tool 4.0 for the new era of manufacturing. *International Journal of Advanced Manufacturing Technology* 92 (5–8): 1893–1900. https://doi.org/10.1007/s00170-017-0300-7.

34 Xu, M., David, J.M., and Kim, S.H. (2018). The fourth industrial revolution: opportunities and challenges. *International Journal of Financial Research* 9 (2): 90–95. https://doi.org/10.5430/ijfr.v9n2p90.

35 Yalcinkaya, E., Maffei, A., and Onori, M. (2017). Application of attribute based access control model for industrial control systems. *International Journal of*

Computer Network and Information Security 9 (2): 12–21. https://doi.org/10 .5815/ijcnis.2017.02.02.

36 Zhang, Y., Qian, C., Lv, J. et al. (2017). Agent and cyber-physical system based self organizing and self-adaptive intelligent shop floor. *IEEE Transactions on Industrial Informatics* 13 (2): 737–747. https://doi.org/10.1109/TII.2016.2618892.

37 Zyubin, V.E. (2007). Hyper-automaton: a model of control algorithms. *Siberian Conference on Control and Communications 2007* (April). IEEE. pp. 51–57. https://doi.org/10.1109/SIBCON.2007.371297.

5

Digital Twins Model of Industrial Production Control Management Using Deep Learning Techniques

Govindaraj Venkatesan[1], Rajendra Narasimman Karthika[2], and Ali Kashif Bashir[3]

[1]Department of Civil Engineering, Saveetha Engineering College, Chennai, Tamil Nadu, India
[2]Department of Information Technology, Saveetha Engineering College, Chennai, Tamil Nadu, India
[3]Manchester Metropolitan University, Manchester, UK

5.1 Introduction

The notion of a "digital twin" has gained popularity in recent years as a powerful tool for boosting industrial production control management. Engineers and operators may build a virtual replica of a real-world production system to model and test various scenarios, identify potential problems, and make educated decisions to increase efficiency and productivity. Deep learning methods, a subset of machine learning, have tremendous potential for improving digital twin (DT) models. Since deep learning algorithms can learn from enormous volumes of data, they may recognize patterns and make predictions that may be difficult or impossible for individuals to make. This can be particularly beneficial for the manufacturing process' scheduling, maintenance, and resource allocation optimization [1, 2].

This research study [3, 4] presents a DT model of industrial production control management that makes use of deep learning techniques for more accurate and successful forecasts. We describe the components of our model's structure, which consists of a deep learning module for predictive modeling, modules for acquiring and processing data, and a simulated version of the real production system. A series of tests were done using real data from an industrial production system to assess the efficacy of our model. We observed that our model was more accurate and efficient in projecting different production aspects when compared to other well-established approaches. Using real data from an industrial production system, we ran a number of tests to assess the efficacy of our model. The proposed model was more accurate and efficient at estimating a variety of

Digital Twins in Industrial Production and Smart Manufacturing:
An Understanding of Principles, Enhancers, and Obstacles, First Edition. Edited by Rajesh Kumar Dhanaraj,
Balamurugan Balusamy, Prithi Samuel, Ali Kashif Bashir, and Seifedine Kadry.

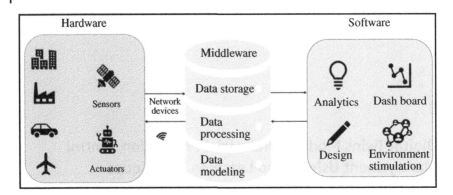

Figure 5.1 Digital twin system.

production measures when the proposed algorithm is compared its performance to state-of-the-art algorithm [5].

Many different technologies that generate and utilize enormous volumes of data are used in manufacturing [6]. Although this data is helpful for making decisions about production lines, it also creates problems for information management, reuse, integration, and design. In this article, we will look at an information management strategy that makes use of the notion of a "digital twin," or a virtual replica of a physical object. We demonstrate how DTs can be used to collect, store, and communicate data from sensors on production lines. We also offer case studies that show the outcomes of our strategy for boosting manufacturing productivity and efficiency. The Digital Twin system is shown in Figure 5.1. To complete tasks, people use sensor-equipped instruments and machinery. Sensors can see and hear what is going on like small eyes and ears. The sensors can provide information about the machine's speed, material consumption, and potential issues. Employees value this knowledge greatly since it enables them to make wiser decisions and develop their companies. But, occasionally, the amount or complexity of the information might be overwhelming. This aids in the quicker, less expensive, and better creation of things. This study emphasizes how deep learning approaches may improve production control management in industrial settings, which is an essential contribution to the increasing body of research on the DT concept [7, 8].

5.2 Literature Review

As Industry 4.0 develops, controls become more complicated and efficient, and diagnostics and maintenance become necessary. As a result, maintenance and proper health care must be discussed in literature and business. Utilizing blockchain technology to build DTs, convert actual spaces into virtual ones, more accurately and effectively assess the state of equipment, and forecast trends. Complex machinery frequently carries out crucial tasks, therefore, the outcome can be unusual when it is not done [9]. By integrating the DT throughout the product lifecycle and management process, the problem of upgrading and replacing the product can be successfully overcome [10, 11].

A customized approach is an effective technique to minimize costs while boosting the system's usability since the DT is more than just a cheaper reproduction of experimental measurements. Combining different care approaches is a wise solution, but in order to create the best treatment plans, it must be carefully analyzed. This publication [12] provides examples to illustrate the modeling and assessment of care methodologies that can be used in DT applications. This study examines the characteristics and theoretical foundations of green building maintenance services (GBMS), and it provides a DT based GBMS framework to help the team address the difficulties of insufficient automation and green resource management (DT-GBMS) clinics. DT technology is a system that delivers real-time intelligent defect detection. Real-world equipment and virtual environments can match thanks to DT technology, allowing for optimization and centralized monitoring. For clever and effective diagnosis, intelligent fault detection offers a mechanism to distinguish between sound and broken machine signals [13]. The research on monitoring and forecasting the usage of the DT concept in manufacturing is first summarized and presented in this article. After that, the gaps in the literature are found and analyzed, and the development of the DT idea is summarized. Applications for business, such as the area of maintenance forecasting is in progress right now [13]. This article provides details on the initial data use, explains how the data were used, extracted, and discussed, and it does so in the context of previous research and DT applications. The industrial revolution is shown in Figure 5.2 [14].

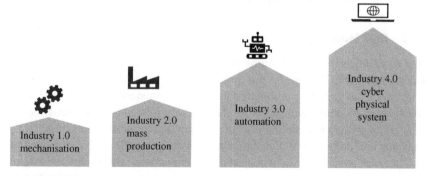

Figure 5.2 Industrial revolution.

The new algorithm that is suggested is validated using real-time sensor data. The information is gathered by sensors positioned on the body and utilized to spot problems before they cause damage to the system. Simulation Based Digital Twin (SBDTs), also known as DTs based on online first-law simulation models, offer several benefits over DTs based on data-driven models. Data aggregation for predictive maintenance (PdM) with ML and sophisticated analytics provides precise diagnosis and prediction as well as the best possible treatment options [15]. To avoid significant and unforeseen machine failures and production losses, PdM is crucial for machines involved in complex processes.

In this research, a low-cost DT guided error compensation system for robotic arms is described. This system is utilized to provide DT models with accurate data about the robotic arms. The inaccuracy lowers the performance of the robotic arm and necessitates intelligent maintenance [16].

In this article, DT and Its Industrial Internet of Things (IIoT) contexts are used to demonstrate a PdM tool for electronic items. Computer simulations in this instance revealed a relative inaccuracy of less than 4% for conductivity analysis and less than 10% for temperature. The most popular drives in industry are asynchronous motors, hence methods for forecasting failures and lowering maintenance are of interest [17].

5.3 Analysis of the Field's Current State

A DT is a physical representation that can be used for simulation, optimization, and prediction. Because they may increase the effectiveness, consistency, and adaptability of the production process, they have become essential for business management. On the other hand, developing and maintaining DTs requires a significant quantity of data and processing power, which presents challenges for data modeling and ML. To address these issues and put smart DTs into production, researchers have developed a variety of strategies recently.

5.3.1 Data Modeling

Data modeling is the act of defining and organizing the data that is used to create and maintain DTs. To capture the traits and behavior of the system, data modeling comprises choosing pertinent data, patterns, models, and schemas. Additionally, data modeling will guarantee the accuracy, reliability, and security of the data. Ontology-based modeling uses the representation of information and relationships in a domain to define data and their meaning. Graph-based modeling in a network model represents objects and connections with nodes and edges. Heavy modeling uses compressed or streamlined data to reduce complexity and data quantity.

5.3.2 Machine Learning

ML is the process of learning from data and utilizing algorithms and models to make predictions or judgments. By performing activities including monitoring, diagnosis, prognosis, optimization, control, and learning, ML can help DTs function and be intelligent. ML techniques such as deep learning, reinforcement learning, and federated learning were used to train DTs. Deep learning learns nuanced patterns and features from massive volumes of data using numerous layers of ANNs. In order to determine the best policies or actions to take in response to environmental feedback, reinforcement learning uses trial-and-error techniques. To train ML models on decentralized data sources while protecting privacy and security, federated learning takes advantage of distributed computing.

The challenges faced are:

- **Data Integration**: Data integration is the process of combining disparate data sources into a single, homogenous data model that adheres to a shared semantics and schema. In order to access the integrated data, it is necessary to Extract, Transfer and Load (ETL) data from a variety of sources, address data quality and integrity issues, and provide a consistent query interface. Data integration is required for the creation of complete and accurate DTs that mirror the state and dynamics of physical systems today. Data integration must also contend with issues including heterogeneity, incompleteness, inconsistency, ambiguity, and scalability. In order to integrate data for DTs, more effective and efficient procedures and technologies are needed.

- **Model Validation**: Model validation is the process of assessing the accuracy and dependability of models used to create and manage DTs. Comparing model outputs to actual measurements or observations from physical systems or other truth sources constitutes model validation. DTs and their applications' quality and effectiveness depend on model validation. On the other side, there are many challenges with model validation, such as a lack of ground truth, complexity, variability, and flexibility. As a result, more stringent and reliable processes and criteria are needed to validate DT models (Table 5.1).

- **Human–Machine Interaction**: Interaction between humans and machines that occurs spontaneously and intuitively is known as human–machine interaction. For DTs to be useful and efficient in industrial production control management, human–machine interaction is essential. Designing and putting into practice user interfaces, visualizations, feedback mechanisms, dialogue systems, and decision support systems that enable information and knowledge exchange between humans and machines is known as "human–machine interaction." However, issues with usability, accessibility, reliability, comprehensibility, and ethics come up in human–machine interaction. DTs need more user-centric and humanware approaches and technology to promote human–machine interaction. DT is depicted in Figure 5.3.

Table 5.1 Digital twin in data analytics and IIoT/IoT.

Data analytics	IIoT/IoT
IT infrastructure	IT infrastructure
Security	Security
Data	Data
Trust	Trust
Privacy	Privacy
Expectations	Expectations
NA	Connectivity

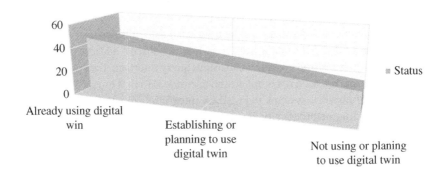

Figure 5.3 Estimated percentage of industrial equipment manufacturing companies that use or plan to use digital twins.

5.4 Identification of Gaps and Research Opportunities

5.4.1 Data Quality

One of the most difficult aspects of using DTs with deep learning is ensuring that the correct data is utilized to train the system. It takes a lot of labor to collect, arrange, and clean up data to guarantee that it is trustworthy and correct.

5.4.2 Interoperability

Because the DT depends on data from several sources, it is crucial to guarantee that the data is interoperable across various platforms and systems. A key area of research to encourage collaboration could be the creation of common data exchange protocols and interfaces.

5.4.3 Model Explainability

Although deep learning approaches have the ability to significantly improve the performance and accuracy of DT models, they are usually regarded as "black box" models, making it difficult to comprehend how the prediction was generated. Deep learning models' applicability and reliability can be improved by creating ways for describing and explaining them.

5.4.4 Scalability

It can take a lot of processing power and storage to create and run DT models for manufacturing processes because they might be huge and complicated. Wider adoption can be facilitated by research into scalable algorithms and infrastructure development.

5.4.5 Ethics and Privacy

The ethics and privacy of using DTs, which rely on so much data, must be taken into account. A key area of research will be creating a framework for data management, privacy, and openness.

5.4.6 Real-time Decision-making

By providing precise and timely information, DTs enable real-time decision-making. To create algorithms that can handle and evaluate data in real time and interface with current systems, research is necessary.

5.5 Methodology

The DT model of production control management is an innovative way for optimizing production operations and increasing efficiency in industrial settings. Deep learning techniques are used in this model. The simulation, monitoring, and optimization paradigm is built around the concept of "digital twins," which entails making a virtual replica of a real system. The proposed model is made up of several pieces, including:

5.5.1 Data Acquisition

In order to put the DTs' paradigm into practice, data must first be collected from a variety of sources, including sensors, control systems, and manufacturing machinery. Real-time data collection includes information on temperature, pressure, flow rate, vibration, and energy usage. The body's DT is built using data gathered over time from a range of sources, including sensors, manufacturing machinery, and control systems.

5.5.2 Data Preprocessing

The gathered data is preprocessed to eliminate noise and outliers and to format it in a way that the deep learning algorithms can use it. The data must first be processed after it is received in order to remove noise, inconsistencies, and missing values as well as to change the vacuum algorithms into a format that can be used for the learning process. The following methods for data preparation are frequently employed:

- **Data Cleaning**: Noise and outliers must be eliminated in order to ensure that the data is reliable and correct.
- **Data Normalization**: To enhance the performance of deep learning models, this entails scaling data into a multi-level process.
- **Feature Selection**: This involves selecting the most important aspects of the data in order to simplify and improve the accuracy of deep learning models.
- **Data Transformation**: This includes transforming the data into a format that deep learning models can use, such as changing categorical variables into numerical values.
- **Deep Learning Techniques**: Convolutional neural networks (CNNs) and recurrent neural networks (RNNs) are two forms of deep learning models that are trained on preprocessed data to foresee system behavior and improve production processes.
- **Performance Evaluation Metrics**: Accuracy, precision, recall, and F1 score are among the metrics that are used to gauge how well the deep learning models are working.
- **Real-time Decision-making**: For real-time decision-making, such as forecasting equipment failures, optimizing production schedules, and allocating resources, trained deep learning models are used.

The concept provides numerous benefits over the production management system, including accuracy, efficiency, prompt monitoring and decision making, and reduced time and maintenance costs. Numerous businesses, including manufacturing, energy production, and transportation can use the notion. A deep learning–based DT model for manufacturing management is generally a smart technique to boost productivity and efficiency in corporate operations. The model's validity and viability, as well as concerns with data quality, interaction, and model interpretation, all require further study.

5.6 Deep Learning Techniques for Predictive Modeling

"Digital twins" are virtual replicas of real-world assets, processes, and systems that may be used to replicate and improve their performance. In the context of industrial production control management, DTs can be used to mimic and optimize supply chain operations, equipment performance, and manufacturing processes.

When using deep learning approaches such as neural networks, DT predictive modeling has been demonstrated to be quite accurate. A DT of an industrial production control management system may employ one or more of the following deep learning approaches for predictive modeling.

5.6.1 Recurrent Neural Networks (RNNs)

In industrial production control management systems, which are popular, time-series data is often modeled using RNNs. RNNs are able to produce precise predictions based on historical data and capture the dynamic relationships between variables over time.

5.6.2 Convolutional Neural Networks (CNNs)

CNNs are widely used for image recognition tasks, but they can also be used to analyze data from other sources, such as sensor data from industrial equipment. By identifying patterns in sensor data, CNNs can be used to predict equipment failures or identify irregularities in the production process.

5.6.3 Auto Encoders

In order to learn to extract significant features from data without the aid of labeled training data, auto encoders are neural networks that can be utilized for unsupervised learning. High-dimensional sensor data can be made easier to model and analyze by using auto encoders to reduce its dimensionality.

5.6.4 Generative Adversarial Networks (GANs)

An advanced neural network termed a GAN is able to generate fresh data that is similar to the training data. In the context of a DT, GANs might be used to generate simulated data that could be used to test how well the production control management system functions in various scenarios.

5.6.5 Long Short-Term Memory (LSTM) Networks

An RNN type called LSTM is used to manage long-term dependencies in time-series data. LSTMs are helpful for forecasting longer term events like the need for equipment maintenance or shifts in manufacturing demand.

5.7 Performance and Evaluation Metrics

There are a number of metrics that can be used to gauge how well a DTs model of industrial production control management prediction model is doing.

5.7.1 Mean Absolute Error (MAE)

The MAE metric computes the average absolute difference between the expected and actual values. This statistic is useful for assessing the overall accuracy of the model's predictions.

5.7.2 Root Mean Squared Error (RMSE)

RMSE is a metric that calculates the square root of a number to return to the original unit of measurement. It measures the average squared difference between the expected and actual values. This statistic, which places more emphasis on larger errors than MAE, is helpful for assessing the overall accuracy of the model's predictions.

5.7.3 Coefficient of Determination (R-squared or R^2)

The R-squared metric (predicted values) measures the degree of variance in the dependent variable (actual values) explained by the independent variable. The effectiveness of the model's fit to the data can be determined using this statistic.

5.7.4 Precision, Recall, and F1-Score

Precision is the percentage of correct predictions for each positive prediction, whereas recall is the proportion of correct predictions for each positive prediction. Because it finds a compromise between accuracy and recall, the F1 score provides a fair assessment of sample performance. These measures can be used to evaluate how effectively a binary distribution function model works.

5.7.5 Curve and Area Under the Curve (AUC)

The ROC curve, which measures specificity as the proportion of accurate negative predictions among all genuine negatives and sensitivity as the percentage of accurate positive predictions among all real positives, graphically illustrates the sensitivity-specificity trade-off.

5.8 Background Study

5.8.1 Predictive Maintenance in Industrial Systems

Due to its potential to streamline maintenance processes and cut downtime, PdM has drawn a lot of interest from the industrial sector. Traditional methods, including planned maintenance or reactive repairs, are frequently expensive and ineffective. To anticipate equipment faults and schedule maintenance tasks in advance, PdM uses cutting-edge analytics and real-time data. Cost reductions, increased operational effectiveness, and increased equipment reliability are all made possible by this strategy.

5.8.2 Industrial Systems with Digital Twins

A potent instrument in the industrial sector is the DT. A DT is a virtual representation of a real-world system or item that duplicates its behavior, collects data in real time, and allows for predictive analytics. DTs give important insights into system performance by bridging the physical and digital spheres, facilitating improved planning, optimization, and preventive maintenance. They make it possible to monitor, stimulate, and analyze industrial systems in real time, which increases productivity and decreases downtime.

5.8.3 Predictive Maintenance Digital Twins

PdM in industrial systems has a lot of potential when deep learning techniques are combined with DTs. It is now possible to continually evaluate sensor data and offer pro-active maintenance recommendations by fusing the real-time monitoring capabilities of DTs with the predictive capacity of deep learning models. The DT serves as a platform for data integration, model deployment, and decision support, and deep learning algorithms enhance the accuracy and timeliness of maintenance predictions.

5.8.4 Deep Learning and Digital Twins

PdM in industrial systems has a lot of potential when deep learning techniques are combined with DTs. It is now possible to continually evaluate sensor data and offer proactive maintenance recommendations by fusing the real-time monitoring capabilities of DTs with the predictive capacity of deep learning models.

5.8.5 Related Research

Deep learning and DTs have both been used in research to investigate the use of PdM. The development of DT frameworks, investigation of appropriate deep learning architectures, optimization of feature engineering approaches, and assessment of the effectiveness of PdM systems have been the key areas of research. These studies have shown the potential advantages of combining deep learning and DTs, including enhanced maintenance effectiveness, less downtime, and cost savings.

5.8.6 Research Gap

DTs and deep learning for PdM have been studied, but additional study is still needed to confirm these techniques in real industrial systems. Further research is needed to determine the efficacy and scalability of deep learning-based DT frameworks in various industrial contexts, the integration of domain knowledge and uncertainty quantification, and the generalizability of the suggested approaches across various systems.

5.8.7 Discussions and Analysis of Digital Twins

There are several other industries within this one, including those that deal with digital design and manufacturing, micro-nano fabrication, composite material molding, conventional and nonconventional machining, and precision and ultra-precision machining. Big data-related tools and methods must be used in each of these fields. In order to optimize designs, improve quality, lower costs, and enhance product performance, firms need this information to define the manufacturing line. The sharing, reuse, generation, and administration of this knowledge are still difficult. Businesses can improve their performance by using knowledge management to learn from their data, modify their processes, and create better products. The DT idea is one of the most essential tools for knowledge management in manufacturing. A computer clone or DT that can imitate and quantify the depth of a physical item. A DT is a computer-generated version of a real-world object, feature, operation, process, or service that is linked to the real thing via sensors and data flow.

Sensors are tools that gauge the physical properties of things like temperature, pressure, motion, or vibration. Production line health, performance indicators, defect indicators, and other information are brought via sensors. It can offer information that is close to real-time, such as a continuous stream of data delivered from the sensor to the communication network's digital counterpart. Rich information and opportunity for analysis and decision-making can be found in data streams. DTs can help collect, store, and communicate information from sensor data in production lines using models, algorithms, and software.

Models can be used to better understand how a process or system operates and how it responds to changes in parameters or input. A computer is instructed by a collection of rules or instructions known as an algorithm on how to complete a task or address an issue. Algorithms aid in the processing, analysis, and generation of outputs or solutions from data. A computer or other device can run a collection of programs or apps under the name "software." The software can help with model and algorithm implementation and offer a user interface for interaction and visualization.

DTs can carry out functions including monitoring, diagnosis, prediction, optimization, control, and learning using models, algorithms, and software. By boosting capacity, these resources can assist businesses in gaining knowledge from big data on the manufacturing line and enhancing the production process. Surveys are another method that DTs can employ to let consumers access and use the data they store.

Semantic inquiry is a method of posing inquiries or requesting data using everyday language or concepts that have specific meanings in a given situation. Inference rules and key inquiries on knowledge graphs can be performed using query-based semantic query techniques. A knowledge map is a graph-based model that uses nodes and edges to represent the relationships between things in a domain.

Nodes stand in for entities like thoughts, things, events, or truths. Edges depict relationships such as those between things, traits, operations, or outcomes. Key questions are those that have a number of requirements that must all be met in order to receive an answer. Rules known as inference rules enable new facts to be inferred logically from preexisting ones. According to the user's inquiry, a sentence-based query type can leverage key queries and inference rules to give pertinent data from image recognition.

5.9 Limitations and Future Research

The limitations of the suggested model and deep learning methods in general while deep learning methods have produced promising results in the development of predictive models for the DT model of industrial production control management, there are still some restrictions to take into account.

5.9.1   Data Availability and Quality

Deep learning models require a large amount of high-quality data to be trained successfully. In contrast, due to sensor noise or missing values, data might be scarce and of low quality in industrial settings. Additionally, data collected in one environment cannot be representative of another, which could lead to overfitting or generalization issues.

5.9.2   Interpretable Predictions

Deep learning models are sometimes regarded as "black boxes," which makes it difficult to comprehend how the model generates its predictions. This could be a drawback in industrial settings where stakeholders must comprehend the logic underlying the model's predictions in order to take the proper action.

5.9.3   Model Complexity and Training Time

Deep learning models can be quite complicated, which makes them computationally expensive and necessitates a large investment in time and materials during training. This might be a drawback in industrial environments where quick responses and real-time predictions are crucial.

5.10   Identification of Future Research Opportunities and Directions

5.10.1   Integration of Multiple Data Sources

To create more complete models for DTs, future study may investigate the integration of various data sources, such as text, image, and audio data. For instance, natural language processing techniques may be used to analyze text data from maintenance logs or repair reports, or computer vision techniques may be used to analyze images from sensors or cameras.

5.10.2   Transfer Learning

Deep learning models can be quite complicated, which makes them computationally expensive and necessitates a large investment in time and materials during training. This might be a drawback in industrial environments where quick responses and real-time predictions are crucial.

5.10.3   Explainable AI

Models are developed in the developing field of explainable AI that provide comprehensive justifications for the predictions they make. Transfer learning is a technique for improving a model's performance on a new task by using models that have already been taught. Future studies might look towards the creation of explainable AI methods for DTs so that stakeholders can more easily comprehend the justification for the predictions made by the model.

5.10.4   Human-in-the-loop Approaches

Human input is incorporated into the decision-making process of the model in "human-in-the-loop" techniques. The usage of human-in-the-loop methods for DTs, wherein operators or maintenance staff might provide feedback on the model's predictions and gradually improve its performance, could be explored in future study.

5.10.5   Interoperability and Standardization

Interoperability and standardization will be more crucial as the use of DTs spreads. Future studies could examine the creation of common frameworks and standards for DTs, facilitating increased cooperation and knowledge exchange between academics and industry professionals. In conclusion, the field of the DT model of industrial production control management using deep learning techniques offers many exciting opportunities for future research, and scholars and practitioners should keep looking into novel methods and strategies to enhance the effectiveness and dependability of these models.

5.11   Applications and Benefits

5.11.1   Use of Digital Twins for Business Management Using Deep Learning

Designing the product design and improvement using simulation and reality. For instance, create the aluminum alloy flange using Workbench and SolidWorks software based on the DT model.

Utilizing real-time data and feedback, the monitors and regulates the production process and product quality. For instance, gathering information from the body and updating the DT using sensors and IoT devices. Based on previous data and ML, it forecasts and eliminates product risks and failures. As an illustration, interior product problems can be identified by utilizing deep learning models and ultrasonic examination. Based on health management and diagnostics, manage and enhance the performance and dependability of products. Using a DT model, for instance, to analyze product wear and tear and offer prompt maintenance advice. The efficiency and quality of product design and manufacture can be improved by reducing the cost and time spent on physical prototypes and testing. Using data-driven and intelligence-driven methods, NetSuite can increase the accuracy and reliability of product prediction and prevention, which will enhance the performance and lifespan of repair and maintenance goods by offering quick and individualized solutions. By facilitating sharing and agreement on products, it ensures the integration and cooperation of many stakeholders in the product industry. DT models that come to mind as examples include: (i) a digital replica of a car that models the aerodynamics, fuel efficiency, safety, and environmental effects of various driving scenarios; (ii) the DT of the wind turbine can monitor its performance, power production, weather, and fault using real-time data from sensors and IoT devices; and (iii) a digital replica of the human body that uses technology and personal data to track vital signs, health, medical history, and lifestyle. The expected usage rate for industrial equipment manufacturing enterprises is shown in Figure 5.3.

5.12 Results and Discussions

The DT approach for industrial production management was used and assessed in factory production lines using deep learning techniques. The model is used to forecast equipment breakdown and enhance production planning. It is trained using data gathered from sensors placed on production lines. The findings demonstrate that the model can accurately, precisely, and reliably forecast device failure using F1 scores. Real-time decision-making within the model boosts effectiveness while decreasing maintenance time and expense.

Discussing the findings demonstrates the model's benefits in boosting business productivity and efficiency. Increased productivity and better performance may result from the model's capacity to anticipate equipment breakdown and shorten manufacturing times. Real-time decision-making in the model provides quick response to unanticipated events, minimizing downtime and production disruptions. The discussion does, however, clearly address the flaws and restrictions of the suggested paradigm. A significant difficulty is the caliber and consistency of the data gathered by the sensors, which has a huge direct impact on the precision and dependability of the prediction model. Another difficulty is the model's deep learning algorithms' complexity, which will necessitate specialized knowledge and expensive processing power to implement and maintain.

Overall, the findings and discussion show how the suggested DT model for industrial management has the potential to boost production and efficiency in fields using deep learning. More study is required to verify the model's viability and efficacy as well as to solve problems with data quality, interaction, and model interpretation.

Some of the ongoing discussions on this topic are:

- DT ideas and traits, as well as how they differ from other ideas like simulation, virtual reality, and cyber-physical systems.
- The actual idea behind DTs, how they are used, and how other technologies like Big Data, AI, and the IoT affect them.
- The energy production, health care, smart cities, etc. The advantages and difficulties of using DTs in different applications and disciplines, include.
- Aspects of DTs that are ethical and social, such as data privacy, security, ownership, governance, and accountability.
- How to assess the correctness, dependability, and efficiency of DTs as well as their performance.

5.13 Potential Domains for Digital Twins

5.13.1 Manufacturing

By providing simulation, monitoring, control, prediction, and improvement, DTs can aid in the optimization of industrial product and system design, manufacture, and operation. They can also make it easier for many stakeholders to work together on supply chain and product life cycle integration. The various domains that use DT technology are shown in Figure 5.4.

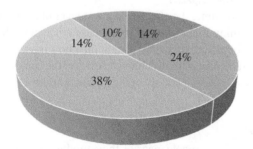

▪ Energy ▪ Automative ▪ Manufacturing ▪ Supply chain ▪ Healthcare

Figure 5.4 Pie chart of different domains that use digital twin.

5.13.2 Energy

By facilitating real-time data analysis, optimization, and decision-making, DTs can assist increase the efficiency and reliability of energy generation, transmission, and consumption. By promoting renewable energy sources and smart grids, they can also aid in lowering the negative effects energy systems have on the environment and their carbon imprint.

5.13.3 Healthcare

By enabling personalized care, precision medicine, and preventative medicine, DTs can improve the diagnosis, treatment, and prevention of diseases and disorders. By making telemedicine and remote monitoring possible, they can also aid in enhancing the quality and accessibility of healthcare services.

5.13.4 Supply Chain

Supply chain managers can use DTs to design and test various shipping, distribution, and inventory management scenarios. They can also enhance network optimization, storage, and delivery protection.

5.13.5 Automotive

5.13.5.1 Product Testing
A product's DT aids in evaluating its quality and performance by testing out different raw materials and components to improve design and boost product performance. For instance, the DT of a new tire can be accurately modeled, tested in various weather conditions, and improved based on the results.

5.13.5.2 Adding Manufacturing Capacity
Businesses can utilize DTs to mimic the effects and advantages of new equipment on production before installing them in order to increase efficiency. The virtual model should take into account, and then offer insight into, how new technologies might boost productivity. It incorporates the characteristics of the company's products, the information it possesses, the history of the production time, the required machines, and so on.

5.13.5.3 Employee Training
Companies can create a digital doppelganger of their production infrastructure so that staff can be trained remotely without having to upgrade their hardware.

For instance, to speed up the process, a manufacturing business in Europe may instruct Mexican personnel on the DT of the facility before installing it in Mexico. hiring new employees and recognizing their need.

5.13.5.4 Predictive Maintenance

DTs of the production equipment and machinery can be used to enhance the health of production lines and companies. In this case, the DT must make use of real-time data gathered from IoT devices and sensors throughout the production process to identify recurring mistakes.

5.13.5.5 Sales

One of the upcoming implications of DT technology is sales, where consumers can voice their opinions before the goods are even delivered to the retailer. Companies may let customers evaluate the vehicle, spot new features, and contrast it with an earlier model thanks to the car's DT.

5.13.5.6 3D: Car Design and Product Development

Through 3D visualization, which also avoids the delays associated with traditional 3D car rendering software, global collaboration is made feasible.

5.13.5.7 Human–machine Interfaces (HMI)

Design interactive 2D and 3D user interfaces for in-car entertainment (IE) systems and digital cockpits.

5.14 Conclusion

In conclusion, the suggested deep learning–based DT model for corporate management provides a useful solution to boost organizational productivity and efficiency. To forecast equipment breakdown, maximize production efficiency, and facilitate prompt decision-making, the model makes use of sensor data and deep learning techniques.

The plant's production line design model has been used and evaluated, and results indicate that it is useful for forecasting equipment failure and enhancing output. Rapid response to unanticipated events is made possible by the model's real-time decision-making, reducing downtime and production interruptions. The quality and consistency of the data acquired by the sensors as well as the complexity of the deep learning techniques employed in the model are issues that still need to be resolved even if the model has a lot of potential.

References

1 Guc, F. and Chen, Y. (2022). Smart Predictive Maintenance Enabled by Digital Twins and Smart Big Data: A New Framework. https://doi.org/10.1109/dtpi55838.2022.9998937.

2 Poletikin, A., Promyslov, V., and Semenkov, K. (2021). Digital Twins as a Key Technology for Ensuring Effective Modernization of Complex Industrial Systems. https://doi.org/10.1109/rusautocon52004.2021.953 7443.

3 Batteux, M., Khebbache, S., and Seo, S.-S. (2022). Assessment of Maintenance Strategies for Digital Twin Applications. https://doi.org/10.1109/icsrs56243.2022 .10067492.

4 Wang, W., Hu, H.M., Zhang, J., and Hu, Z. (2020). Digital Twin-based Framework for Green Building Maintenance System. https://doi.org/10.1109/ieem45057.2020.9309951.

5 Bhatti, G. and Singh, R.P. (2021). Intelligent Fault Diagnosis Mechanism for Industrial Robot Actuators using Digital Twin Technology. https://doi.org/10.1109/iprecon52453.2021.9641000.

6 Aivaliotis, P., Georgoulias, K., and Alexopoulos, K. (2019). Using Digital Twin for Maintenance Applications in Manufacturing: State of the Art and Gap Analysis. https://doi.org/10.1109/ice.2019.8792613.

7 Buchner, A., Micheli, G., and Gottwald, J. et al. (2022). Human-centered Augmented Reality Guidance for Industrial Maintenance with Digital Twins: A Use-Case Driven Pilot Study. https://doi.org/10.1109/ismar-adjunct57072.2022.00024.

8 Siddiqui, M., Kahandawa, G., and Hewawasam, H.S. (2023). Artificial Intelligence Enabled Digital Twin For Predictive Maintenance in Industrial Automation System: A Novel Framework and Case Study. https://doi.org/10.1109/icm54990.2023.10101971.

9 Chen, Q., Zhu, Z.-W., Si, S., and Cai, Z. (2020). Intelligent Maintenance of Complex Equipment Based on Blockchain and Digital Twin Technologies. https://doi.org/10.1109/ieem45057.2020.9309898.

10 Karhela, T. and Vyatkin, V. (2018). Automatic Generation of a Simulation-Based Digital Twin of an Industrial Process Plant. https://doi.org/10.1109/iecon.2018.8591464.

11 Mulubrhan, F. and Saptaji, K. (2022). Digital Twin Enabled Industry 4.0 Predictive Maintenance Under Reliability-Centered Strategy. https://doi.org/10.1109/iceeict53079.2022.9768590.

12 Santos, J.F., Tshoombe, B.K., Santos, L.H.B. et al. (2022). Digital twin-based monitoring system of induction motors using IoT sensors and thermo-magnetic finite element analysis. *IEEE Access* 11: 1682–1693. https://doi.org/10.1109/access.2022.3232063.

13 Söderberg, R., Wärmefjord, K., Carlson, J.S., and Lindkvist, L. (2017). Toward a digital twin for realtime geometry assurance in individualized production. *CIRP Annals* 66 (1): 137–140.

14 Wu, Z., Chen, S., Han, J. et al. (2022). A low-cost digital twin-driven positioning error compensation method for industrial robotic arm. *IEEE Sensors Journal* 22 (23): 22885–22893. https://doi.org/10.1109/jsen.2022.3213428.

15 Schluse, M. and Rossmann, J. (eds.) (2016). From simulation to experimentable digital twins. Simulation-based development and operation of complex technical systems. *Systems Engineering (ISSE), 2016 IEEE International Symposium on IEEE.*

16 Brenner, B. and Hummel, V. (2017). Digital twin as enabler for an innovative digital shopfloor management system in the ESB logistics learning factory at Reutlingen University. *Procedia Manufacturing* 9: 198–205.

17 Grieves, M. and Vickers, J. (2017). Digital Twin: Mitigating Unpredictable, Undesirable Emergent Behavior in Complex Systems. In: Kahlen, J., Flumerfelt, S., Alves, A. (eds) Transdisciplinary Perspectives on Complex Systems. Springer, Cham. https://doi.org/10.1007/978-3-319-38756-7_4.

6

Digital Twin for Sustainable Development of Intelligent Manufacturing

Kuldeep Singh Kaswan[1], Jagjit Singh Dhatterwal[2], Sumit Singh Dhanda[3], Balamurugan Balusamy[4], and Azween Abdullah[5]

[1]Department of Computer Science & Engineering, Galgotias University, Greater Noida, India
[2]Department of Artificial Intelligence & Data Science, Koneru Lakshmaiah Education Foundation, Vaddeswaram, Andhra Pradesh, India
[3]School of Computer Science & Engineering, IILM University, Greater Noida, UP, India
[4]Shiv Nadar Institute of Eminence, Delhi, India
[5]Faculty of Applied Science and Technology, Perdana University, Kuala Lumpur, Malaysia

6.1 Introduction to Digital Twin Environment

The idea of a computerized twin has earned critical consideration and respect as of late, arising as an amazing asset for improving the capacities of different businesses. A virtual replica or representation of a real-world entity, system, or procedure is known as a digital twin. It includes not only the physical characteristics of the entity it represents but also its conduct and constant information. Through sensors, Web of Things (WoT) gadgets, and other data sources, this digital replica is connected to its physical counterpart, allowing for the exchange of information and offering a comprehensive comprehension of the physical entity's performance [1].

The digital twin environment bridges the physical and virtual worlds and offers numerous advantages in various fields. Concerning assembling, computerized twins have arisen as a groundbreaking innovation, upsetting customary creation processes and empowering insightful assembling frameworks.

Manufacturers can create virtual models in a digital twin environment that replicate their actual production systems, including machines, equipment, and entire production lines. Ongoing information is gathered from sensors implanted in the actual resources, and this information is utilized to refresh and synchronize the computerized twin, making a constant portrayal of the actual framework.

Digital Twins in Industrial Production and Smart Manufacturing:
An Understanding of Principles, Enhancers, and Obstacles, First Edition. Edited by Rajesh Kumar Dhanaraj, Balamurugan Balusamy, Prithi Samuel, Ali Kashif Bashir, and Seifedine Kadry.

Manufacturing processes can be continuously monitored, analyzed, and improved thanks to this synced connection between the real and virtual worlds [2].

The computerized twin climate offers a few benefits for smart assembling. First and foremost, it enables manufacturers to identify inefficiencies, bottlenecks, and areas for improvement by providing them with a deeper comprehension of their production systems. Continuous information from the actual resources can be utilized to mimic and break down various situations, empowering prescient upkeep, process advancement, and execution improvement.

Additionally, the digital twin environment facilitates the implementation of data-driven decision-making processes. To make better decisions, allocate resources more effectively, and increase production efficiency, manufacturers can use the insights provided by the digital twin. They can cost-effectively minimize risks and optimize their operations by simulating and testing various strategies and configurations in the virtual environment.

The environment of digital twins has the potential for long-term growth, which is yet another significant advantage. By observing and investigating continuous information, producers can distinguish open doors for energy improvement, squander decrease, and asset effectiveness. The digital twin enables the execution of sustainable practices, such as predictive energy management, intelligent supply chain optimization, and waste reduction strategies. Sustainable development and environmental goals are both aided by this incorporation of sustainability considerations into the manufacturing process [3].

6.1.1 Digital Twin and Its Role in Intelligent Manufacturing

A virtual replica or representation of a real-world entity, system, or procedure is known as a digital twin. It includes not only the physical characteristics of the entity it represents but also its behavior and real-time data. It enables a synchronized connection among the real and digital worlds because it is created and updated using real-time data from connected sensors, Internet of Things (IoT) devices, and other data sources.

Job of Advanced Twin in Shrewd Assembling

Digital twins are driving intelligent manufacturing systems in a significant way. Digital twins play a key role in intelligent manufacturing, and their advantages include the following:

- Data Collection and Monitoring in Real Time: Digital twins make it easier to keep an eye on manufacturing processes and physical assets all the time.

Sensors and IoT gadgets gather continuous information, which is utilized to refresh and synchronize the advanced twin. Proactive maintenance and performance optimization are made possible by this data collection, which gives manufacturers a comprehensive understanding of the performance, behavior, and condition of their assets [4].

- Predictive Repairs: Digital twins are able to anticipate and pinpoint potential equipment failures or maintenance requirements by analyzing real-time data from the physical assets. By taking a proactive approach to maintenance, unplanned downtime can be reduced, maintenance schedules can be improved, and equipment lifespan can be extended. Digital twins enable predictive maintenance, which boosts equipment efficiency and lowers maintenance costs.

- Optimization and Simulation: Manufacturers can simulate and optimize a variety of scenarios and configurations in a virtual environment that is provided by digital twins. Before putting the most efficient and effective strategies into action in the real world, manufacturers can test a variety of strategies and process parameters in the digital twin. This capability of simulation aids in the improvement of product quality, the reduction of energy and waste, and the optimization of production procedures.

- Support for Real-Time Decisions: Digital twins' real-time data collection and analysis can provide useful insights for decision-making. Utilizing this data, manufacturers can quickly respond to shifting market conditions or production demands, optimize resource allocation, and make data-driven decisions. In manufacturing, real-time decision support made possible by digital twins increases operational efficacy, adaptability, and responsiveness [5].

- Innovation and Constant Improvement: In manufacturing, digital twins facilitate innovation and continuous improvement. By observing and dissecting the exhibition information of actual resources, makers can distinguish regions for development, execute changes, and assess the effect of those adjustments of the virtual climate. The manufacturing processes' efficiency, quality, and innovation are all aided by this iterative improvement and innovation process.

- Sustainable Production: Digital twins help make manufacturing more environmentally friendly. By incorporating ongoing information and investigation, makers can upgrade asset use, decrease energy utilization, limit squander age, and work on ecological execution. Digital twins make sustainable practices like energy management, waste reduction strategies, and intelligent supply chain optimization possible, which makes manufacturing operations more sustainable.

6.1.2 Sustainable Development in Manufacturing

The integration of environmental, social, and economic considerations into production processes and practices is referred to as sustainable development in manufacturing. The goal is to promote long-term economic viability while minimizing negative effects on the environment, society, and generations to come. The idea emphasizes the necessity of a comprehensive strategy that strikes a balance between economic expansion, social well-being, and environmental stewardship. The digital twin environment has a lot of potential for supporting manufacturing's sustainable development. Digital twins enable manufacturers to optimize processes, improve resource efficiency, reduce waste, and minimize their impact on the environment by utilizing advanced analytics and real-time data. An overview of the most important aspects of sustainable development in the digital twin environment is as follows:

- Efficiency of Resources: Advanced twins empower producers to enhance asset use by giving continuous experiences into energy utilization, material use, and asset designation. By dissecting the information gathered from the advanced twin, makers can distinguish failures, carry out energy-saving measures, and streamline creation cycles to limit asset squander [6].
- Reduced Waste: By monitoring and analyzing production processes in real time, the digital twin environment helps reduce waste. Manufacturers can improve process efficiency, reduce scrap and rework, and optimize workflows by locating inefficiencies and bottlenecks in the virtual environment. Cost savings, enhanced product quality, and reduced environmental impact are the outcomes of this waste reduction.
- Control of the Energy: By enabling real-time energy consumption monitoring, analysis, and optimization, digital twins support sustainable energy management practices. Manufacturers can identify energy-intensive areas, implement energy-saving measures, and optimize production schedules to integrate energy data from the physical assets into the digital twin in order to cut down on energy use and emissions of greenhouse gases (GHGs).
- Analyses of the Effects on the Environment: Assessment and reduction of manufacturing processes' environmental impact can be accomplished with the help of digital twins. Manufacturers are able to evaluate the environmental consequences of their decisions and identify opportunities for improvement by simulating and analyzing various scenarios in the virtual environment. Sustainable practices can be implemented and environmental risks reduced with this proactive approach [7].
- Optimization of the Supply Chain: Manufacturers can better manage logistics, transportation, and inventory with the help of the digital twin environment,

which gives them a complete picture of the supply chain. Manufacturers can reduce carbon emissions, reduce waste produced throughout the supply chain, and improve overall sustainability by optimizing processes in the virtual supply chain.

- Assessment of the Lifecycle: By providing a comprehensive view of the product lifecycle—from design and manufacturing to use and disposal—digital twins can aid lifecycle assessment. Manufacturers can identify opportunities for improvement, such as eco-design, product reuse, and recycling strategies, by integrating data on materials, energy consumption, and environmental impact throughout the product lifecycle.
- Sustainable Change: By allowing manufacturers to test and evaluate new technologies, materials, and processes in the virtual environment prior to implementing them in the real world, the digital twin environment encourages sustainable innovation. This makes it possible to evaluate their potential benefits, feasibility, and impact on the environment, encouraging the creation and adoption of sustainable innovations [8].

6.1.3 Measurable Examination Approach in Evaluating the Effect of Computerized Twin on Reasonable Turn of Events

The quantitative factual investigation approach is utilized to assess the effect of computerized twin innovation on feasible assembling advancement. The objective of this strategy, which makes use of statistical techniques and the collection of pertinent data, is to examine the connections, trends, and correlations that exist between the adoption of digital twins and sustainability indicators. The following is a description of the statistical analysis technique used to evaluate the impact of a digital twin on sustainable development.

- Collection of Data: The collection of a comprehensive dataset containing data on digital twin adoption and relevant sustainability indicators is the first step in the statistical analysis approach. This information can be gathered from different businesses or assembling associations that have carried out advanced twin innovation in their activities. Variables related to resource efficiency, waste reduction, energy consumption, production efficiency, and other sustainability metrics ought to be included in the dataset.
- Clear Insights: The collected data are summarized and described using descriptive statistics. This includes determining each variable's mean, median, standard deviation, and range. Graphic insights give an outline of the information conveyance and feature any critical patterns or examples [9].
- Analysis of Correlations: The correlation method is used to assess the degree and direction of the link between digital twins' deployment and environmentally

friendly metrics. This study analyzes if these variables are statistically significantly connected. Correlation coefficients, such as the correlation coefficient calculated by Pearson, can be used to quantify the degree of linkage between variables, indicating the magnitude and course of the connection. Controlling for other relevant variables, regression analysis is used to determine the impact of digital twin adoption on sustainability indicators. In order to determine how each independent variable affects the dependent variable (the sustainability indicator), multiple regression analysis can be used. This analysis assists in determining the specific contribution that the use of digital twins makes to the outcomes of sustainable development.

- Testing the Hypothesis: Speculation testing is utilized to approve the factual meaning of the connections recognized in relationship and relapse examinations. The null hypothesis, in which there is no connection between adoption of digital twins and sustainability indicators, and the alternative hypothesis, in which there is a significant connection, are two examples of hypotheses that are put forth. The significance of the results and their statistical validity are assessed using statistical tests like *t*-tests and analysis of variance (ANOVA).

- Visualization of Data: The statistical results are represented visually using data visualization methods like graphs, charts, and plots. Stakeholders and decision-makers will find the analysis' relationships, trends, and patterns easier to understand thanks to these visualizations which aid in their presentation.

- Conclusion and Interpretation: Interpreting the statistical results and drawing meaningful conclusions is the final step. The examination helps in grasping the effect of computerized twin reception on manageable advancement results in assembling. In light of the factual discoveries, analysts, policymakers, and producers can go with informed choices, foster techniques, and execute measures to advance reasonable improvement through computerized twin usage [10].

6.2 Statistical Analysis Methods

- Association Examination: The strength and course of the connection between two not entirely set in stone through relationship examination. It recognizes whether there is an immense connection between modernized twin gathering and reasonability markers. The degree of the factors' relationship can be measured using Spearman's position relationship coefficient or Pearson's connection coefficient.

- Analyses of Regression: The relationship between a dependent variable (the sustainability indicator) and one or more independent variables, such as digital twin adoption and other relevant factors, is the subject of regression analysis. Multiple independent variables can be included in multiple regression analysis to examine how they affect the dependent variable as a whole. It aids in determining the specific contribution that the use of digital twins makes to the outcomes of sustainable development.

- Testing the Hypothesis: To confirm the statistical significance of the relationships found in correlation and regression analyses, hypothesis testing is used. To determine the significance of the results, statistical tests like *t*-tests and ANOVA are used to formulate null and alternative hypotheses. Hypothesis testing aids in determining the statistical validity of the observed relationships.

- Time Series Examination: When the data collected represents observations over time, time series analysis can be used. It distinguishes patterns, examples, and irregularity in the information. The utilization of time series methods like remarkable smoothing or AutoRegressive Integrated Moving Average (ARIMA) can be utilized to take a gander at what the reception of computerized twins means for signs of practical improvement over the long haul.

- Analyses of Clusters: Bunch examination is utilized to recognize gatherings or groups inside a dataset in light of likenesses or dissimilarities between perceptions. It can be used to sort manufacturing systems or businesses by how much digital twin use they have and see if some clusters have better sustainability results than others [11].

- Investigations of Difference (ANOVA): ANOVA is utilized to analyze the method for different classes or gatherings. It can be used to determine whether manufacturing systems with varying levels of digital twin adoption have significant differences in sustainability indicators. ANOVA assists in determining whether digital twin technology has a statistically significant effect on the outcomes of sustainable development across various groups.

- Test of Chi-Square: The chi-square test is used to determine whether categorical variables are independent or related. It can be used to investigate the connection between the adoption of a digital twin—a categorical variable—and sustainability outcomes in various manufacturing contexts, such as reductions in waste or improvements in energy efficiency.

6.2.1 Selection of Appropriate Statistical Techniques for the Analysis

The selection of the appropriate statistical methods for the analysis is influenced by the nature of the data, the nature of the relationship being studied, and the specific research question. Here are a few elements to consider while picking measurable methods for examining the effect of computerized twin on economic turn of events.

- Goals of the Study: Define the specific questions you want to answer and the goals of your research in a clear way. Find out if you want to learn about relationships, make predictions, compare groups, or see patterns over time. The statistical methods that best answer your research questions will be chosen based on your objectives.

- Data Formats: Consider the sorts of information you have gathered or approach. Determine whether your data is categorical, continuous, or time series. This will assist you with distinguishing suitable factual techniques that can deal with the particular information types and backing significant investigation.

- Evaluation of Relationship: Consider the idea of the relationship you need to survey between computerized twin reception and manageability markers. Are you interested in correlating the relationship's strength and direction (correlation analysis)? Do you want to use regression analysis to evaluate the impact of using digital twins while controlling for other factors? Are you using ANOVA or chi-square to compare groups or categories? Pick statistical methods that are compatible with the relationship you want to investigate.

- Sample Size: Think about the size of your example. While larger sample sizes may permit more robust analysis, certain statistical techniques may not be applicable to small sample sizes. Regression analysis, for example, requires a sufficient sample size to support statistical inference and provide accurate estimates.

- Assumptions: Learn about the underlying assumptions of various statistical methods. Regression analysis, for instance, relies on the assumption that residuals are independent, normal, and linear. Check to see if your data conforms to these presumptions or if you need to think about other methods that are better suited to non-normal or nonlinear data.

- Time Reliance: Assuming that you have time-series information or information gathered throughout various time spans, think about utilizing time series examination methods to distinguish patterns, examples, and irregularity. Time series analysis techniques like exponential smoothing or ARIMA can help you understand how the use of digital twins affects sustainability indicators over time.

- Complex Connections: Advanced methods like structural equation modeling (SEM) or machine learning algorithms like random forests and neural networks may be useful if your research involves intricate relationships or interactions between multiple variables. These procedures can deal with non-straight connections, multivariate investigation, and give more complete experiences into the effect of advanced twin on supportable improvement [12].

6.2.2 Description of the Variables Considered in the Analysis

The specific research context and goals may influence the variables that are taken into consideration in the analysis of the impact of the digital twin on sustainable development. However, the following are some typical variables that could be taken into account in such an analysis:

- Digital Twin Adoption: This variable indicates the degree to which manufacturing processes incorporate digital twins or make use of them. It can be expressed as a continuous variable indicating the degree of integration of digital twin technology or as a scale of low, medium, or high adoption.
- Indicators of Sustainability: The various facets of manufacturing's sustainable development are captured by these variables. Economic, social, and environmental indicators are all options. Consumption of energy, production of waste, emissions of GHGs, water use, social impact, economic performance, and resource efficiency are all examples of sustainability indicators.

6.2.3 Statistical Models Used to Evaluate the Relationship Between Digital Twin and Sustainable Development in Intelligent Manufacturing

A few factual models can be utilized to assess the connection between computerized twin and reasonable improvement in well-informed fabricating. The research objectives, characteristics of the data, and nature of the relationship being investigated all influence the choice of model. An overview of some of the most frequently used statistical models in this context is provided below:

- Linear Regression: The statistical model known as linear regression is used to investigate the connection that exists between a single independent variable and a dependent variable. While controlling for other factors in the evaluation of the impact of digital twins on sustainable development, linear regression can be utilized to examine how the adoption of digital twins affects sustainability indicators. The model gauges the extent and bearing of the relationship and can recognize critical indicators.
- Various Relapses: By simultaneously considering multiple independent variables, multiple regression extends linear regression. The collective impact of digital twin adoption and other relevant variables on sustainability indicators can be evaluated using this model. It sheds light on the overall connection between the digital twin and outcomes of sustainable development, as well as the distinct contribution that each predictor variable makes [13].

6.3 Impact on Resource Efficiency

Intelligent manufacturing that makes use of digital twin technology may have a significant impact on resource efficiency and contribute to sustainable development. Digital twin environments can improve resource efficiency in intelligent manufacturing in the following ways:

- Optimal Use of Resources: Real-time monitoring and analysis are made possible by the virtual representation that digital twins provide of actual manufacturing systems. By incorporating information from different sensors and sources, advanced twins can improve the distribution of assets like energy, natural substances, and hardware. They can distinguish failures, limit squander, and further develop mostly asset use.
- Predictive Repairs: Advanced twins work with prescient support by consistently checking gear and frameworks progressively. By breaking down execution information, they can foresee gear disappointments or upkeep needs before they happen. Unplanned downtime is minimized, resource waste associated with repair and replacement is reduced, and maintenance schedules are optimized with this proactive approach.
- Optimization and simulation: Before implementing them in the real world, manufacturers can simulate and improve production processes in digital twin environments. Manufacturers can streamline workflows, identify areas for improvement, and reduce resource consumption through virtual testing and analysis. By enhancing simulations in the virtual environment, they can reduce the number of experimental iterations and resource-intensive testing in the real world.
- Energy Productivity: Digital twins make it possible to analyze and monitor energy usage in great detail across manufacturing systems. They can follow energy utilization designs, recognize energy-escalated cycles or hardware, and recommend energy-saving measures. By optimizing energy use, digital twins aid in the reduction of carbon emissions and improvement of manufacturing operations' overall energy efficiency [14].
- Optimization of the Supply Chain: Advanced twin innovation can reach out past the limits of individual assembling frameworks and incorporate the whole production network. By making a virtual portrayal of the production network, computerized twins can enhance operations, stock administration, and transportation, prompting decreased asset utilization, limited squander, and further developed supportability.
- Assessment of the Lifecycle: By tracking and analyzing the environmental impact of products throughout their entire lifecycle, digital twins can support

lifecycle assessment. Manufacturers can identify opportunities for improvement, develop greener alternatives, and make informed decisions to improve resource efficiency by taking into consideration factors like the selection of materials, manufacturing processes, product use, and disposal.

6.3.1 Statistical Assessment of Resource Consumption Reduction Achieved Through the Implementation of a Digital Twin

The following steps are typically followed to carry out a statistical analysis of the resource consumption reduction achieved using a digital twin:

- Specific Metrics: Decide the particular measurements that will be utilized to gauge asset utilization decrease. This could incorporate energy use, material waste, water utilization, or some other applicable markers in light of the assembling system and objectives.
- Basic Statistics: Before using the digital twin, collect information about how much you use of the resources. This establishes the starting point for resource consumption and serves as a baseline for comparison.
- Execution Stage: In the manufacturing setting, use the digital twin system. To accurately capture information about resource consumption, make sure that data collection mechanisms are in place. Keep track of and record the amount of resources used during the implementation phase.
- Information Examination: Evaluate the resource savings achieved by implementing the digital twin by analyzing the collected data. This involves contrasting the utilization of resources during the implementation phase with the earlier-collected baseline data.
- Factual Procedures: Determine the significance of the resource consumption reduction by employing appropriate statistical methods. Statistical techniques like t-tests, ANOVA, regression analysis, and other relevant ones may be utilized, depending on the data distribution and the goals of the research [15].
- Hypothesis Testing and Confidence Intervals: Estimate the range of resource consumption reduction achieved by calculating confidence intervals. Test your hypotheses to see if the observed decrease in resource consumption is statistically significant.
- Interpretation: To comprehend the impact of the digital twin implementation on resource consumption reduction, interpret the statistical results. This includes assessing the reduction's magnitude, statistical significance, and implications for resource efficiency and sustainable development.
- Restrictions and Suggestions: The availability of data, sample size, or other potential confounding factors should all be discussed in relation to the statistical evaluation. Based on the findings, make suggestions for areas of improvement or focus.

6.3.2 Analysis of Energy Efficiency Improvements in Manufacturing Processes

Dissecting energy productivity enhancements in assembling processes includes evaluating the energy utilization designs when executing energy-saving measures. An analysis of this kind typically involves the following steps:

- Specific Metrics: Determine the metrics that will be used to evaluate improvements in energy efficiency. Energy intensity and any other relevant indicators are examples of common metrics.
- Basic Statistics: Before implementing energy-saving measures, gather data on the manufacturing processes' energy consumption. This provides insight into the initial energy performance and a baseline for comparison.
- Audit of Energy: Conduct an energy audit to find opportunities to save energy. This entails examining the systems, equipment, and manufacturing processes for inefficiencies and potential enhancements. On-site inspections, data collection, and patterns of energy use analysis may all be part of an energy audit.
- Efforts to Save Energy: Based on the findings of the energy audit, take measures to save energy. Upgrades to equipment, process optimization, energy management systems, employee training on energy efficiency practices, and other initiatives to save energy could be some of these measures.
- Post-Execution Information Assortment: Collect data on the amount of energy used in the same manufacturing processes following the implementation of energy-saving measures. Ensure that energy usage data is captured using precise measurement mechanisms.
- Information Examination: Examine the energy utilization information when the execution of energy-saving measures. Determine the amount of energy saved the improvement in percentage, or any other relevant indicators.
- Measurable Appraisal: Utilize statistical methods to evaluate the significance of improvements in energy efficiency. Factual strategies, for example, t-tests, ANOVA, relapse investigation, or other fitting methods, can be utilized to assess the extent and measurable meaning of the enhancements.
- Interpretation: Learn how energy-saving measures affect energy efficiency by interpreting the findings. Assess the rate improvement, outright energy reserve funds, recompense periods, or profit from venture (return for money invested) for the executed measures. Consider the energy efficiency enhancements' viability, scalability, and long-term advantages.
- Recommendations and Reporting: Set up an extensive report summing up the energy productivity examination, including the technique, discoveries, and suggestions for additional enhancements. Identify areas where additional energy-saving measures can be implemented or suggest future R&D projects [16].

6.3.3 Quantification of Waste Reduction and Improved Recycling Rates

Assessing the amount of waste generated and recycled before and after implementing initiatives to reduce waste and increase recycling rates is necessary for quantifying these results. A general approach to quantifying these enhancements is as follows:

- Specific Metrics: Make a list of the metrics that will be used to measure decreased waste and increased recycling rates. Normal measurements incorporate all out squander created (in weight or volume), squander per unit of creation, reusing rate (level of waste reused), or some other important markers well defined for the business or waste streams.
- Basic Statistics: Before implementing initiatives to reduce waste and recycle, gather information on recycling rates and the amount of waste produced. This helps establish the initial waste generation and recycling performance and provides a baseline against which to compare.
- Squander Review: To find opportunities for reducing waste and recycling, conduct a waste audit. Investigate the kinds of waste created, their sources, and removal techniques. Find opportunities to reduce, reuse, recycle, or recover waste. Prioritize efforts to reduce waste by taking into consideration the entire waste management hierarchy (reduce, reuse, recycle).
- Initiatives for Reducing Waste and Recycling It: Based on the waste audit's findings, implement initiatives to reduce waste and recycle. Changes to processes, substitution of materials, waste segregation, employee training, partnerships with recycling facilities, and other waste management practices are all examples of these initiatives [17].
- Post-Execution Information Assortment: In the wake of carrying out squander decrease and reusing drives, gather information on squander age and reusing rates utilizing suitable estimation and global positioning frameworks. Make certain that the volumes of recycling and waste are accurately measured and recorded.
- Information Examination: Before and after implementing initiatives to reduce waste, examine data on waste generation and recycling. Compute the waste decrease accomplished, reusing rate improvement, or other pertinent pointers.
- Measurable Appraisal: Utilize statistical methods to evaluate the significance of reduced waste and increased recycling rates. Factual strategies, for example, t-tests, ANOVA, relapse investigation, or other fitting methods can be utilized to assess the extent and measurable meaning of the enhancements.
- Interpretation: To comprehend the impact of initiatives to reduce waste and recycle, interpret the results. Analyze the reduction in total waste, the rise in recycling rates, or the percentage of waste produced that has been reduced. Take

into consideration the improvements' effects on sustainability, cost savings, and the environment.
- Recommendations and Reporting: Prepare a comprehensive report detailing the methodology, findings, and suggestions for further improvement of the waste reduction and improved recycling rates analysis. Identify areas where additional initiatives to reduce waste or recycle can be implemented, or make suggestions for future research and development efforts.

6.4 Environmental Impact Assessment

Ecological Impact Assessment (EIA) in a computerized twin climate for the economic improvement of keen assembling includes assessing the potential natural impacts related with the execution and activity of advanced twin innovation. In this context, the essential steps for conducting an EIA are as follows:

- Scoping: Take into account the particular digital twin implementation as well as the manufacturing processes that are associated with it when defining the EIA's scope and boundaries. Determine the potential effects and aspects of the environment that need to be evaluated, including energy consumption, carbon emissions, the production of waste, water consumption, and other relevant factors.
- Collecting Initial Information: Collect pertinent data on the manufacturing processes' current environmental performance, such as resource consumption, emissions, waste generation, and other environmental indicators. This lays out a pattern against which the effects of the computerized twin execution can measure up.
- Influence Distinguishing Proof: Determine and evaluate the potential effects that the implementation of the digital twin might have on the surrounding environment. Take into account both direct and indirect effects, such as changes in energy consumption or material usage, as well as supply chain, transportation, and product lifecycle effects.
- Impact Evaluation: Assess the recognized effects quantitatively or subjectively. Analyzing the baseline data and determining how the use of the digital twin technology might affect resource efficiency, energy use, emissions, waste management, and other environmental factors are all part of this. To estimate the impact's magnitude and significance, make use of the appropriate tools and methods for environmental assessment.
- Enhancement and Mitigation Measures: Identify and suggest means of mitigating the potential negative effects on the environment. This could incorporate energy-saving drives, squander decrease techniques, enhancement of material

use, or whatever other measures that can further develop asset effectiveness and limit natural mischief. Additionally, look for opportunities to improve environmental performance by implementing the digital twin, such as optimizing supply chain management or enhancing monitoring capabilities.

- Management and Monitoring: Create a plan for monitoring and management to keep track of and control the effects on the environment throughout the digital twin's implementation and operation. This incorporates laying out execution markers, carrying out observing frameworks, and guaranteeing that suitable measures are set up to address any unanticipated ecological dangers or effects.
- Communicating and Reporting: The methodology, baseline data, affect assessment results, mitigation measures, and monitoring plan should all be included in a comprehensive report detailing the EIA's findings. To promote transparency and gather feedback, communicate the proposed strategies and the impacts on the environment to stakeholders like manufacturers, regulators, and the public.
- Continual Development: Continue to monitor, evaluate, and improve environmental performance as the digital twin implementation progresses. Review and update the EIA on a regular basis to incorporate new data, technological advancements, and shifting sustainability objectives [18].

6.4.1 Statistical Evaluation of the Digital Twin's Impact on Greenhouse Gas Emissions

The following steps are typically followed when conducting a statistical analysis of the digital twin's impact on GHG emissions:

- Specific Metrics: Determine the metrics that will be used to measure GHG emissions, such as total CO_2 equivalent (CO_2e), nitrous oxide (N_2O), and carbon dioxide (CO_2). Based on the manufacturing processes and objectives, select the specific emissions indicators.
- Basic Statistics: Before using the digital twin, gather information about GHG emissions. This helps establish the initial GHG emissions associated with the manufacturing processes and serves as a baseline for comparison.
- Advanced Twin Execution: In the manufacturing environment, implement the digital twin system and ensure that accurate data collection mechanisms are in place to capture information about GHG emissions. Throughout the implementation phase, keep track of and document GHG emissions [19].
- Information Examination: Dissect the gathered information to evaluate the effect of the advanced twin execution on GHG emanations. Compare the baseline data collected earlier with the GHG emissions during the implementation phase.
- Factual Procedures: Apply proper measurable procedures to assess the meaning of the computerized twin's effect on GHG discharges. The magnitude and

statistical significance of the changes or reductions in emissions can be evaluated using statistical methods like t-tests, ANOVA, regression analysis, or other relevant methods.

- Hypothesis Testing and Confidence Intervals: To estimate the range of GHG emissions reductions achieved, calculate confidence intervals. Lead speculation testing to decide if the noticed changes in GHG discharges are measurably critical.
- Interpretation: To comprehend the impact of the digital twin implementation on GHG emissions, interpret the statistical results. Analyze the extent of the reductions in emissions, their statistical significance, and the implications for the objectives of reducing climate change and promoting sustainability.
- Restrictions and Suggestions: The availability of the data, the size of the sample, or any potential factors that could cause confusion should all be discussed in relation to the statistical evaluation's restrictions. Based on the findings, make suggestions for areas of improvement or focus [20].

6.4.2 Evaluation of Water and Air Contamination Decrease Accomplished Through Wise Assembling with a Computerized Twin

Surveying the decrease of water and air contamination accomplished through shrewd assembling with a computerized twin includes assessing the progressions in poison emanations and release of foreign substances. Some methods carrying out such an assessment are as follows:

- Specific Metrics: Determine the metrics that will be used to evaluate the reduction in air and water pollution. For water contamination, measurements can incorporate poison focuses (e.g., substance oxygen interest, all out suspended solids) or explicit toxins (e.g., weighty metals, natural builds). Pollutant emissions can be included in metrics for air pollution.
- Basic Statistics: Before using a digital twin to implement intelligent manufacturing, gather information on air and water pollution levels. This provides insight into the initial levels of pollution associated with the manufacturing processes and serves as a baseline for comparison [21].
- Perceptive Assembling Execution: Utilize the digital twin system to support intelligent manufacturing practices. To accurately collect relevant information about pollution, make sure that data collection mechanisms are in place. Throughout the implementation phase, monitor and record wastewater characteristics, pollutant emissions, and other relevant parameters.
- Information Examination: Evaluate the water and air pollution reduction achieved by intelligent manufacturing by analyzing the collected data. Compare the levels of pollutants in the implementation phase to the earlier baseline data.

- Measurable Appraisal: Utilize statistical methods to assess the significance of the reduction in pollution. The magnitude and statistical significance of the reductions in pollution can be evaluated using statistical techniques such as t-tests, ANOVA, regression analysis, or any other appropriate method.
- Interpretation: Decipher the factual outcomes to grasp the effect of insightful assembling with a computerized twin on water and air contamination. Assess the extent of contamination decrease, their factual importance, and the ramifications for natural manageability [22].
- Recommendations and Reporting: The assessment's methodology, baseline data, results of the data analysis, and recommendations for further improvement are all included in a comprehensive report. Identify areas where additional measures for reducing pollution can be taken, or make suggestions for future research and development projects.

6.4.3 Analysis of the Carbon Footprint and Life Cycle Assessments (LCAs) of Products Manufactured Using the Digital Twin Environment

The evaluation of the GHG to conduct a life cycle assessment (LCA), emissions and environmental effects throughout the product's life cycle are required and analyzing the carbon footprint of products produced in a digital twin environment. Here are some ways to deal:

- Limits of the System and Scope: Characterize the degree and framework limits of the investigation, taking into account the whole life pattern of the items, including unrefined substance extraction, fabricating processes, item use, and end-of-life stages. Choose the functional unit that represents the particular product quantity or function under study for comparison.
- Collection of Data: Collect data on the amount of energy used at each stage of the product life cycle, material inputs, emissions, and other relevant parameters. This includes information about how to extract raw materials, make things, transport them, use products, and get rid of or recycle old ones. Guarantee exact and delegate information assortment to give a thorough image of the item's natural effects.
- Analysis of Carbon Footprint: Determine the product's carbon footprint by quantifying the emissions of GHGs at each stage of the life cycle. This includes both direct and indirect emissions, such as those produced by burning fossil fuels. Utilize established emission factors and methods, such as those provided by industry-specific guidelines or international standards (such as the ISO 14040 series) [23].
- Assessment of Life Cycles (LCA): To evaluate the environmental effects beyond the carbon footprint, conduct a comprehensive LCA. Assess other effect classes,

like asset exhaustion, water use, air contamination, and wastage, utilizing proper effect appraisal techniques (e.g., ReCiPe, Eco-Marker, and Effect 2002+). To estimate the product's potential impact on the environment, take into account its life cycle stages and affect pathways.

- Hotspot Evaluation and Interpretation: The hotspots or stages of the life cycle that contribute the most to the product's environmental impacts can be identified by interpreting the results of the carbon footprint and LCA analyses. This can assist with focusing on progress amazing open doors and illuminate decision-production for feasible item plan, material determination, process advancement, and store network the board.
- Recommendations and Reporting: The methodology, data sources, analysis results, and suggestions for minimizing environmental impacts and reducing carbon footprint should all be documented in a comprehensive report. Give experiences into potential improvement systems, for example, utilizing energy-proficient assembling processes, advancing material decisions, executing reusing projects, or decreasing item bundling [24].
- Continual Development: Throughout the product's life cycle, make use of the analysis's findings to steer efforts toward continuous improvement. Screen the execution of suggested measures and track progress after some time. Update the carbon footprint and LCA analysis on a regular basis to take into account changes in materials, technologies, and processes.

6.5 Economic Benefits

In intelligent manufacturing, the digital twin environment offers numerous economic advantages. Manufacturing businesses and organizations benefit from these advantages because they foster long-term growth. Here are a few key monetary advantages related with the execution of a computerized twin climate:

- Diminished Expenses: Using digital twin technology, businesses can optimize their manufacturing processes to cut costs. Associations can distinguish areas of shortcoming and waste by reenacting and investigating different situations utilizing a virtual imitation of the genuine creation framework. Designated upgrades in asset usage, creation arranging, and production network the board are thus conceivable, which eventually brings about a decrease in functional expenses.
- Increased Efficiency: Manufacturing operations can be monitored, analyzed, and improved in real time in the digital twin environment. It aids businesses in locating bottlenecks, streamlining processes, and increasing overall productivity. Manufacturers can improve overall production efficiency, reduce

downtime, and optimize equipment performance through data-driven insights and predictive analytics [25].

- Improved Quality Control: Advanced twin innovation takes into consideration constant observing and examination of assembling cycles and item execution. Organizations are able to spot and address quality issues in real time by incorporating quality control measures into the digital twin environment. Product quality and customer satisfaction rise because of this proactive approach to quality control's reduction of defects, rework, and customer complaints.
- Quicker Time to Market: Organizations can streamline product development and innovation cycles with the help of the digital twin environment. By carefully mimicking item plans and directing virtual testing, producers can lessen the time and cost related with physical prototyping. New product time-to-market is shortened as a result, giving businesses a competitive advantage and increasing revenue potential.
- Optimized Asset Administration: By providing insights into asset performance, condition monitoring, and predictive maintenance, digital twin technology makes asset management easier. Associations can advance upkeep plans, limit impromptu personal time, and broaden the life expectancy of basic hardware. As a result, assets are used more efficiently, maintenance costs are lower, and operations are more reliable.
- Enhanced Control of the Supply Chain: The environment of the digital twin enables improved supply chain collaboration and coordination. Organizations can improve production and distribution coordination, synchronized planning, and real-time visibility by integrating digital twins of various stakeholders. This upgrades store network responsiveness, diminishes stock levels, and enhances strategies, bringing about cost investment funds and further developed consumer loyalty [26].
- New Opportunities for Business: Manufacturers may be able to benefit from new business opportunities after adopting a digital twin environment. Organizations can provide value-added services like predictive maintenance, performance optimization, and product customization by making use of the data and insights provided by the digital twin. The organization's economic expansion and sustainability are aided by these new revenue streams.

6.5.1 Statistical Analysis of Cost Savings Achieved Through the Implementation of a Digital Twin

To lead a measurable investigation of cost reserve funds accomplished through the execution of a computerized twin, you would ordinarily follow these means:

- Define the Variable for Saving Money: The first step is to define the variable that represents the cost savings made possible by using a digital twin. This could be a particular metric, like a reduction in total costs, operational cost savings, or cost per unit savings. Define the measurement unit and any particular calculations involved in a clear manner.

- Collection of Data: Accumulate the vital information connected with cost reserve funds from the execution of a computerized twin. This might include information about operations, financial records, and other things that are important. Make certain that the data are reliable, accurate, and cover a suitable period.
- Information Preprocessing: To ensure that the data is suitable for analysis, clean and preprocess it. This may entail dealing with missing data, removing outliers, and, if necessary, transforming the data (such as a logarithmic transformation) [27].
- Analyses Descriptive of: Analyze the cost savings data using descriptive analysis to summarize and investigate it. To comprehend the central tendency and dispersion of the achieved cost savings, compute summary statistics such as mean, median, standard deviation, and range.
- Formulation of the Hypothesis: Create a hypothesis to determine the significance of the cost savings realized by using a digital twin. The invalid speculation (H0) could express that there is no tremendous contrast in cost reserve funds between the pre-execution and post-execution stages, while the elective speculation (H1) could propose that cost reserve funds are altogether higher in the wake of carrying out the computerized twin.
- Choosing the Statistical Tests: In view of the idea of the information and the inquiry being explored, pick a suitable factual test. If the data have two groups, such as pre-implementation and post-implementation, and a normal distribution, a paired t-test can be used. An investigation of fluctuation (ANOVA) or nonparametric tests like Kruskal-Wallis can be utilized when different gatherings or variables are involved.
- Lead Fact Check: Apply the picked measurable test to the information on cost reserve funds. A p-worth and test measurement will be given by the test. The test estimation gauges the strength of confirmation against the invalid hypothesis, while the p-regard exhibits the probability of seeing the data if the invalid hypothesis is substantial.
- Interpretation: Decipher the measurable test's discoveries and assess them. You can dismiss the invalid speculation and reason that there is a huge contrast in cost reserve funds among when carrying out the computerized twin if the p-esteem falls beneath the foreordained importance level (e.g., 0.05). The investigation does not give adequate proof to dismiss the invalid speculation in the event that the p-esteem is not huge.
- Responsiveness Examination: Perform sensitivity analysis to determine the results' robustness. In order to confirm the results of the statistical

analysis, this may entail evaluating the effects of various assumptions or data variations.

- Reporting: Including the statistical test used, the results, and the interpretation, summarize the results of the statistical analysis. Include any additional insights or suggestions derived from the analysis, as well as the savings realized through the use of a digital twin [28].

6.5.2 Evaluation of Productivity Improvements and Reduced Downtime

The following steps can be taken to evaluate the productivity gains and decreased downtime brought about by the implementation of a digital twin:

- Define the Variables of Productivity and Idle time: Define the variables that represent decreased downtime and increased productivity clearly. Efficiency can be estimated utilizing measurements, for example, yield per unit of time, units delivered, or work efficiency. The total amount of time that production is halted or that equipment is not functioning as intended is known as downtime.
- Collection of Data: Before and after implementing a digital twin, collect data on productivity and downtime. This might incorporate creation records, gear logs, support logs, and other significant information. Check to see that the data is reliable, accurate, and covers a suitable period.
- Information Preprocessing: To ensure that the data is suitable for analysis, clean and preprocess it. This may necessitate transforming the data, removing outliers, and handling missing data.
- Descriptive Analyses: Summarize and investigate the data on productivity and downtime using descriptive analysis. Compute synopsis insights like mean, middle, standard deviation, and reach to grasp the focal inclination and scattering of the factors. To identify any apparent changes, compare the periods prior to and following implementation.
- Choosing the Statistical Tests: In view of the idea of the information and the inquiry being explored, pick a suitable factual test. A matched *t*-test can be utilized to decide the meaning of the distinctions in the event that you have matched information, for example, efficiency and free time estimations taken preceding and following the execution of the computerized twin. Contingent upon the conveyance of the information, you should utilize an examination of fluctuation (ANOVA) or nonparametric tests assuming that there are different gatherings or elements included.
- Lead Fact Check: Apply the picked measurable test to the information in regards to efficiency and margin time. A p-worth and test measurement will be given by the test. The p-esteem shows the likelihood of noticing the information in the

event that the invalid speculation is valid, while the test measurement estimates the strength of proof against the invalid theory.

- Interpretation: Decipher the measurable test's discoveries and assess them. If the p-value falls below the expected significance level (e.g., 0.05), you can disregard the false hypothesis and infer that the advanced twin results in significant efficiency enhancements and shorter margin time. The investigation does not give adequate proof to dismiss the invalid speculation in the event that the p-esteem is not huge.
- Responsiveness Examination: Perform sensitivity analysis to determine the results' robustness. In order to confirm the results of the statistical analysis, this may entail evaluating the effects of various assumptions or data variations.
- Reporting: Including the statistical test used, the results, and the interpretation, summarize the results of the statistical analysis. Include any additional insights or suggestions derived from the analysis, as well as the enhancements in productivity and reductions in downtime attributable to the use of a digital twin.

6.6 Social and Human Factors

6.6.1 Analysis of the Impact of Digital Twin on Workforce Safety and Ergonomics

In order to assess the impact of the digital twin on the safety and ergonomics of the workforce, it is necessary to evaluate the modifications and enhancements to safety measures and ergonomic factors that have resulted from its implementation. You can carry out such an analysis by following these steps:

- Characterize Security and Ergonomics Factors: Define the variables that represent ergonomics and worker safety in a clear way. Metrics like the number of accidents, incident rates, near-miss incidents, and safety compliance rates are examples of safety variables. Physical strain, repetitive motion, and ergonomic risk assessments are examples of ergonomic variables.
- Collection of Data: Before and after using the digital twin, collect information about ergonomics and worker safety. Incident reports, records of safety inspections, ergonomic assessments, employee feedback, and other relevant data are examples of this. Check to see that the data is reliable, accurate, and covers a suitable period.
- Information Preprocessing: To ensure that the data is suitable for analysis, clean and preprocess it.
- Analyses Descriptive Of: Lead clear examination to sum up and investigate the security and ergonomics information. Determine relevant metrics and

summary statistics like incident rates and average ergonomic risk scores. To identify any apparent changes, compare the periods prior to and following implementation.

- Choosing the Statistical Tests: In view of the idea of the information and the inquiry being explored, pick a suitable factual test. A matched t-test or a nonparametric comparable can be used to examine the significance of the differences if you have matched data (such as ergonomics and health estimates from the computerized twin). Contingent upon the conveyance of the information, you should utilize an examination of fluctuation (ANOVA) or nonparametric tests assuming that there are different gatherings or elements included.
- Lead Factual Test: Apply the selected statistical test to the ergonomics and safety data. A p-value and test statistic will be provided by the test. The test measurement estimates the strength of proof against the invalid speculation, while the p-esteem demonstrates the likelihood of noticing the information in the event that the invalid theory is valid.
- Interpretation: Decipher the measurable test's discoveries and assess them. You can dismiss the invalid speculation and infer that the computerized twin has altogether further developed ergonomics and laborer wellbeing if the p-esteem falls beneath the foreordained importance level (e.g., 0.05). The investigation does not give adequate proof to dismiss the invalid speculation in the event that the p-esteem is not huge.
- Responsiveness Assessment: Perform awareness investigation to decide the outcomes' heartiness. This may involve evaluating the effects of various assumptions or data variations to confirm the statistical analysis's findings.
- Reporting: Counting the measurable test utilized, the outcomes, and the understanding, sum up the consequences of the factual examination. Explain how the advanced twin affected ergonomics and worker safety, as well as any additional information or suggestions based on the study.

6.6.2 Assessment of Job Creation and Employment Opportunities Resulting from the Adoption of the Digital Twin Environment

The implementation of a digital twin environment has the potential to have a significant impact on employment in a variety of industries and to create significant job opportunities. The following is an evaluation of the opportunities for employment that have arisen because of the adoption of the digital twin environment:

- Advanced Twin Turn of Events and Execution: Data scientists, software developers, system engineers, and domain experts are among the skilled professionals needed to create and implement digital twin technologies. Digital twin systems must be designed, developed, and integrated into existing infrastructure and processes by these specialists. During this phase, crucial job roles include digital twin architects, modelers, and analysts.

- Analytical and Management of Data: The vast amount of data generated by digital twins must be gathered, stored, processed, and analyzed. Data engineers, database administrators, and data analysts are all required to manage the flow of data and guarantee its quality, security, and accessibility. In order to gain useful insights and guide decision-making based on the data, experts in data science and analytics are also required.

- Optimizing and Maintaining: The real-time monitoring, predictive maintenance, and optimization of physical assets are made possible by digital twins. Asset management, condition monitoring, and maintenance optimization specialists are in high demand as a result. These jobs include examining information from the computerized twin to distinguish possible issues, plan support exercises, and improve resource execution.

- Cybersecurity: Cybersecurity becomes a major concern in the digital twin environment because of the increased connectivity and data exchange. System administrators, security analysts, and ethical hackers are among the cybersecurity specialists required to safeguard digital twin systems from potential vulnerabilities and threats.

- Support and Instruction: As associations embrace computerized twins, there is a requirement for preparing and support experts who can teach clients on the functionalities and uses of the innovation. These experts give specialized help, investigating, and client backing to guarantee the compelling usage of computerized twin frameworks.

- Services for Consulting and Advice: The use of digital twin technology frequently necessitates expertise and strategic planning. Organizations can get help understanding the potential benefits, determining their readiness, and creating adoption roadmaps from digital twin implementation consulting firms and advisory services. These jobs include giving direction, best practices, and proposals for effective execution. Innovation and entrepreneurship: Entrepreneurs and innovators now have more chances to create cutting-edge applications, tools, and services thanks to the widespread use of digital twins. Digital twin-related specialized solutions, customization, and integration services can be provided by small businesses and startups.

6.7 Case Studies and Results

6.7.1 Presentation of Real-world Case Studies Demonstrating the Statistical Analysis Outcomes

- Production-Based Predictive Maintenance: Study of Case: Digital twin technology was utilized by General Electric (GE) to optimize gas turbine maintenance. They were able to anticipate equipment failures and carry out preventative maintenance by collecting real-time sensor data from the turbines and employing statistical analysis tools. Reduced downtime, increased dependability, and significant cost savings were the outcomes of this strategy.
- Building Energy Optimization: Contextual Investigation: The Edge Building, Amsterdam, known as the most environmentally friendly office building in the world, uses a digital twin system to cut down on energy use. The building's energy systems were dynamically modified to achieve maximum efficiency through statistical analysis of sensor data, weather patterns, and occupancy data. The building's remarkable energy performance resulted in a significant decrease in its environmental impact.
- Clinical Customized Care: Investigation of Case: Disease treatment at the Memorial Sloan Kettering Cancer Center (MSKCC). MSKCC improved personalized cancer treatment by utilizing cutting-edge twin technology. Strategies of factual examination were utilized to distinguish designs and foresee therapy reactions overwhelmingly of patient information, which included genomics, clinical history, and treatment results. Oncologists were able to tailor treatment plans to each individual patient as a result, which led to better outcomes for patients and fewer side effects.
- Metropolitan Traffic Care: Investigation of Case: Singapore's strategic country drive employed advanced twin technology for traffic management and urban planning. By combining data from sensors, traffic cameras, and public transportation systems, statistical analysis was used to improve transportation efficiency, reduce congestion, and optimize traffic flow. Improved personal satisfaction for residents and greater metropolitan versatility were the outcomes of this information-driven strategy.
- Optimization of the Supply Chain: Study of Case: Walmart's supply chain operations were improved by utilizing digital twin technology. By breaking down verifiable deals' information, weather conditions, and stock levels, measurable examination strategies were applied to anticipate customer interest and advance item dispersion. Walmart and its customers both benefited from this

because it resulted in improved inventory management, fewer stockouts, and reduced costs.

6.7.2 Conversation of the Discoveries and Their Suggestions for Practical Advancement in Wise Assembling

The use of advanced twin innovation in canny assembling has uncovered huge outcomes that have critical repercussions for feasible turn of events. The following is a discussion of these findings and their implications:

- Streamlining of Assets: Computerized twin technology makes it possible to continuously check and improve assembling processes. Through statistical analysis of the data from the digital twin, manufacturers are able to find inefficiencies, reduce waste, and make the most of resources. Thus, producing tasks adversely affect the climate and utilize energy, materials, and different assets.
- Productivity of Energy: The computerized twin's information can be investigated measurably to recognize energy-concentrated cycles and gear. Makers can upgrade energy use, decrease ozone depleting substance discharges, and advance energy effectiveness by dissecting examples of energy utilization and distinguishing regions for development. This promotes long-term growth by lowering manufacturing's carbon footprint.
- Diminished Squander: Makers can gather and break down information on item quality, deformity rates, and creation varieties in the computerized twin climate. Utilizing statistical analysis techniques, manufacturers are able to identify the underlying causes of defects, reduce waste in the manufacturing process, and implement corrective measures. Diminished squander lessens natural effect, increments item quality, and diminishes scrap rates.
- Predictive Repairs: Manufacturers are able to anticipate equipment failures, identify maintenance requirements, and schedule proactive maintenance activities through statistical analysis of the digital twin's sensor data. This strategy extends the useful life of machinery and equipment while also reducing unplanned downtime and the necessity of emergency repairs. Manufacturers have the ability to reduce resource consumption, minimize waste, and enhance sustainability as a whole by optimizing maintenance procedures.
- Evaluation of the Lifecycle: Information can be gathered and examined all through an item's lifecycle, from plan and assembling to utilization and removal, on account of computerized twin innovation. The energy efficiency of manufacturing processes, the environmental impact of various product design choices, and opportunities for improvement can all be identified through statistical analysis. Producers can settle on very much educated decisions that focus on maintainability and backing the round economy because of this lifecycle evaluation system.

- Uprightness in the Store Network: The store network is made more apparent and straightforward thanks to computerized twin innovation. Authentic assessment of stock organization data, including getting materials, transportation, and arranged tasks, can recognize disappointments and important entryways for upgrade. Producers can advance maintainable practices all through the whole worth chain and decrease fossil fuel byproducts as well as waste by improving cycles in the production network.

6.7.3 Comparison of Results with Industry Benchmarks and Standards

It is necessary to take into consideration specific industry benchmarks and standards that are relevant to the utilization of computerized twin innovation in savvy fabricating to make an immediate correlation of the outcomes to those benchmarks and principles. I do not, however, have access to the most recent industry-specific benchmarks and standards beyond my September 2021 knowledge cutoff because I am an AI language model. As a result, I am unable to offer specific, current comparisons.

However, I am able to provide a general structure for comparing outcomes to industry standards and benchmarks. When conducting such a comparison, consider the following steps:

- Recognize Applicable Benchmarks and Norms: Find industry-specific benchmarks and standards for the use of digital twin technology in intelligent manufacturing by conducting research. Metrics for environmental performance, standards for energy efficiency, targets for reducing waste, and other relevant indicators established by industry associations, regulatory bodies, or international standards organizations could be examples of these.
- Collect Performance Metrics and Data: Collect information from your statistical analysis results and the implementation of your digital twin. Key performance indicators (KPIs) such as energy consumption, waste generation, defect rates, resource utilization, and any other metrics related to operational efficiency and sustainability may be included in this.
- Compare the Results to the Standards: Compare the data and performance metrics you have gathered to the established standards. Check how your implementation of a digital twin and the results of your statistical analysis compare to industry standards or goals. Find out whether you meet, exceed, or fall short of the benchmarks.
- Analyze Variations and Find Opportunities for Improvement: Examine any varieties or inconsistencies between your outcomes and the benchmarks/principles. Determine the areas in which your performance falls short of expectations

set by the industry. You can use this analysis to find specific opportunities for improvement in your digital twin environment and statistical analysis procedures.

- Do the Right Thing: Develop and put into action corrective measures based on the analysis to fill in any gaps or areas that need work. Implementing industry-recommended best practices, enhancing statistical analysis methods, or improving data collection procedures are all examples of this.
- Ceaselessly Screen and Survey: Create a procedure for evaluating your digital twin's performance against benchmarks and standards on an ongoing basis. Assess your progress on a regular basis, make any necessary adjustments to your strategies, and keep up with any changes to industry benchmarks and standards.

6.8 Limitations and Future Directions

6.8.1 Identification of Limitations in the Statistical Analysis and Data Collection Process

Although statistical analysis and data collection procedures are potent instruments for gaining insights and making educated choices, they do have some drawbacks. Here are a few normal restrictions to consider:

- Quality and Availability of the Data: The reliability and accuracy of statistical analysis can be significantly influenced by the data's quality and availability. Inaccuracies, missing values, and other problems with the data can lead to incorrect conclusions and biased outcomes. The statistical models' accuracy may be harmed if there is not enough relevant data available to cover all possible scenarios.
- Examining Inclination: While gathering information for investigation, there is a gamble of inspecting inclination, which happens when the chosen test does not precisely address the whole populace. Results may be skewed or erroneous because of this. It is critical to ensure that the example is illustrative of the populace under study and sufficiently catches its variety and inconstancy.
- Simplifications and Assumptions: In order to model complex systems or phenomena, statistical analysis frequently relies on assumptions and simplifying the data. In the real world, these assumptions might not always hold true, which could lead to bias or errors in the analysis. In light of the data and the issue at hand, it is essential to be aware of the assumptions made and evaluate their validity.
- Causation Versus Correlation: Measurable investigation can distinguish connections between factors, yet it cannot be guaranteed to lay out causation. While

a measurable relationship might exist between two factors, it does not suggest that one variable straightforwardly causes changes in the other. Care ought to be taken to try not to make causal cases exclusively founded on measurable connections.

- Problems with Interpretation: The outcomes of statistical analysis can sometimes be difficult to understand. Incorrect conclusions or poor decision-making can result from interpreting statistical results incorrectly or failing to take into account the larger context. It is critical to interpret the results of statistical analysis in the right context and to seek advice from experts when necessary.
- Changing Dynamics of Data: Changes in external factors, technological advancements, or shifts in business operations can alter data patterns and relationships over time. As the underlying data dynamics change, statistical models and analysis results may become less accurate or relevant. Models and analyses need to be updated and reevaluated on a regular basis to be accurate and keep up with the changing conditions.
- Ethical Issues to Consider: Ethical guidelines should be followed when collecting and analyzing data to ensure privacy, confidentiality, and informed consent. It is fundamental for handle touchy information fittingly, safeguard individual data, and agree with pertinent guidelines and rules.

6.8.2 Suggestions for Future Research and Improvements in the Assessment Methods

- Reconciliation of Ongoing Information: Digital twin environments for intelligent manufacturing can capture real-time data from a variety of sensors, machines, and processes. Future exploration can zero in on utilizing this information to foster appraisal techniques that give constant execution criticism, empowering persistent improvement and streamlining of assembling processes.
- Reenactment Based Evaluations: The manufacturing processes can be simulated in a way that is both realistic and dynamic in digital twin environments. The development of simulation-based assessment methods that enable the evaluation of various scenarios, decision-making abilities, and the impact of various strategies on productivity and sustainability can be the subject of future research.
- Performance Evaluation Using Predictive Analytics: Predictive analytics methods can be used to evaluate and predict the performance of manufacturing processes by analyzing historical and current data in digital twin environments. Identifying potential bottlenecks, optimizing resource allocation, and enhancing overall sustainability can all benefit from this.

- Metrics for Sustainability Integration: Intelligent manufacturing relies heavily on sustainable development. In the context of digital twin environments, sustainability metrics may be the focus of future research. This could include assessing the natural effect, energy utilization, squander decrease, and asset proficiency of assembling processes.
- Team-based and Collaborative Assessments: Diverse teams and stakeholders frequently collaborate on intelligent manufacturing projects. Methods for evaluating individuals' individual performance as well as their capacity to collaborate, communicate, and work effectively in a group in the digital twin environment can be the subject of future research.
- Appraisal of Human Elements: Future exploration can zero in on assessing the ergonomic plan of workstations, human-robot joint effort, and the effect of human variables on in general execution and security in the computerized twin climate on the grounds that smart assembling includes communications among people and machines.
- Evaluation of Long-term Performance: The environments of digital twins offer opportunities for manufacturing processes' long-term performance evaluation. Methods for assessing the sustainability and effectiveness of intelligent manufacturing systems over long periods of time can be developed in future research, allowing for continuous improvement and decision-making.
- Questions of Morality: With the rising utilization of computerized reasoning and mechanization in canny assembling, moral contemplations become fundamental. Assessment techniques that evaluate ethical decision-making, privacy concerns, and the responsible use of technology in the digital twin environment for sustainable development may be the subject of future research.

6.8.3 Challenges and Opportunities for the Digital Twin Environment

6.8.3.1 Challenges

- Interoperability and Integration of Data: Advanced twin conditions depend on the combination of information from different sources, including sensors, machines, and frameworks. Guaranteeing consistent information mix and interoperability can be a huge test, particularly while managing inheritance frameworks or heterogeneous information designs.
- Privacy and Security of Data: Digital twin environments process many data, including sensitive data about manufacturing operations and processes. This data must be safeguarded against unauthorized access, cyber threats, and privacy violations.
- Complexity and Scalability: It can be challenging to scale digital twin environments to handle large-scale operations as manufacturing systems become more complex and interconnected. It can be technically challenging to ensure that

the digital twin can accurately represent the physical system's complexity and maintain real-time synchronization.

- Infrastructure and Cost Requirements: Digital twin environments for intelligent manufacturing necessitate significant investments in connectivity, infrastructure, software, and hardware. The expense of setting up and keeping up with such a climate can be a hindrance, particularly for little and Small and Medium-sized Enterprises (SMEs).
- Workforce Readiness and Skills: A skilled workforce that is capable of designing, putting these systems into action, and operating them is necessary for maximizing the potential of digital twin environments. There might be a deficiency of ability with the essential skill, and associations need to put resources into preparing and upskilling their labor force.

6.8.3.2 Opportunities

- Processes for Manufacturing Optimization: Digital twin environments, making it possible to optimize and fine-tune processes, provide a virtual representation of the manufacturing system. This may result in increased sustainability, decreased waste, and improved resource efficiency.
- Predictive Maintenance and Resource Management: By analyzing real-time data from digital twin environments, manufacturers can forecast maintenance needs, detect anomalies, and optimize asset utilization. This approach supports sustainability goals by minimizing downtime, reducing maintenance costs, and extending the lifespan of equipment.
- Control and Monitoring in Real Time: Manufacturing operations can be monitored and controlled in real time thanks to digital twin environments. Proactive decision-making, swift response to disruptions, and process optimization for sustainability and energy efficiency are all made easier by this.
- Product Development and Virtual Prototyping: Products can be tested and prototyped virtually in digital twin environments prior to production. This contributes to sustainable design practices by reducing the need for physical prototypes, minimizing waste, and speeding up the product development cycle.
- Sharing of Information and Collaboration: Digital twin environments make it easier for designers, engineers, operators, and suppliers to work together with other stakeholders. Knowledge sharing, creativity, and collective efforts toward sustainable intelligent manufacturing development are all aided by this.
- Lifecycle Appraisal and Round Economy: Computerized twin conditions can uphold lifecycle appraisal by catching information overall lifecycle of items, including their plan, assembling, use, and removal. Recycling, remanufacturing, and minimizing environmental impact are just a few of the principles of the circular economy that can be implemented by manufacturers thanks to this.

- Continuous Advancement and Enhancement: Digital twin environments provide a feedback loop for manufacturing processes' continuous improvement and optimization. Manufacturers can identify bottlenecks, inefficiencies, and areas for improvement through data and performance metrics analysis, resulting in increased productivity and sustainability.

6.9 Conclusion

In conclusion, intelligent manufacturing's sustainable development greatly depends on the digital twin environment. The digital twin environment makes it possible for optimization, predictive maintenance, real-time monitoring, virtual prototyping, and integration of real-time data by creating virtual replicas of actual systems. These capacities add to asset effectiveness, decreased squander, further developed item plan, and improved generally manageability. However, in order to fully utilize the digital twin environment's potential, a number of obstacles must be overcome. These difficulties incorporate information mix, security, versatility, cost, and labor force preparation. Infrastructure investments, data interoperability, cybersecurity measures, workforce training, and other measures are all necessary for overcoming these obstacles. The digital twin environment offers many opportunities, despite the difficulties. Among the most important opportunities are manufacturing process optimization, predictive maintenance, real-time monitoring and control, virtual prototyping, collaboration, lifecycle assessment, and continuous improvement. Utilizing these valuable open doors can prompt improved efficiency, diminished ecological effect, and the reception of roundabout economy standards. Stakeholders must work together, share information, and give priority to investments in technology, infrastructure, and skill development if intelligent manufacturing is to develop sustainably in the digital twin environment. Manufacturers can unlock the full potential of this technology and pave the way for a manufacturing industry that is more sustainable in the future by embracing the digital twin paradigm and addressing the obstacles.

References

1 Dhatterwal, J.S., Kaswan, K.S., and Jaglan, V. (2023). 4 Data fabric technologies and their innovative applications. In: *Data Fabric Architectures: Web-Driven Applications*, 61–89. Berlin, Boston: De Gruyter.

2 Singh, H. and Kaswan, K.S. (2016). Clinical decision support systems for heart disease using data mining approach. *International Journal of Computer Science and Software Engineering* 5 (2): 19.

3 Kaswan, K.S., Dhatterwal, J.S., and Kumar, A. (2023). *Swarm Intelligence: An Approach from Natural to Artificial.* John Wiley & Sons.

4 Berwal, P., Dhatterwal, J.S., Kaswan, K.S., and Kant, S. (2022). *Computer Applications in Engineering and Management.* CRC Press.

5 (a) Siddhartha, G.S., Lella, Y.S., Mandava, V.S. et al. (2023, May). Machine learning based case-based reasoning techniques using in medical diagnosis systems. In: *2023 International Conference on Disruptive Technologies (ICDT),* 359–363. IEEE. (b) Yan, H., Yu, P., and Long, D. (2019). Study on deep unsupervised learning optimization algorithm based on cloud computing. In: *2019 International Conference on Intelligent Transportation, Big Data & Smart City (ICITBS),* 679–681.

6 Dhatterwal, J.S., Naruka, M.S., and Kaswan, K.S. (2023, January). Multi-agent system based medical diagnosis using particle swarm optimization in healthcare. In: *2023 International Conference on Artificial Intelligence and Smart Communication (AISC),* 889–893. IEEE.

7 Altintas, Y., Yang, J., and Kilic, Z.M. (2019). Virtual prediction and constraint of contour errors induced by cutting force disturbances on multi-axis CNC machine tools. *CIRP Annals* 68 (1): 377–380.

8 Zheng, P., Wang, H.H., Sang, Z.Q. et al. (2018). Smart manufacturing systems for Industry 4.0: conceptual framework, scenarios, and future perspectives. *Frontiers of Mechanical Engineering* 13 (2): 137–150.

9 Kaswan, K.S., Naruka, M.S., and Dhatterwal, J.S. (2023, January). Enhancing effective learning capability of SOAR agent based episodic memory. In: *2023 International Conference on Artificial Intelligence and Smart Communication (AISC),* 898–902. IEEE.

10 (a) Tao, F., Qi, Q.L., Liu, A. et al. (2018). Data-driven smart manufacturing. *Journal of Manufacturing Systems* 48 (13): 157–169. (b)Zhong, R.Y., Xu, X., Klotz, E. et al. (2017). Intelligent manufacturing in the context of Industry 4.0: a review. *Engineering* 3 (5): 616–630.

11 Yadav, A. and Jayswal, S.C. (2017). Modelling of flexible manufacturing system: a review. *International Journal of Production Research* 56 (7): 2464–2487.

12 Kaswan, K.S., Gautam, R., and Dhatterwal, J.S. (2022). Introduction to DSS system for smart cities. In: *Decision Support Systems for Smart City Applications* (ed. L. Gaur, V. Agarwal, and P. Chatterjee), 53–76. Academic Press.

13 Bouzary, H. and Frank Chen, F. (2018). Service optimal selection and composition in cloud manufacturing: a comprehensive survey. *International Journal of Advanced Manufacturing Technology* 97 (1/4): 795–808.

14 Bouzary, H., Chen, F.F., and Krishnaiyer, K. (2018). Service matching and selection in cloud manufacturing: a state-of-the-art review. *Procedia Manufacturing* 26: 1128–1136.

15 Chithaluru, P., Fadi, A.T., Kumar, M., and Stephan, T. (2023). Computational intelligence inspired adaptive opportunistic clustering approach for industrial IoT networks. *IEEE Internet of Things Journal* 13: 156–163.

16 Dhatterwal, J.S., Kaswan, K.S., and Ojha, R.P. (2022). The role of multiagent system in industry 4.0. In: *A Roadmap for Enabling Industry 4.0 by Artificial Intelligence* (ed. J.M. Chatterjee, H. Garg, and R.N. Thakur), 227–246. Wiley.

17 Bauer, D., Stock, D., and Bauernhansl, T. (2017). Movement towards service-orientation and app-orientation in manufacturing it. *Procedia CIRP* 62: 199–204.

18 Giret, A., Garcia, E., and Botti, V. (2016). An engineering framework for service-oriented intelligent manufacturing systems. *Computers in Industry* 81: 116–127.

19 Dhatterwal, J.S., Kaswan, K.S., and Kumar, N. (2023). Telemedicine-based development of M-health informatics using AI. In: *Deep Learning for Healthcare Decision Making* (ed. S. Sanchez-Martinez), 159. Wiley.

20 Jain, A., Singh, J., Kumar, S. et al. (2022). Improved recurrent neural network schema for validating digital signatures in VANET. *Mathematics* 10 (20): 3895.

21 Cao, Y., Wang, S.L., Kang, L. et al. (2015). Study on machining service modes and resource selection strategies in cloud manufacturing. *International Journal of Advanced Manufacturing Technology* 81 (1/4): 597–613.

22 He, B., Zhang, D., Gu, Z.C. et al. (2020). Skeleton model-based product low carbon design optimization. *Journal of Cleaner Production* 67: 56–78. https://doi.org/10.1016/j.jclepro.2020.121687.

23 Yadav, A., Chithaluru, P., Singh, A. et al. (2022). An enhanced feed-forward back propagation Levenberg–Marquardt algorithm for suspended sediment yield modeling. *Water* 14 (22): 3714.

24 Grossmann, I.E. (2019). Optimization and management in manufacturing engineering: resource collaborative optimization and management through the internet of things. *Optimization Methods and Software* 34 (1): 220–223.

25 Pradhan, A.K., Das, K., Mishra, D., and Chithaluru, P. (2023). Optimizing CNN-LSTM hybrid classifier using HCA for biomedical image classification. *Expert Systems* 56: e13235.

26 Soni, R., Bhatia, M., and Singh, T. (2019). Digital twin: intersection of mind and machine. *International Journal of Intelligent Internet of Things* 2 (3): 667–670.

27 Chithaluru, P. and Prakash, R. (2020). Organization security policies and their after effects. In: *Information Security and Optimization* (ed. K.C. Kam), 43–60. Chapman and Hall/CRC.

28 Zheng, P., Lin, Y., Chen, C.H. et al. (2018). Smart, connected open architecture product: an IT-driven co-creation paradigm with lifecycle personalization concerns. *International Journal of Production Research* 57 (8): 2571–2584.

7

Digital Twins in Flexible Industrial Production and Smart Manufacturing: Case Study on Intelligent Logistics and Supply Chain Management

S. Jeyalakshmi[1], A. Prasanth[2], M. Yogeshwari[1], and Ahmed A. Elngar[3]

[1]*Associate Professor, Department of Information Technology, Vels Institute of Science, Technology & Advanced Studies, Chennai, India*
[2]*Associate Professor, Department of Computer Science and Engineering, Vel Tech Rangarajan Dr. Sagunthala R&D Institute of Science and Technology, Chennai, Tamil Nadu, India*
[3]*Faculty of Computers and Artificial Intelligence, Beni-Suef University, Beni-Suef City, Egypt*

7.1 Introduction

Intelligent manufacturing is one of the fundamental pillars of the Fourth Industrial Revolution, and DT technology is a key enabler of this by mapping physical data into cyberspace and allowing for the manipulation of real products via the study and exploration of information models. This theory was first proposed in the American Apollo space program. Some first conceived of this idea of product life cycle management course, combined with a 3D digital twin model theory of the physical entity, information entity, and communication, therefore, creating mapping links between physics and information. Over the years, the DT has been the subject of increasing study and development in both the classroom and the business world [1].

The idea of a DT as a revolutionary technology with far-reaching consequences for the economy's future is catching hold. A DT is a replica of a physical system or process that may be used for simulation, prediction, and optimization. Data-driven operation monitoring and optimization, new product and service development, and a broader range of value generation and business model options are all possible with the help of DT and smart algorithms [2].

To construct a DT of anything is to make a replica of it in a digital environment, either as a single model or as a collection of individual things. By putting the model through its paces, finally learn what to know about the physical world and figure out how to make it work for us. As a result, DT technology is the most effective means to achieve comprehensive informatization and industrialization [3].

Digital Twins in Industrial Production and Smart Manufacturing:
An Understanding of Principles, Enhancers, and Obstacles, First Edition. Edited by Rajesh Kumar Dhanaraj, Balamurugan Balusamy, Prithi Samuel, Ali Kashif Bashir, and Seifedine Kadry.

Industry 4.0 refers to the widespread adoption of advanced technological systems across the entire product lifecycle, from initial concept to final disposal. This includes all production process phases, from design and development to production and distribution. The integration of engineering, manufacturing, logistics, services, and marketing is expected to lead to Industry 4.0. Beyond automation, it focuses on intelligent, self-adaptive industrial processes and real-time communication. It also aims to provide a path toward more agile, efficient, and high-quality manufacturing [4].

Industry 4.0 is a movement that promotes next-generation intelligent manufacturing as a means to improve the manufacturing sector by increasing adaptability, facilitating rapid design iterations, using digital information technology, and educating and training a more flexible technical labor force. Examples of cutting-edge manufacturing technology include the cloud, Internet of Things (IoT), and cyber-physical systems [5]. To enable dependable communication systems, IoT has been included in several applications [6]. For IoT-based applications, network durability is seen as an important attribute [7].

7.2 Related Works

Barykin et al. [8] have suggested that the whole suite of technologies required to build a supply chain DT model consists of simulation modeling, optimization, as well as information analytics. This model may be utilized for risk management, and it is connected to a continuous stream of digital information. According to strategy affect output, it is useful to run simulations and make plans in advance. Changes introduced into the DT model of the supply chain help shed light on the dynamics of the physical supply chain.

Damjanovic-Behrendt and Behrendt [9] have discussed the widespread use of cyber-physical systems (CPSs) and DTs to automate operations in industries as diverse as manufacturing and agriculture are predicted to cause radical shifts in established business practices. Smart manufacturing can better monitor execution in simulated lifecycle processes as well as the insights necessary for making educated choices and forecasts, managing assets, and performing preventative maintenance. However, numerous computational and network issues relating to the design, operation, and administration of CPS- and DT-based complex systems remain unresolved.

Moshood et al. [10] have mentioned that the logistics industry is a potential application area for DT. The resulting research might be utilized to identify further inconsistencies in DT approaches to logistics and manufacturing. Before committing to a large-scale DT deployment, enterprises must achieve a proper technical and digital maturity level. While many businesses may still see DT as a distant strategic development objective, it is to their advantage to lay the groundwork for their eventual adoption by designing their digital infrastructure with DT in mind.

Leng et al. [11] have analyzed the use of DT technology to enhance SMS design using the Web of Science database. They also presented a function-structure-behavior-control-intelligence-performance (FSBCIP) framework that can analyze the critical steps of DT-SMSD against the background of smart manufacturing system design (SMSD) definitions and the worth of DT-SMSD. The potential of DT to aid in key areas of the SMSD design process. Although additional blueprint models have been added to the DT-SMSD method, such as a configuration, motion, control, and optimization (CMCO) designs architecture or DT shop floor, it was evident that further research and development is needed in this area.

The industrial Internet of Things (IIoT), multi-domain physical-chemical modeling, virtual reality (VR), data analytics, industrial artificial intelligence (AI), blockchain, and cloud computing are all inspected as possible pillars of the DT-SMSD and offered design instances spanning several layers of a production system and multiple manufacturing paradigms. We also proposed four areas of study for DT-SMSD moving forward, based on our findings from the studies and conversations. This research aims to shed new light on major industrial difficulties associated with the development of new SMSs in the wake of the arrival of Industry 4.0.

Chen et al. [12] have discussed process analytical technology (PAT) methods, data management systems, unit operations, flowsheet models, system analyses, and integration strategies that are just some of the recent advances in pharmaceutical production. However, restriction in PAT accuracy, real-time model calculation, model maintenance capacity, real-time data connection, or confidentiality and security of data issues are inhibiting their complete integration. Solutions to these problems are proposed in a variety of forms.

Some of the problems that have hampered PAT attempts thus far may be solvable with the aid of novel approaches like near-infrared spectroscopy (NIRS), in-line ultraviolet spectroscopy (UVS), iterative optimization technologies, and other offline adaptive methods. Full integration and automation of DT also require establishing the data flow from the virtual components to a physical plant. To accomplish process optimization within the design area, the virtual plants should be able to make adjustments to the system settings and direct the physical plant. All of these parts should have proper physical and digital safeguards in place.

Marmolejo-Saucedo [13] has explained how DT may be used as a decision support tool in the supply chain. The goal is to boost product and process visibility by disseminating this data to all parties involved in the supply chain. The suggested instrument combines dynamic simulation approaches with what-if multi-scenario analysis models for facility placement and linear mixed-integer optimization. To put the DT through its paces, it was decided to use a case study from the pharmaceutical business.

Madni et al. [14] have mentioned the earliest implementations of DT into the supply chain in literature survey. Its novel contribution is to see the difficulty of designing a supply network as a problem best tackled within the context of the DT. Experts estimated that 85% of commercial IT infrastructures worldwide using DT can be estimated by the exponential growth of IoT.

7.3 Case Study

The concept of DT was first introduced in 2002 by Michael Grieves at the University of Michigan [15]. Since then the concept has evolved with the advent of IoT. DT is a dynamic virtual model of a system, process, or service. Thus, a product, factory, or business service can have DT. With this advent, the DT technology has become cost-effective to implement and is gaining increasing acceptance in the IIoT community, which tends to focus on large, complex, capital-intensive equipment. At the same time, the aerospace and defense industry, which continues to invest in Industry 4.0, has started to invest in DT technology [16].

DT is at the forefront of the Industry 4.0 revolution facilitated through advanced data analytics and IoT connectivity which has increased the volume of data usable from manufacturing, healthcare, and smart city environments [17, 18]. IoT is evolving to be a giant within the electronics industry in recent times. Due to its expansive nature, a consensus can be made that there must be no single definition of what IoT encompasses [19]. The application of DT technology in logistics and distribution is one of the future directions of DT, which realizes the accurate distribution of goods through the control of the distribution process of mobile robots [20, 21].

7.3.1 Digital Twin

The use of DT technologies has been more popular in recent years since it has greatly benefited the expansion and modernization of businesses via increased connectivity and digitalization. However, there are still problems in integrating DT components, exchanging data in real time, and restoring reality. DT are computer simulations that faithfully recreate the actual object's or process's characteristics and behavior. The purpose of this analysis is to improve its efficiency by looking at how it has performed or acted in the past. Application of the notion of the DT to commodities, technologies, and even complete corporate ecosystems holds the potential for shedding light on the past, improving the present, and predicting the future performance of various sorts of industries.

DT are revolutionizing the supply chain industry by providing a wide range of tools for improving teamwork, making better decisions with data, and streamlining operations. In this essay, the author suggests creating a digital duplicate of a pharmaceutical corporation to better comprehend the case study. This technology, which is built on simulators, solvers, as well as information analytics tools, couples several tasks in a crucial interface for the business (Figure 7.1).

Recently, DT technology has received a lot of attention. Gartner, the industry-leading IT research and advisory group, has consistently ranked DT as one of the top ten most important technological developments since 2016. The world's largest manufacturer of military hardware, Lockheed Martin, has identified six new technologies in the aerospace and defense sectors, the first of which is the DT. China's research and development consortium, the China Association for Science

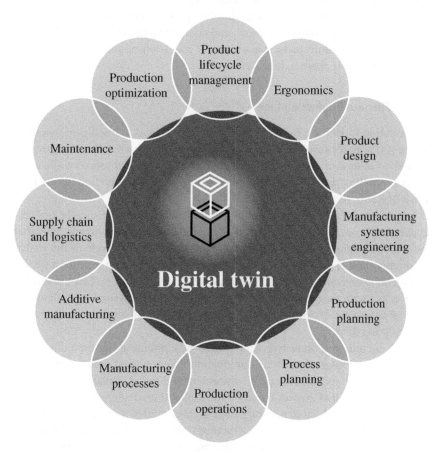

Figure 7.1 Applications of DT.

and Intelligent Manufacturing, named DT innovative manufacturing assembly as one of the top 10 scientific and technological achievements in intelligent manufacturing in 2017. Different components are added at various points in their product life cycle, and DT technology is present throughout the whole cycle. As a result, there are several flavors of DT performance. DT applications from three different angles are: product development, production, and maintenance.

The term "digital twin" is central to the Industry 4.0 paradigm shift. To better comprehend, foresee, and enhance the efficiency characteristics of its physical counterpart, it creates a virtual replica of the asset, product, or process. With the development of IoT, DT can continually gather data from sensors and exchange data with its physical counterpart. To practice for future missions, NASA created an early DT that mimics Apollo 13's environment and behavior. The whole Space Center, however, is now under the watchful eye of DT. Some have characterized DT as a multi-scale, probabilistic simulation of complex products that use cutting-edge physical models, sensors, and components to mimic the behavior of their real-world counterparts. Algorithms, a collection of analytics, and a data model are the three main parts of DT.

7.3.2 Flexible Production Line

A simulation model of a flexible production line utilized in intelligent manufacturing is first necessary for a DT system. This model allows for the input, export, upload, or release of data in a full-fidelity three-dimensional format while accurately simulating the real production line using a line of production information service system, twin data, and so on.

Second, production data and a sensing virtual model are built using the real production line's needs for people, materials, equipment, the environment, and other aspects. A dynamic model database covering the entire life cycle of the virtual production line is produced by integrating the manufacturing execution system (MES), Mx data collection (MDC), enterprise resource planning (ERP), and other service data with the production line's information service system. This allows for the real-time transformation, specification, association, analysis, and integration of perceptual data.

Third, include the derivative data created by combining the time dimension data with knowledge about a real production line, a virtual production line, or the information service system, as shown in Figure 7.2. Creating a dynamic, symbiotic model that evolves based on the digital thread and offering assistance and services for production activities such as collaborative resource management, production planning, manufacturing methods, and so on. By enabling interaction between the actual and virtual worlds, model-based twin data drives virtual manufacturing lines.

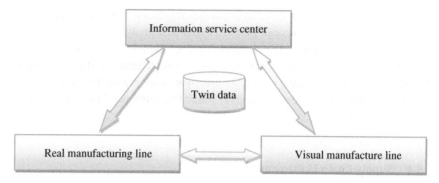

Figure 7.2 Digital twin interaction.

The lightweight model's testing has come to an end. The experiment makes use of Unity 3D as the rendering engine to compare the original and the lightweight version3. When comparing two models, take into account their sizes after export, the number of faces that are visible in Unity 3D, the number of fixed points, and so on.

7.3.3 Smart Manufacturing

The digitization of manufacturing processes is a growing trend. To fully embrace the benefits of Industry 4.0, it is essential to adopt a smart or digital factory. This advanced manufacturing approach will empower business with increased efficiency, optimized processes, and improved productivity.

All aspects of manufacturing are changing as a result of digitalization, with more opportunities for interaction between humans and machines, as well as between machines themselves and between manufacturing or business functions. To boost productivity and efficiency in the workplace, manufacturers of the future will use cutting-edge technology like DT and augmented reality (AR). The increased product variety, decreased manufacturing costs, or reduced carbon emissions necessitate greater production flexibility in the Smart Factories of the future. This flexibility is made possible by the automation of production processes and the automatic transmission of data throughout the manufacturing chain.

Many technologies contribute to the success of Industry 4.0 and Smart Manufacturing. These include AI, cloud computing, edge computing, big data and analytics, smart sensors, CPSs, DT, and IoT. As an illustration, the IoT makes it easier to produce and collect sensor data in real time, which may then be sent digitally in the context of digital manufacturing. Data mining and value creation via knowledge discovery is made possible by AI technologies, with additional help provided by edge and cloud computing. Big data technologies have the potential to increase the efficiency of manufacturing systems by facilitating faster decision-making. This is because these technologies can perform systematic analyses of a broad variety of data gathered throughout the complete product lifetime. Many authors agree that CPS is a game-changing technology that adds intelligence to traditional production methods. Combining computer principles with physical processes, CPS improves manufacturing features such as reliability, interoperability, predictability, and tracking.

7.3.4 Supply Chain Management

To build a supply chain that can last, you need to strike a balance between consistency and adaptability. These two aspects of the supply chain have the potential to cushion the blow of unforeseen events and must be included in preparations. In today's digital world, antiquated methods of logistics no longer make sense. Better, more cutting-edge answers may be found thanks to decentralized communication.

The development of a "DT" or digital representation of a physical supply chain, is predicated on the integration of simulation, optimization, as well as data analytics. Developing a digital duplicate has the potential to make supply networks more resilient and long-lasting in the face of failure.

Visibility in the supply chain is a subfield of supply chain management. The practice of increasing supply chain transparency to boost internal decision-making and productivity. Supply chain visibility studies have come into the academic spotlight in the last several years. As businesses expand their supply chains to meet rising consumer demand, they face new problems brought on by globalization's megatrends. More automation for managing end-to-end supply chain processes is now achievable with cutting-edge technologies like DT, AI, IoT, and robotic process automation.

Strategic planning relies heavily on simulation and optimization, two fundamental technologies. In situations when quick action is required, such as when dangers arise, the quality of judgments taken is critically dependent on the accessibility of important data in the moment made. The Internet has made it possible to gather vast volumes of information on supply chains, including the likelihood of supply failure and details about individual suppliers' resources and efficiency in producing goods (Figure 7.3).

Supply chain management technologies now allow for the pinpointing of vital nodes and the provision of early warnings of events with potentially catastrophic consequences for the supply chain.

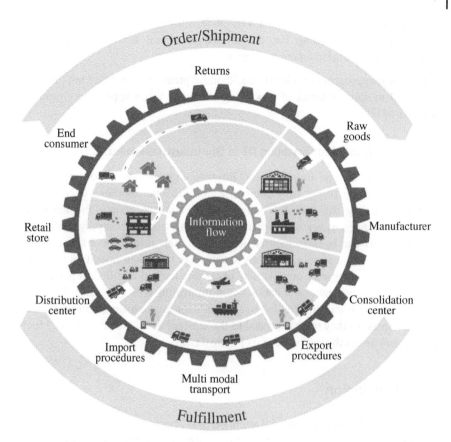

Figure 7.3 Supply chain management.

7.3.5 Predictive Analytics for Industry 4.0

Using both historical and real-time data, predictive analytics allows for a wide range of potential outcomes to be determined. It uses methods like clustering and association rules to find and record connections in data.

- Moving averages—predicts future values of the outcome variable(s) by identifying their past trends; and
- Linear regression—uses the connections between the explanatory variables and the outcome variable(s) to make predictions.

Predictive analytics fall into one of two camps, depending on the methods used to make predictions.

- Methods of regression analysis (for instance, multinomial logit models), and
- ML methods (including but not limited to neural networks, supervised/ unsupervised/reinforcement learning).

Statistical approaches provide the backbone of predictive analytics, but they typically fail to scale in terms of processing efficiency when applied to enormous datasets of DTs.

7.3.6 The Technology Enabling DT in Our Industrial Production

Anything that can be linked to the Internet and does so is considered part of the IoT. Many things may communicate with one another via embedded systems, opening the door to communication between people and machines [22]. Efficiency gains across the board, from smart houses to supply chains, are possible if IoT can be implemented according to new principles of transparency and flexibility. In a supply chain, early warnings to human operators might be based on data analyzed from a variety of sensors located in different places. For IoT devices to give any useful data for communicating with one other, they must be equipped with sensors. The IoT may include sensors of many shapes and sizes. The information they offer is crucial to the concept of DT, since a DT may then review and assess the true state of its physical twin.

7.4 Conclusion

Many different sectors pay special attention to DT because of their significance in the growing convergence of manufacturing's digital and physical infrastructures. In a fully functional DT, the physical and virtual pieces, as well as the interconnected data transmission routes, play crucial roles. Many businesses have found uses for DT since the advent of IoT technology, but the pharmaceutical and biopharmaceutical industries have lagged. As the complexity and breadth of DT applications vary greatly, so do the potential benefits. In this chapter, we investigated the role of DT in adaptive manufacturing and smart production. We also included a discussion of a case study involving intelligent logistics as well as supply chain management.

References

1 Fan, Y., Yang, J., Chen, J. et al. (2021). A digital-twin visualized architecture for flexible manufacturing system. *Journal of Manufacturing Systems* 60: 176–201.

2 Lu, Y., Liu, C., Kevin, I. et al. (2020). Digital twin-driven smart manufacturing: connotation, reference model, applications and research issues. *Robotics and Computer-Integrated Manufacturing* 61: 101837.

3 Yu-ming, Q. and San-peng, D. (2020). Research on intelligent manufacturing flexible production line system based on digital twin. *2020 35th Youth Academic Annual Conference of Chinese Association of Automation (YAC)*. IEEE. pp. 854–862.

4 Cinar, Z.M., Nuhu, A.A., Zeeshan, Q., and Korhan, O. (2020). DT for Industry4.0: a review. *Industrial Engineering in the Digital Disruption Era: Selected papers from the Global Joint Conference on Industrial Engineering and Its Application Areas, GJCIE 2019* (2–3 September 2019). Gazimagusa, North Cyprus, Turkey: Springer International Publishing. pp. 193–203.

5 He, B. and Bai, K.J. (2021). Digital twin-based sustainable intelligent manufacturing: a review. *Advances in Manufacturing* 9: 1–21.

6 Prasanth, A., Sabeena, G., and Sowndarya, N.P. (2023). An artificial intelligence approach for energy-aware intrusion detection and secure routing in the internet of things-enabled wireless sensor networks. *Concurrency, and Computation: Practice and Experience* 1–21.

7 Prasanth, A. and Jayachitra, S. (2020). A novel multi-objective optimization strategy for enhancing quality of service in IoT enabled WSN applications. *Peer-to-Peer Networking and Applications* 13: 1905–1920.

8 Barykin, S.Y., Bochkarev, A.A., Kalinina, O.V., and Yadykin, V.K. (2020). The concept for a supply chain digital twin. *International Journal of Mathematical, Engineering and Management Sciences* 5 (6): 1498.

9 Damjanovic-Behrendt, V. and Behrendt, W. (2019). An open-source approach to the design and implementation of DT for smart manufacturing. *International Journal of Computer Integrated Manufacturing* 32 (4–5): 366–384.

10 Moshood, T.D., Nawanir, G., Sorooshian, S., and Okfalisa, O. (2021). DT drove supply chain visibility within logistics: a new paradigm for future logistics. *Applied System Innovation* 4 (2): 29.

11 Leng, J., Wang, D., Shen, W. et al. (2021). DT-based smart manufacturing system design in Industry4.0: a review. *Journal of Manufacturing Systems* 60: 119–137.

12 Chen, Y., Yang, O., Sampat, C. et al. (2020). DT in pharmaceutical and biopharmaceutical manufacturing: a literature review. *Processes* 8 (9): 1088.

13 Marmolejo-Saucedo, J.A. (2020). Design and development of DT: a case study in supply chains. *Mobile Networks and Applications* 25 (6): 2141–2160.

14 Madni, A.M., Madni, C.C., and Lucero, S.D. (2019). Leveraging digital twin technology in model-based systems engineering. *Systems* 7 (1): 7.

15 Grieves, M. (2014). *Digital Twin: Manufacturing Excellence Through Virtual Factory Replication: A White Paper*. Melbourne, FL, USA: Michael Grieves, LLC.

16 Matthews, S. (2018). *Designing Better Machines: The Evolution of the Digital Twin Explained*. Hanover, Germany: Keynote Delivered at Hannover Messe.

17 Fuller, A., Fan, Z., Day, C., and Barlow, C. (2020). Digital twin: enabling technologies, challenges, and open research. *IEEE Access* 8: 108952–108971.

18 Bilberg, A. and Malik, A.A. (2019). Digital twin driven human–robot collaborative assembly. *CIRP Annals* 68 (1): 499–502.

19 Bhatti, G., Mohan, H., and Singh, R.R. (2021). Towards the future of smart electric vehicles: digital twin technology. *Renewable and Sustainable Energy Reviews* 141: 110801.

20 Liu, X., Jiang, D., Tao, B. et al. (2022). Genetic algorithm-based trajectory optimization for digital twin robots. *Frontiers in Bioengineering and Biotechnology* 9: 1433.

21 Wu, J., Yang, Y., Cheng, X.U.N., et al. (2020). The development of digital twin technology review. *2020 Chinese Automation Congress (CAC)*. IEEE. pp. 4901–4906.

22 Xia, F., Yang, L.T., Wang, L., and Vinel, A. (2012). Internet of Things. *International Journal of Communication Systems* 25 (9): 1101.

8

Applications and Use Cases of Digital Twins in Industry: 3D Graphics, Visualization, Modeling, Printing, and Reality Platforms

Suganthi Selvakumar[1], Saranya Jayaraman[1], Selvakumar Varadarajan Subramani[1], Sheena Christabel Pravin[2], Gayathri Radhakrishnan[1], and Ponmurugan Karuppiah[3]

[1]Department of Electronics and Communication Engineering, Rajalakshmi Engineering College, Anna University, Chennai, India
[2]School of Electronics Engineering, VIT University, Chennai Campus, India
[3]Department of Botany and Microbiology, King Saud University, Riyadh, Saudi Arabia

8.1 Introduction

A new technology called digital twin (DT) has recently acquired prominence in the age of smart systems and Internet of Things (IoT). This idea came into existence in the 1960s when NASA built a model illustrating the physical characteristics of its spacecraft. Real-time data processing, running simulations in a virtual environment, and conducting experiments in a physical setting are the three ways through which physical and virtual components can interact [1]. This ongoing connection between the two supports prompt process intervention and action execution in the physical assets. Although there are numerous uses of DT across many industries, the following industry areas are the focus of this chapter: textiles, food products, electronics, telecommunications, and smart systems.

The booming of computer-aided designing and manufacturing technology in the 1960s and 1970s triggered the evolution of DTs. In order to replicate the behavior of physical items and systems and to test different design concepts, engineers and designers were able to develop virtual representations of those objects and systems. However, the phrase "digital twin" was not first used until the early 2000s. Dr. Michael Grieves, a professor at the University of Michigan, came up with the phrase while collaborating with NASA to create a computer simulation of the space shuttle. A DT, according to Grieves, is a virtual version of a physical system, process, or object that may be used to mimic and study its behavior in real time [1].

Digital Twins in Industrial Production and Smart Manufacturing:
An Understanding of Principles, Enhancers, and Obstacles, First Edition. Edited by Rajesh Kumar Dhanaraj,
Balamurugan Balusamy, Prithi Samuel, Ali Kashif Bashir, and Seifedine Kadry.
© 2024 The Institute of Electrical and Electronics Engineers, Inc. Published 2024 by John Wiley & Sons, Inc.

DTs were first used in the aerospace sector, where they were employed to build virtual versions of planes and spacecraft. These models were used to test the vehicle's performance in various circumstances and mimic various events. Engineers were able to identify and address possible problems before they materialized in the real world, saving time and money. The automotive sector also used DTs, using them to replicate the behavior of engines and other components [2]. Engineers were able to test several design concepts and improve the vehicle's performance as a result. Through the use of DTs, producers were also able to track the performance of their goods in real time, facilitating preventive maintenance and minimizing downtime.

The aerospace sector was one of the primary segments that employed DTs in 2005. The virtual replication of passenger flights and spacecraft using DTs allowed for the simulation of numerous situations and the testing of the vehicle's performance in diverse environments. This conserved engineers' time and money through identification and addressing of any issues before they manifested in the real world [3].

In 2005, the automotive industry started utilizing DTs as well. They helped engineers evaluate various design concepts and improve the performance of the car by simulating the behavior of engines and other components. Through the use of DTs, producers were also able to track the performance of their goods in real time, facilitating preventive maintenance and minimizing downtime.

In 2005, DTs were also applied in the healthcare sector. In order to replicate various treatments and procedures, they were used to develop digital models of human organs and systems. This enabled medical professionals and researchers to test novel treatments and create improved medical equipment [4].

In order to produce virtual prototypes of things, DTs were utilized in the product design process in 2006. This allowed designers a chance to experiment with different design ideas and model how the product would act in various scenarios. They may then anticipate possible problems and make design modifications prior to the fabrication of actual prototypes, cutting costs, and speeding up the time to market. Industrial and manufacturing industries have employed DTs to forecast maintenance requirements and minimize downtime. Engineers could track the operation of physical assets like equipment, machinery, and infrastructure in real time and forecast when maintenance was necessary to prevent expensive breakdowns by building virtual models of those assets.

The performance of power plants and other energy assets was optimized in the energy sector using DTs [5]. Engineers might simulate different situations and test various operational techniques to optimize energy production and lower costs by developing virtual models of energy systems.

To mimic training scenarios for soldiers and pilots, DTs were employed in the military and aviation sectors. Pilots and soldiers could practice in a simulated environment that closely resembled the real thing by building virtual representations of planes and other equipment.

The implementation of DTs grew in 2008, and the technology was progressing. Buildings and its systems, including lighting, security, and heating, ventilation, and air conditioning (HVAC), were virtually modeled using DTs.

These models were applied to enhance occupant comfort, lower energy usage, and optimize building efficiency. Virtual models of supply chain systems, including factories, warehouses, and distribution centers, were made using DTs.

These models were used to simulate various scenarios and test various operational methods in order to enhance the performance, lower costs, and increase efficiency of the supply chain. To manage and monitor physical assets including machinery, equipment, and infrastructure, DTs were employed. Engineers might forecast maintenance requirements, optimize operation, and monitor performance in real time by building virtual models of these assets.

Virtual representations of the human body, including its organs and systems, were made using DTs. These models were used to test new medical gadgets, create new medications, and replicate various medical procedures and treatments. Virtual representations of environmental systems, including rivers, oceans, and weather patterns, have been made using DTs [6]. These models were applied to simulate various scenarios, forecast environmental effects, and create mitigation plans.

DTs were employed in the year 2020 to generate virtual representations of entire cities, complete with structures, roads, utilities, and transportation systems. These models were used to simulate various scenarios, enhance governmental processes, and provide better services to citizens. Virtual models of autonomous vehicles, such as cars, trucks, and drones, were made using DTs. These models were used to test new software and algorithms, simulate various driving situations, and enhance vehicle performance.

Energy systems, such as power grids, renewable energy sources, and storage systems, were virtually modeled using DTs. These models were applied to grid reliability enhancement, scenario simulation, and energy generation and consumption optimization. Patients' medical histories, genetic makeup, and treatment regimens were all included in virtual patient models made using DTs. These models were used to test new medications and medical equipment, replicate various medical procedures and treatments, and personalize healthcare. Buildings, bridges, and tunnels all had virtual replicas made using DTs for construction. The design process was optimized using these models, which also increased construction efficiency and decreased costs [7].

DT usage has risen recently owing to the explosion of IoT. IoT devices produce trillions of data, which are used to model physical systems in a digital format that is incredibly exact and thorough. These models can be used to streamline processes, boost efficiency, and reduce costs [8].

As per the report from Global Market Insight, the US\$8 billion market for DTs will touch a 25% compound annual growth rate (CAGR) between 2023 and 2032. Eventually, the market for DTs is anticipated to expand by around US\$32 billion by 2026, according to the latest report by global technology research [9]. Additionally, a 2022 estimate claims that by 2028, about 60% of CEOs from a variety of industries intend to use DTs in their operations.

DTs are employed in many different applications nowadays, such as asset management, predictive maintenance, and design and testing of the products. In order to model and test various situations, such as traffic flow, energy consumption, and environmental impact, they are also used to build virtual replicas of cities.

Sensors and DTs are frequently used to simulate a real system or object electronically. Sensors are tools that can monitor changes in the physical environment, including vibration, pressure, temperature, and humidity. To gather information about their performance and behavior, they might be integrated with machinery, tools, and infrastructure.

For instance, sensors can be employed in a manufacturing facility to keep an eye on the functioning of the equipment on the assembly line. To construct a DT of the manufacturing line, data from the sensors can be integrated with information from other sources, such as maintenance logs and historical performance data. To predict and identify problems and improve performance, the DT can be used to mimic various settings, such as altering the production rate or modifying the machine settings.

Sensors can be employed in the transportation sector to track the performance of moving vehicles. A DT of the vehicle and its surroundings can be created using the sensor data in conjunction with information from other sources, such as weather and traffic data. To find possible problems and improve performance, the DT can be utilized to mimic various scenarios, such as changing the route or the speed [10].

In the field of medicine, sensors can be employed to continuously monitor patients' vital signs. To build a DT of the patient, data from the sensors can be coupled with information from other sources, such as genomic and medical records. The DT can be used to fine-tune treatment strategies and forecast how a disease will develop. All of these instances include the role of DTs to produce a footprint of a real-world equipment or item whose behavior can be improved. In order to monitor a physical object or system in real time, sensors are a crucial part of the process.

These DTs are developed using engineering simulations and models, not sensor data, as is the case with sensor-based DTs. Physical systems are designed, prototyped, and tested using non-sensor-based DTs. They may also be employed to assess the effects of adjustments and enhance system performance [11]. However, non-sensor-based DTs might not be a perfect representation of how physical systems behave in actual environments.

In IoT communication, data is collected from physical entities using the sensors. IoT data transmissions are utilized to replicate physical objects digitally and also evaluate, adjust, and optimize. Further, IoT enables continuous updation of information and hence is suitable for the DT applications. The device configuration and maintenance history are easily procured by DTs using data from IoT sensors that are installed in the object [12]. The construction industry deploys customized modeling for its various datasets. Building information modeling (BIM), the erstwhile methodology of this industry, can be replaced by DTs as virtual representations. BIM is the digital copy of a building with its structural and functional details. It offers a repository of details about a structure or project, such as geometry, positioning, location data, numbers, and characteristics of construction materials. Construction companies or their customers can monitor projects in real-time using DTs, which uses sensors to deliver real-time data as opposed to BIM, which only provides static data [13]. Construction teams may monitor the work being done, see possible issues, and modify plans using DTs, ensuring that projects are finished safely, promptly, affordably, and to the acclaimed standard of quality. Additionally, DT methods in the building sector improvise resource planning and logistics, safety monitoring, and the tracking of other resources (such as materials, labor, and equipment). Sensors and DTs are frequently used to simulate a real system or object virtually. Sensors are tools that can monitor changes in the physical environment, including vibration, pressure, temperature, and humidity. To gather information about their performance and behavior, they might be integrated in machinery, tools, and infrastructure [14].

A DT of the actual object or system being monitored can be produced when sensor data is accepted with data from other sources, such as history records or weather data. The real-time simulation of the actual object's behavior, the detection of possible problems, and performance enhancement can all be done with this DT.

For instance, sensors can be employed in a manufacturing facility to keep an eye on the functioning of the equipment on the assembly line. To construct a DT of the manufacturing line, data from the sensors can be integrated with information from other sources, such as maintenance logs and historical performance data. To find possible problems and improve performance, the DT can be deployed to mimic various scenarios, such as altering the production rate or modifying the machine settings [15].

Sensors can be employed in the transportation sector to track the performance of moving vehicles. A DT of the vehicle and its surroundings can be created by fusing data from the sensors with data from other sources, such as weather and traffic data [16]. Additionally, the DT appends various phenomena, including changing the route or the pace.

In the medical field, sensors can be employed to continuously monitor patients' vital signs. To build a DT of the patient, data from the sensors can be integrated with information from other sources, including genomic and medical records [17]. The DT can be deployed to improve treatment strategies and forecast how a disease will develop. All of these instances include the role of DTs in replication of a real-world system or item that can be deployed to imitate its behavior. In order to monitor a physical object or system in real time, sensors are a crucial part of the process.

In smart manufacturing, DTs are deployed to aid machine tools and also perform supervising and fault-detection independently. Futuristically, the twin must dynamically capture the real-time sensor signals [18] and datasets stored in clouds, and simultaneously machine-learn the necessary knowledge from the historical sensor signal datasets. An advanced twin development of this capacity has not yet been thoroughly experimented. This gap may be sealed by tracking the DT for smart machinery and generation of sensor signals.

To construct and manipulate the DT the digital systems, one for construction and another for adaptation respectively, should be used. Along with real-time answers and delay-related computational architectures, the literature also includes the modular architectures of both systems. Additionally, the systems are implemented on a platform that is based on Java TM. The smart manufacturing industry makes efficient use of Digital Twin Construction System (DTCS) and Digital Twin Adaptation System (DTAS) by first calculating milling torque before creating intelligent machine tools.

To recapitulate the applications, DTs are actually the digital representations of physical world systems, situations, or things that can be utilized to fine-tune processes, boost efficiency, and cut costs. Creation of a digital representation of a physical system through combination of data from many sources, such as sensors, IoT devices, and others is the emerging trend. Numerous industries, including manufacturing, healthcare, architecture, and engineering, among others, can benefit by using DTs. They can be employed to mimic system changes, monitor and improve processes, and offer important insights into effectiveness and performance. Data security, interoperability, and scalability are just a few of the issues that need to be resolved because the DT is still in the nascent stage as far as these applications are concerned.

8.2 Digital Twins in Microelectronic Manufacturing

The affordable employment of DT technology in manufacturing technologies envisages a DT protocol to ensure an efficient methodology for DT strategies.

8.2.1 System Development Life Cycle

The development of design, development, verification, validation, deployment, and maintenance solutions must follow best practices. Therefore, a DT lifecycle ensures to deliver a degree of guaranteed capacity over time in a given context, effective DT solutions today either implicitly or explicitly [8]. There is lack of reusability as the DTs are meant for one-time usage. The evaluation of DTs' viability in industrial control and automation processes is still in its infancy. As a result, the creation and maintenance of DT solutions lacks systematic approaches to enable common software attributes like reuse, interoperability, and so on. Numerous methods for systematic design do exist in manufacturing. Predominantly used method called system development life cycle (SDLC) explains the plethora of activities or stages undergone by a system right from envisioning to commissioning and testing [9–12], which is depicted in Figure 8.1. Typically, SDLC encompasses six key stages, namely: (1) the planning stage, involving determination of the need to develop a new system or software for achieving the organization's milestones and targets; (2) the analysis stage, which includes agglomeration of the above modules and verification of functionality needed to address the same; (3) the

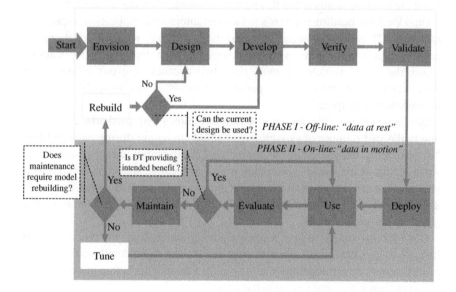

Figure 8.1 Stages of system development life cycle.

design stage, which grabs the outcomes of the analysis stage and evolves them into a theoretical model with strategies for implementation; (4) the development stage, which involves realizing the design through appropriate technologies such as machine learning algorithms; (5) the testing stage, where the implemented system is scrutinized to confirm its realization; and finally (6) the deployment stage, where the DT is delivered to the customer after authorization [9–12].

8.2.2 Digital Twins in Semiconductor Industry

DTs can also make semiconductor designers more aware of potential issues that they may not otherwise notice until much later in the overall process. That is one of the reasons why Siemens collaborated with global semiconductor IP company Arm to develop a DT.

The work in this area relates to more connected vehicles. The DT allows simulating and verifying subsystem and system-on-chip (SoC) designs to gain new insights into how those chips will perform within a car long before the automobile goes into production. For example, the DTs can show an integrated circuit's (IC) estimated power and performance metrics, which helps the designers to understand the working status of the IC and where trouble shooting is required.

Dipti Vachani, heading the Arm's IoT and its automotive applications especially, has stated "Developing future transportation solutions requires collaboration across complex ecosystems. Arm technology has been deployed in applications across the whole vehicle for more than two decades, and our collaboration with Siemens redefines what is possible in terms of safety-capable, scalable heterogeneous compute."

Designing chips for the automotive industry comes with particular challenges due to the vast number of variables a car could encounter during its useful life. "In developing and validating an automotive SoC, the input [designers must deal with] is 'the whole world'—including whatever weather and road conditions a vehicle must operate in. And the output that must be verified is that their new SoC/vehicle is not running over anyone," according to David Fritz, manager—the global technology Siemens.

However, using a DT makes semiconductor systems designers more aware of how those variables could affect their chips. Having those details earlier in the process allows design team members to make adjustments before getting too far ahead of themselves and having to backtrack.

8.2.3 Aiding Production Decisions

Designers might review a DT's features before deciding which techniques to use when finishing a printed circuit board with a soldering iron. Lead-free soldering emerged as the industry standard in 2006. People can handle it with a conventional iron, but they must know that this technique requires higher temperatures than other soldering techniques.

However, designers may choose automated options, such as miniwave soldering. It produces high-quality solder joints and is an easily reproducible method. Soldering procedure decisions do not come up in the earliest design phases. However, examining a DT allows thinking ahead about how to meet production deadlines without encountering slowdowns.

DTs also let designers to estimate a semiconductor's lifecycle from design stages, rather than postponing the design validation until production begins in a manufacturing plant. This capability can unlock new insights that enhance all stages of creation for a semiconductor system.

Anne Asensio, holding the position of vice president—design of the prestigious Dassault Systèmes, a French software company sees DTs as wholly uprooting current design processes, saying, "We are at the very beginning of a huge change that may transform the way we design. It's a huge revolution for the future, and we've only seen the first stage so far."

She continued, "This ability to visualize any given element in 3D where, it could be an object, a service, a system or an entire city and play with all the physical rules within the conformity of science is an incredible power. For designers, it is a clear access to design not just the product but its behavior, the experience and its entire life cycle at the front end of the creative process."

8.2.4 Achieving More Balance and Security in the Semiconductor Supply Chain

Statistics from 2019 showed that the United States has an approximately 85% share of the chip design market. However, it has only a 12% share of the semiconductor manufacturing market.

Various efforts are underway to bring more semiconductor manufacturing back to the United States, thereby, securing the country's supply chain in a time when semiconductors are in tremendous demand there and elsewhere. DTs could simultaneously assist designers and manufacturers in overcoming some of the current production issues, particularly until new semiconductor factories get built.

For example, one of Bosch's German semiconductor factories features a DT that aids employees in dealing with challenges related to process updates and building construction. The DT version of the factory contains approximately 500,000 objects, making it a useful resource for ironing out future production plans.

Another ongoing effort to use DTs to deal with supply chain issues comes from the US Department of Defense (DoD). More specifically, it will depend on them to validate chip assemblies or individual devices before putting those components into weapons. The idea is that this project could strengthen the supply chain while also establishing provenance for the semiconductor systems.

If designers and manufacturers more frequently use DTs to collaborate, production levels could remain high while ensuring all parties are on the same page about components in development. Giving everyone access to the same updated and reliable information can reduce many issues that could otherwise slow production or result in missed expectations.

8.2.5 Digital Twins for Semiconductor Designing

It is still not as common to use DTs in semiconductor design as manufacturing. However, this overview shows there are numerous compelling reasons to rely on them throughout that phase. Designers must continually develop new, better chips at least as fast as competing companies in the marketplace. DT technology could help them achieve that feat and notice other advantages that were out of reach before using these innovations.

8.3 Digital Twins in Food Products Manufacturing Industry

"A manufacturing execution system (MES) combined with Advanced Planning and Scheduling (APS) software lets manufacturers optimize their production by knowing how and when to change over equipment efficiently," [4]. "MES allows manufacturers to be proactive, not reactive. Instead of slowing down during disruption, manufacturers can use manufacturing operations data to apply advanced analytics to press forward in their digital transformation." According to Keith Chambers heading Operations Management Software, AVEVA, "digitizing and automating manual processes helps optimize the system, so lines are run in the most efficient order possible." The industry is moving fast, but the consumer is causing that rapid change. To meet the consumers' needs, food and beverage companies must address these changing dynamics in a more agile way (Figure 8.2).

8.3.1 Digital Twin Technology for Data Handling

Digitalization has resulted in marvelous growth of numerous industrial sectors. Food and nutrition industries have explored many innovative ways of adopting DT technology. Although this industry and soft drink companies handle huge amounts of data, they face trouble in retrieval of the data from it. They have resorted to meaningful analysis through the DT methodologies to gain insights from the data rather than simply storing it.

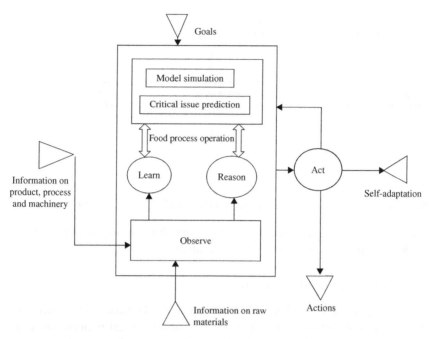

Figure 8.2 Digital twin for process optimization in the food and soft drink industry.

8.3.2 Creation of Digital Footprint in Food Production

A potential method is application of the DT in food and soft drink manufacturing for developing digital versions of them. These digital versions are actually ingredients along with ratios and their variants that can be used to assess product nutritional values to confirm the claims by marketing heads, thereby complying with regulations. A successful model hence includes production necessities that allow rapid change of ingredients catering to processing across different plants and transformable to production requirements specific to the site under consideration (Figure 8.3).

8.3.3 Digital Twin of the Complete Food Factory

A highly appreciable utilization of the DT is in setting up digital models of machines and production lineage. These models may include different levels of replication for different purposes, from high level models for production flow to accurate machine models that can effectively simulate production strategies. The models also have a completely accurate depiction of manufacturing capabilities that enables process optimization using simulated, enhanced, and digitally

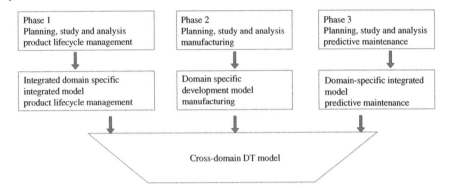

Figure 8.3 Digital twin for plant control. Source: Syda Productions/Adobe Stock.

validated DTs. For improvisation in production planning and plant design, the models articulate manufacturing engineers and planners.

8.3.4 Connectivity Establishment

Connecting digital and real portfolios with the IIoT DTs makes them much more useful than static models. Modeling and simulation of plant performance is an age-old strategy of food and soft drink companies. In addition, production control and monitoring systems are hardcore technologies for them. Besides, every model being important, real transformation happens by updating the DT using data collected from the IIoT for improvised physical twin performance.

8.3.5 Utilization of IIoT Information

The DT is advantageous because of its connecting capability between the real and the digital environment. This provides access to real-world performance data from sensors on production and packaging equipment along with the IIoT and provides easier data interpretation.

This connectivity enables businesses to track actual performance in order to improve the well-being and working of a piece of equipment or an entire facility and to learn more about how the physical assets are actually performing [19]. Additionally, it aids in supply chain and market forecasting for businesses.

8.3.6 Identification of Discrepancies and Anomalies

Manufacturers' major learning starts from viewing the differences between simulated results obtained using the DT and real-world data. Realization and analysis of these discrepancies helps to rectify anomalies in the plant leading to diagnosis and restoration of ideal performance.

8.3.7 Enhancement of Quality and Productivity

Companies can monitor production as a whole in addition to monitoring a specific piece of equipment, food, or beverage. For instance, IIoT data can be shown on a dashboard to highlight important production information like cycle times or yield, so operators can spot problems right away. In order for plant operators to respond and address problems as they arise, if not earlier, the IIoT platform can identify discrepancies and send out alerts.

IIoT data can also give management insight to benchmark manufacturing lines, plants, and regions for ongoing improvement. Additionally, by comparing actual specifications with anticipated batch specifications, it helps reduce the gap between R&D and the plant.

8.3.8 Production Parameters Mapped with Quality Outcomes

Lab results and production equipment data combined can help shed light on potential quality drifts that result in scrap or rework. The makers can pinpoint and address the real causes by comparing the results with factors that may be affecting the output.

8.3.9 Optimized Supply Chain and Manufacturing Network

The benefits of DTs extend beyond businesses. Comparing production across many facilities helps increase the productivity of the businesses. Through the IIoT, corporate headquarters or a center of excellence get data from factories via the cloud. In order to identify regional variations, this makes it possible to analyze product data across the network, whether it consists of their own factories or contract manufacturers. Additionally, this research may provide new information that may be used to pinpoint issues or exchange best practices in order to guarantee global product consistency and high productivity.

8.3.10 Closure of Lack of Information in the Supply Chain

Supply chain demand is forecasted by digital firms. For instance, they can interact with retailers on marketing and packaging while also closely collaborating with ingredient suppliers on R&D. Alternately, vendors can utilize advanced natural ingredient supplies proactively. This enables formulators to alter recipes in response to dynamic data changes such as humidity, seasonal availability, or changes in component ratios. As a result, digitalization enables businesses to exchange information to improve quality, efficiency, and agility.

8.4 Digital Twins for Process Optimization in Textile Industry

In the fashion industry, for example, the DT of an apparel can contain all the information including design origin and material to be used, to details about the marketing campaign that the garment has been displayed. Simulations using the DT enable creation of accurate 3D models, completely removing the need to create prototypes. Thus, the overhead on the waste associated with prototype creation is eliminated (Figure 8.4).

Further, simulations can also be employed to detect any supply chain shortages in advance, which further reduces waste that would typically be produced in the testing phase of production.

The DTs can be leveraged to aid sustainability in the fashion industry by helping manufacturers make decisions on clearance of products as they near the end of their lifecycles.

As products reach their expiry and need to receive final processing, DTs can be utilized for decision making regarding excess materials and deciding the harvest time of materials. Again, this articulates in reducing unnecessary waste.

Finally, in addition to innovation of the sampling process, DTs can also be used to help resolve fashion world's sizing problem.

An issue that many consumers face when buying clothing is the non-uniformity amongst sizing. For example, consumers could purchase three pairs of dresses from three different brands, and would embark on a completely different size at each store.

A major issue in this is that the customers tend to return the clothing, for reasons such as ill-fitting clothes that prevailed as the number one reason till 2021. Such mass garment returns heavily demand optimization of the fashion industry's carbon footprint.

Figure 8.4 Digital twins in fashion designing.

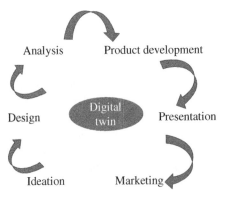

With DTs however, global and standardized sizing database can be created to assure garments are as per universal sizing requirements.

This ensures reduction of waste, helps minimize the number of garment returns, and improves consumers' satisfaction when purchasing clothing from brands which utilize the universal sizing data. Also, DTs can be deployed in the form of 3D printing on textiles for implementing sustainability in the fashion and textile industry.

Technologies that transform the apparel value chain can be explained as below.

8.4.1 Blockchain

By integrating block chain solutions, brands can track the manufacturing of their products right from the raw material stage to the end product. Blockchain promotes communication between those involved in the value chain and the retailer, creating trust between all industry actors. Increased communication and tracking can both be used to promote transparency to the brand and the consumer.

Transparency is becoming progressively more important to consumers—in recent surveys over 90% of consumers said that transparency plays an important role in their purchasing decisions, 11% to 94% of respondents said that they will stay loyal to a transparent brand. While for 12% of respondents it is no longer enough for brands to simply state bold claims without evidencing them. Brands are being increasingly called out for greenwashing, and so, through the implementation of blockchain solutions, they can produce evidence to back up their claims. Many fashion brands are now launching pilot programs to trace their supply chains using blockchain technology, including H&M and Kering.

8.4.2 Radio-frequency Identification (RFID)

Radio-frequency identification (RFID) are tags that can be placed onto items and act like digital barcodes enabling each item of clothing to be individually identified using radio frequency waves. RFID tags can help brands and retailers track their products' production line from start to finish, providing valuable insight into the complex processes involved in their supply chain [20]. These tags promote transparency into a product's development, boasting both social and environmental benefits. Transparency as mentioned previously plays an important role in modern-day consumerism. There is an increase in brands now successfully using RFID tags to provide them with crucial information about their business operations, including Adidas, Burberry, Prada, Decathlon, Nike, Tommy Hilfiger, and many more.

8.4.3 Artificial Intelligence (AI)

There is around 30% of overproduction in apparel manufacturing and also about 13% of these clothes are not sold. Artificial intelligence (AI) can help by making accurate predictions of trends, understanding customer behavior, and optimizing workflows and supply chain operations [20]. Brands and retailers can use this knowledge to only produce what is needed, reducing their production rates, whilst simultaneously minimizing wastage. AI can be a powerful tool to not only benefit those in their supply chain on a socially responsible level but also environmentally too. The implementation of AI is now used by big fashion brands including Dior, Zara, H&M, and Nike14. They all use AI in their business models to optimize routine processes and tasks, predict trend forecasts, and amplify the customer experience.

8.4.4 3D Design Software

3D design software can be used by brands and retailers to sample and adapt the design of products until they are ready to go into production, promoting the efficiency of the design process and reducing the amount of labor required during the manufacturing process [21]. 3D design also gives designers the power to experiment with their designs digitally, allowing them more creative scope. 3D design software can benefit workers involved in the supply chain whilst reducing production costs for brands and waste. Fashion brand Finesse is the first AI-led fashion house which uses AI and machine learning (ML) to create looks that the consumer can choose from, in hopes of tackling overproduction and waste.

8.5 Digital Twin for Building Smart Systems

The administration of water distribution networks (WDNs) today places a lot of emphasis on the digital transition and increasing the systems' energy efficiency. In particular, in terms of water and energy management, the implementation of recent technologies may be essential to achieving perennial water networks. Particularly, DT is an apt technology that combines geographic information system (GIS) data with virtual models, algorithms, data acquisition, and information from smart actuators. This study establishes a new method for effectively using DT knowhow in water distribution networks.

8.5.1 Smart Water Grids (SWGs)

Smart water grids (SWGs) are modern methods for managing water that combine information, communication, and technology (ICT) and WDNs. They ensure a safe and effective water supply by taking into account future risks and uncertainties including population expansion, flooding, disasters, and the escalation of the need for water supply catering to different industrial segments [22].

To accomplish sustainable water use, strategic and intelligent management of water resources is necessary. In SWGs, water resource transmission and distribution are monitored and regulated using sensors, meters, digital controls, and analytical tools in order to improve the efficiency of operation controls, automation, and digitalization.

Sustainable and self-sufficient water systems are the goal of smart water management (SWM). Operational optimization follows as a result of this, which reduces leakage and losses, improves water quality, and boosts customer satisfaction. The use of cutting-edge information technologies in SWM systems also guarantees the following advantages: knowledge on water management, early leak detection and effective water loss reduction, monitoring water quality, financial advantages for water and energy conservation, such as a reduction in water bills of up to 30%, increased quality and efficiency.

Digital conversion and the automation allow the transmission and collection of data remotely for the better management of water. This big data has to be leveraged in performing prediction analysis, which thus articulates increased efficacy and efficiency of water management after conglomeration with AI and DT technology. Routine and mandatory tasks like controlling pressure, leakage, and checking the purity of the water should depict the traditional manual methods. Through automation of these tasks, work processes can be restructured leading to reduced response time and effort.

In this study, an optimization process was used as a backup for a leakage control. The maintenance of control valves is calibrated with reference to an efficient water network. The network's hydraulic resolution equations are included along with optimization constraints in a nonlinear programming (NLP) model. The benefits of pressure and leakage reduction are then elucidated. The DT technology helps to improve the water system's combined effectiveness.

8.5.2 Technology for Smart Water Management (SWM)

The three technologies namely smart sensors and pipes, smart water metering and geographic information system (GIS), make up a SWM system:

- Smart sensors and pipes: Using a network of wireless smart sensors, it is possible to watch the pressure in a system and limit water loss [20]. The primary benefit of water loss management over alternative approaches is the network is continuously monitored without the need for local operator involvement. The wireless sensor's low energy consumption allows it to operate for extended periods of time without incurring significant energy expenses.

- Smart water metering: This technology provides for a better balance between ensuring that everyone has access to clean drinking water, a management entity's right to compensation for services provided, and everyone's shared obligation to protect finite water availability [22]. A software that aids in real-time decision-making and supports WDNs to strike a better balance between satisfying demand and increasing productivity. A smart water meter is made up of hardware for water flow monitoring, a transmitter coupled with memory unit for data storage for smart monitoring of water grids.
- GIS facilitates the incorporation of the geographical entities in an orientated model, for enhanced planning and administration. The simulation of actual properties along with geographical information is the uniqueness of GIS.
- The term "supervisory control and data acquisition," or "SCADA," implies usage of memory units including shared and networked computers and servers. It often makes use of a database of previous sensor measurements to centrally monitor geographically scattered assets. The SCADA system design encompasses necessary components and interfaces for convenient supervision.
- GIS data, city-scale reality models, and computer-simulated engineering simulations must all be combined in order to replicate a water system (Figure 8.5).

Additionally, SCADA, and its peripherals continuously update the DT with virtual operational data. A DT system reproduces real-time simulations of the water network to realize smarter water networks without compromising accuracy. This

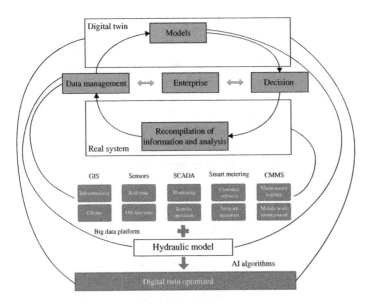

Figure 8.5 Structure of digital twin (DT).

enables analysis over the course of its life-period. To model disruption situations for resilience evaluation purposes, water distribution system DTs are being created. This will enable academics to assess the health and prognosis of assets and develop preventative maintenance methods [23].

It is important to process and correlate significant information in order to build a DT. (i) The GIS, which offers spatial location information; (ii) the sensors, which gather data from the hydraulic network; (iii) SCADA, which keeps track of and manages the acquired data; (iv) in independent metered zones, smart metering aims to regulate network operation and customer service; and (v) the primary data sources used to construct the platform were computerized maintenance management systems (CMMS), which allow one to track and maintain regular maintenance. AI methods and optimization techniques are used to incorporate all the information sources into a hydraulic model.

Traditional models are not successful for long-term because various inputs, such as the settings of pumps and valves, must be accurately calculated. In the presence of an online model, real-time field measurements are used to obtain these inputs rather than estimation. Data from field instruments that are input into the model via SCADA are used to run hydraulic simulations. After that, optimization techniques can be used to change the simulations' output. The set points for pumps or valves, as an example, might be the result of optimizations and communicated back to SCADA. By automating closed-loop operation, the resulting set points allow for the optimal possible operation of the network. On the other hand, an online model would repeat the procedure every hour, enabling the ongoing information on the level of water demand, node pressure, and potential supply outages. Dynamically obtaining information on consumption status, node load, and potential supply disruptions is possible.

The optimization procedures are a critical part of DTs and are useful for a number of tasks, including (i) maintaining the pressure level in the water network by using the best valve setting points to reduce water losses, increase asset life, and reduce energy consumption; (ii) scheduling pumps in the most effective way to reduce running cost; and (iii) developing predictive models based on earlier data.

DT technology offers improved forecasting, preparation, and readiness for future events as well as a better knowledge of the behavior of the water network by providing a complete image of the water infrastructure. The availability of an accurate and comprehensive collection of information facilitates in-line management of assets, the reduction of lifecycle costs, the optimization of asset performance and lifespan, and the enhancement of service levels. Additionally, a DT gives a panoramic view of the entire infrastructure, enabling trouble shooting and model building, ideal maintenance schedule creation, predictive and prescriptive analytics, testing and evaluating various scenarios, and ultimately producing real-time, actionable insights for critical and effective decision making [24].

Better planning and preparation, durability, recovery, and sustainability of the infrastructure, more dependable water service, and forecasting of future renewal needs are some of the longer-term benefits of DT technology. Another benefit is the ongoing reduction in the price of management, maintenance, and emergency services.

Valencia had a digital revolution that led to the creation of thorough real-time dashboards that are accessible every day of the week. This makes it possible to simulate occurrences from the past, present, and future under various operational conditions. The public firm uses DT to combine data from several software systems to manage Oporto's water supply, sewage systems, a couple of effluent treatment plants, water quality maintenance, and storm water drainage. The developed module is utilized to improve services and increase the resilience of the water infrastructure, as well as to predict flooding and issues with water quality [25].

In order to improve installation and sustainable operations, San Diego leverages DT technology for training its water purification operators. The model encompasses all process controls and online instrumentation [26].

8.5.3 Method of Working

The digital profile of the water network duplicates the precise behavior of the physical entity. Digital water is no longer seen as a complex network but as a map of opportunities. The mapping could produce useful outcomes for improved designing.

The two methods that are typically used to stop leaks in water networks are the loss-reduction and utility forecasting models. These two approaches can be included into DTs after going through several stages, as shown in the flowchart in Figure 8.2. The specifics of each stage are described below (Figure 8.6):

Level 1, Step-1.1: To detect anomalies like leaks or unusual water loss, smart sensors are positioned everywhere throughout the water network. They communicate the warning to the DT and transmit the data to the data platform. The data repository for upcoming occurrences is modified as a result, and the virtual model receives the information it requires to analyze and detect problems of this sort in the network.

Level 1, Step-1.2: The virtual system duplicates all the sensor data and identifies the exact location of the leak in the actual network. The variables employed in DT are the fluctuations in flow, pressure, and head losses observed at various network nodes.

Level 1, Step-1.3: In the leakage simulation module, the causes of the leakage, such as flaws in supply points, bursts, or unauthorized connections, are investigated to identify and resolve the issue.

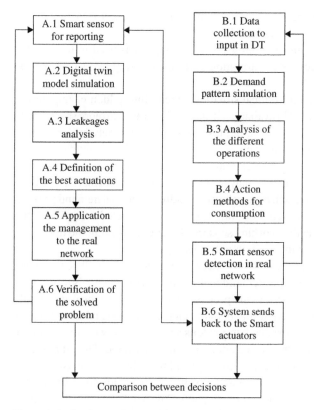

Figure 8.6 Basic methodology for development of DT model.

Level 1, Step-4: To determine the most effective method of reducing intervention water losses, various methods are taken into account and simulated.

Level 1, Step-5: The actuators receive information about the selected answer from the DT model. The type of intervention will determine the complexity and amount of time required to implement the chosen answer.

Level 1, Step-6: Feedbacks are necessary to confirm whether the issue has been correctly solved after the actuators have completed the intervention. These details are then gathered within the data platform and may be very important to increase the effectiveness of the reaction in the event of future anomalies.

The flowchart in Figure 8.2's right branch, which focuses on predicting water usage, has a number of steps as well.

Level 2, Step-2.1: The data platform, in accordance with step 2.1, gathers all the data obtained from necessary metering applications (e.g., SCADA and GIS data on customer locality, networking function, and needed demands, that will be stored to create a data warehoused sample). To facilitate studies in the virtual model, the data are sorted, filtered, and condensed.

Level 2, Step-2.2: The information gathered at step B.1 is used to test the DT model. At this point, various consumption patterns that vary by place and time are discovered.

Level 2, Step-2.3: Different critical situations, including sudden change in the need of water, are simulated in available virtual model so as to explore sharp needy demands. For maintenance of constant pressure and flow, the network processing method under the tested circumstances will be examined for parameter identification for checking impact and regulation to be adapted for future purpose.

Level 2, Step-2.4: Following the analysis shown from the performance of virtual network, the detailed plans will be developed to address potential demand peaks in the actual network. The chosen procedures are created in accordance with all the data obtained from the active working sensors, thereby confirming the upcoming sharp changes and improving water supply effectiveness.

Level 2, Step-2.5: Under the circumstances for verification, intelligent sensors notice abrupt shift in the supply demand pattern, the pertinent data is thoroughly experimented, and updated actions are proposed for maintaining the water supply to be feasibly steady. In the event that demand changes, the virtual model selects the best option from those specified at previous step, in other way, fresh data are gathered when previously undetected anomalies are discovered. Then it has to go back to step 2.1, for creation of a new process to address the new pattern shift.

Level 2, Step-2.6: The results from actual solutions set ready from the fundamental model are sent to actuators for activation. This would be done by defining a best procedure for simulating DT in a better way for predefined performances, which would improvise the device functioning (Figures 8.7 and 8.8).

Digital twin solutions go through various levels of maturity and adoption

	Descriptive analysis	Predictive modeling	Scenario planning and simulations	Operational excellence
Description	Visualize the city in 3D and its changes over time such as new assets, roads and bridges, movement of people, demographic and economic changes	Model, predict and forecast underlying activities in sectors, such as real estate, transport, sustainability and socio-economic sectors	Run multiple what-if scenarios and simulations by pulling levers of change and its effects on the city	Improve operations by making the organization proactive through live command centres and AI powered recommendations backed by real-time data integration

Figure 8.7 Solutions for DT adoption.

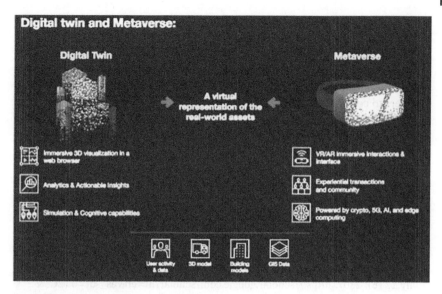

Figure 8.8 DT and metaverse relativity.

8.6 Applications of DT Smart Systems

1. Twins to forecast systems—A DT focused on complicated system prediction to a greater extent.
2. System modeling—It can be designed to simulate the behavior of complex systems.
3. Integration of assets—A DT designed to simplify data extraction in complex systems and support standard data formats.
4. Twins for care—A DT designed to help with purposes involving maintenance.
5. DTs for system visualization—A complicated system can be visualized (for example, in 3D) using a DT.

DTs used to simulate the behavior of (future) goods are known as twins for product simulation.

8.7 Conclusion

DTs and their businesses have increased by 71% during 2020–2022. A DT is being developed by 63% of manufacturers or intends to be developed in the years to come. Interestingly, the primary twin apps used recently are data simulation, interconnecting assets, predicting performance of the system, long term vision, regular maintenance, product specialization and product simulation, as per the latest market survey applicable for next 5 years.

Therefore, creation of digital copies of physical assets has become a common practice of emerging industrial practices. On the other hand, understanding the subtle differences between various DTs and the connecting point spots enables suppliers to concentrate their main aim and end customers to select the projects that will produce the most value.

Undoubtedly, DTs is the most happening technology that leverages the automation processes of industries, thereby enhancing the productivity through reduction of wastage and optimization of design.

References

1 Dembski, F., Wössner, U., Letzgus, M. et al. (2020). Urban digital twins for smart cities and citizens: the case study of Herrenberg, Germany. *Sustainability* 12 (6): 2307. https://scholar.google.com/scholar?hl=en&as_sdt=0%2C5& q=digital+twins+in+3D+graphics+&btnG=#d=gs_qabs&t=1679666355771& u=%23p%3DNcqKPW0KAKUJ.

2 Kuts, V., Otto, T., Tähemaa, T. et al. (2019). Digital twin based synchronised control and simulation of the industrial robotic cell using virtual reality. *Journal of Machine Engineering* 19 (1): 128–145. https://scholar.google.com/scholar? hl=en&as_sdt=0,5&qsp=5&q=digital+twin+virtual+reality&qst=br#d=gs_ qabs&t=1679666506963&u=%23p%3DjM9Qh1xT_LEJ.

3 Schroeder, G., Steinmetz, C., Pereira, C.E. et al. (2016). Visualising the digital twin using web services and augmented reality. *2016 IEEE 14th international conference on industrial informatics (INDIN)*. IEEE. https://scholar .google.com/scholar?hl=en&as_sdt=0%2C5&q=digital+twin+in+printing+ and+reality+platforms+&btnG=#d=gs_qabs&t=1679666576172&u=%23p %3DSENQ64D4EuQJ.

4 Rakib, S., Al Sunny, S.M.N., Liu, X. et al. (2018). MTComm based virtualization and integration of physical machine operations with digital-twins in cyber-physical manufacturing cloud. *2018 5th IEEE International Conference on Cyber Security and Cloud Computing (CSCloud)/2018 4th IEEE International Conference on Edge Computing and Scalable Cloud (EdgeCom)*. IEEE. tps://scholar.google.com/scholar?hl=en&as_sdt=0%2C5&q=digital+twin+in+ printing+and+reality+platforms+&btnG=#d=gs_qabs&t=1679666610989&u= %23p%3D80IF4f-s-UUJ.

5 Osho, J., Hyre, A., Pantelidakis, M. et al. (2022). Four Rs framework for the development of a digital twin: the implementation of representation with a FDM manufacturing machine. *Journal of Manufacturing Systems* 63: 370–380.

https://scholar.google.com/scholar?hl=en&as_sdt=0%2C5&q=digital+twin+in+printing+and+reality+platforms+&btnG=#d=gs_qabs&t=1679666655322&u=%23p%3D4Crpo1F9n4AJ.

6 Židek, K., Pitel, J., Adámek, M. et al. (2020). Digital twin of experimental smart manufacturing assembly system for industry 4.0 concept. *Sustainability* 12 (9): 3658. https://scholar.google.com/scholar?hl=en&as_sdt=0%2C5&q=digital+twin+in+printing+and+reality+platforms+&btnG=#d=gs_qabs&t=1679666710601&u=%23p%3Dsghg2SrYX2kJ.

7 Rakib, S., Nahian Al Sunny, S.M., Liu, X., et al. (2018). MTComm based virtualization and integration of physical machine operations with digital-twins in cyber-physical manufacturing cloud. *2018 5th IEEE International Conference on Cyber Security and Cloud Computing (CSCloud)/2018 4th IEEE International Conference on Edge Computing and Scalable Cloud (EdgeCom)*. IEEE. pp. 46-51. https://scholar.google.com/scholar?hl=en&as_sdt=0%2C5&q=digital+twin+in+3D+printing+and+reality+platforms+&btnG=#d=gs_qabs&t=1679666771317&u=%23p%3D63-sQ7pjJZUJ.

8 Paripooranan, C.S., Abishek, R., Vivek, D.C. et al. (2020). An implementation of AR enabled digital twins for 3-D printing. *2020 IEEE International Symposium on Smart Electronic Systems (iSES) (Formerly iNiS)*. IEEE. https://scholar.google.com/scholar?hl=en&as_sdt=0%2C5&q=digital+twin+in+printing+and+reality+platforms+&btnG=#d=gs_qabs&t=1679666655322&u=%23p%3D4Crpo1F9n4AJ.

9 Li, M. and Wenjie, S. (2021). Application of virtual reality technology and digital twin in digital media communication. *Journal of Intelligent & Fuzzy Systems* 40 (4): 6655–6667. https://content.iospress.com/articles/journal-of-intelligent-and-fuzzy-systems/ifs189501.

10 Jitong, X., Lin, W., and Yan, Z. (2020). Design and development of digital twin system for intelligent motor. *2020 IEEE International Conference on Advances in Electrical Engineering and Computer Applications (AEECA)*. IEEE. https://ieeexplore.ieee.org/abstract/document/9213679.

11 Sepasgozar, S.M.E. (2020). Digital twin and web-based virtual gaming technologies for online education: a case of construction management and engineering. *Applied Sciences* 10 (13): 4678. https://www.mdpi.com/762788.

12 Vidal-Balea, A., Blanco-Novoa, O., Fraga-Lamas, P. et al. (2022). A collaborative industrial augmented reality digital twin: developing the future of shipyard 4.0. *Science and Technologies for Smart Cities: 7th EAI International Conference, SmartCity360°, Virtual Event (2–4 December 2021), Proceedings*. Cham: Springer International Publishing. https://link.springer.com/chapter/10.1007/978-3-030-42416-9_18.

13 Elahi, B. and Tokaldany, S.A. (2021). Application of Internet of Things-aided simulation and digital twin technology in smart manufacturing. In: *Advances*

in Mathematics for Industry 4.0, 335–359. Academic Press https://www.sciencedirect.com/science/article/pii/B9780128189061000152.

14 Zhuang, C., Liu, J., and Xiong, H. (2018). Digital twin-based smart production management and control framework for the complex product assembly shop-floor. *The International Journal of Advanced Manufacturing Technology* 96: 1149–1163. https://link.springer.com/article/10.1007/s00170-018-1617-6.

15 De Giacomo, G., Ghedallia, D., Firmani, D. et al. (2021). IoT-based digital twins orchestration via automated planning for smart manufacturing. *Workshop on Generalization in Planning (GenPlan)*. https://whitemech.github.io/papers/2021/genplan2021dgflmm.pdf.

16 Bradac, Z., Marcon, P., Zezulka, F. et al. (2019). Digital twin and AAS in the industry 4.0 framework. *IOP Conference Series: Materials Science and Engineering. Vol. 618. No. 1*. IOP Publishing. https://www.sciencedirect.com/science/article/pii/S2405896320301798.

17 Tchana, Y., Ducellier, G., and Remy, S. (2019). Designing a unique digital twin for linear infrastructures lifecycle management. *Procedia CIRP* 84: 545–549. https://www.sciencedirect.com/science/article/pii/S221282711930798X.

18 Dani, A.A.H. and Supangkat, S.H. (2022). Combination of digital twin and augmented reality: a literature review. *2022 International Conference on ICT for Smart Society (ICISS)*. IEEE. https://www.hindawi.com/journals/mpe/2022/4361135/.

19 Ma, L. (2022). Design of Chinese opera cultural platform based on digital twins and research on international cultural communication strategies. *Mobile Information Systems* 2022: https://www.hindawi.com/journals/misy/2022/6996377/.

20 Lee, J., Jia, X., Yang, Q. et al. (2021). Collaborative platform for remote manufacturing systems using industrial internet and digital twin in the COVID-19 era. *International Manufacturing Science and Engineering Conference. Vol. 85079*. American Society of Mechanical Engineers. https://asmedigitalcollection.asme.org/MSEC/proceedings-abstract/MSEC2021/1115372.

21 Wenna, W., Weili, D., Changchun, H. et al. (2022). A digital twin for 3D path planning of large-span curved-arm gantry robot. *Robotics and Computer-Integrated Manufacturing* 76: 102330. https://ieeexplore.ieee.org/abstract/document/9730282.

22 Randall, T. and Koech, R. (2019). Smart water metering technology for water management in urban areas analysing water consumption patterns to optimise water conservation. *Water E-Journal* 4: 1–14. [Google Scholar].

23 Bauer, P., Stevens, B., and Hazeleger, W. (2021). A digital twin of Earth for the green transition. *Nature Climate Change* 11: 80–83. [Google Scholar] [Cross-Ref].

24 He, B. and Bai, K.J. (2021). Digital twin-based sustainable intelligent manufacturing: a review. *Advances in Manufacturing* 9: 1–21. [Google Scholar] [CrossRef].

25 Word, D.T. (2020). Oporto Water Utility Leverages Digital Twin for Integrated Management of Urban Water Cycle—Digital Water Works. Available at: https://digitalwaterworks.net/oporto-water-utility-develops-technology-platform-for-integrated-management-of-urban-water-cycle/ (accessed on 1 April 2022).

26 Curl, J.M., Nading, T., Hegger, K. et al. (2019). Digital Twins: The Next Generation of Water Treatment Technology. Available at: https://www.jacobs.com/sites/default/files/2020-01/jacobs-digital-twins-awwa-article.pdf (accessed on 1 April 2022).

9

Cobots in Smart Manufacturing and Production for Industry 5.0

Jeyalakshmi Jeyabalan[1], Eugene Berna[2], Prithi Samuel[3], and Vikneswaran Vijean[4]

[1]*Department of Computer Science and Engineering, Amrita School of Computing, Amrita Vishwa Vidhyapeetham, Chennai, India*
[2]*Department of Artificial Intelligence and Machine Learning, Bannari Amman Institute of Technology, Anna University Erode, Tamil Nadu, India*
[3]*Department of Computational Intelligence, School of Computing, SRM Institute of Science and Technology, Kattankulathur Campus, Chennai, India*
[4]*Faculty of Electronic Engineering & Technology, Universiti Malaysia Perlis, Kampus Alam UniMAP, Arau, Perlis, Malaysia*

9.1 Introduction

A "collaborative robot" (or "cobot") is a robot designed to assist humans in a variety of tasks. Cobot applications are different from those that use industrial robots without human interaction. Lightweight construction materials, softened edges, and natural speed and force restrictions are some of the ways in which cobots might promote safety [1, 2].

Industrial robots and service robots are the two main categories of robots. The robotics and automation domain makes use of industrial robots. Service robots are used in scope of providing useful service to mankind. It can be used for specific purposes pertaining to personal domestic needs or professional purposes. It can be related to medical hospitality or delivery. But either way these robots have not worked with intervention of humans.

Collaborative robots, often known as cobots, are machines programmed to collaborate with humans. Possible applications for collaborative robots include information robots in public spaces (an example of service robots), logistics robots that transport materials within a building, and industrial robots that automate unergonomic tasks such as assisting people with the movement of heavy parts, machine feeding, and assembly operations (cobots) [3].

Digital Twins in Industrial Production and Smart Manufacturing:
An Understanding of Principles, Enhancers, and Obstacles, First Edition. Edited by Rajesh Kumar Dhanaraj, Balamurugan Balusamy, Prithi Samuel, Ali Kashif Bashir, and Seifedine Kadry.
© 2024 The Institute of Electrical and Electronics Engineers, Inc. Published 2024 by John Wiley & Sons, Inc.

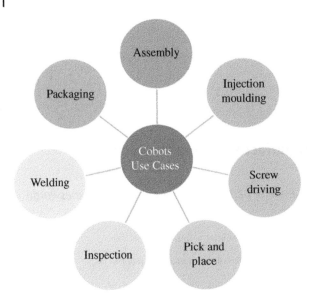

Figure 9.1 Cobots use cases in manufacturing.

They have various scopes of collaborations for humans and robots.

- Coexistence
- Sequential collaboration
- Cooperation
- Responsive collaboration

Cobots are used in the manufacturing industry for various use cases as shown in Figure 9.1. They are used through the process of a product evolution, from assembling using pick and place robots to packaging robots.

Cobots are getting widespread attention because of the expected value of market space in the near future. The projected market reach would grow several folds as shown in Table 9.1 in a span of just six years. The key industries that are expected to be the key focus deployment areas are shown in the table. It implies that the incorporation is pervasive in nature. The key aspects serving as drivers are technical advancements like cloud, data analytics, pervasive automation, need for cheaper labor, critical applications needing workplace safety and so on [4].

Now, robots built for teamwork can operate within a few inches of a human worker. When creators of cobots applied AI to their models, the ensuing improvements were even more striking.

What about robots, do they have AI? AI, however, makes it possible for robots to learn and improve their intelligence [5].

The terms "robot" and "artificial intelligence" are often used interchangeably, despite the fact that they refer to two quite different things. Robots are physical objects, while AI is software.

Table 9.1 Industrial scope for cobots.

Aspect	Details
Market Size	- Valued at around US\$ 710 million in 2020 - Projected to reach US\$ 12.3 billion by 2026 - Expected CAGR of over 45% during the forecast period
Key Industries	Manufacturing, Healthcare, Logistics and Warehousing, Retail and Hospitality, Agriculture
Market Drivers	- Advancements in technology - Increasing labor costs - Emphasis on workplace safety - Flexibility and ease of use - Need for agility and adaptability
Key Players	Universal Robots, ABB Ltd., FANUC Corporation, KUKA AG, Rethink Robotics, Yaskawa Electric Corp.

AI is similar to a computer program that exhibits some of the behaviors connected to human intelligence, such as learning, planning, reasoning, knowledge sharing, problem solving, and more; a robot is a machine designed to carry out one or more simple to complex tasks automatically with the utmost speed and accuracy. AI refers to the study of how machines may reason and act in ways akin to humans. The most advanced robots have AI that can detect and respond to changes in their environment.

Robots with state-of-the-art AI can now work in tandem with humans in ways never before possible in the creative process. In actuality, AI systems are supposed to have capabilities that are radically distinct from those of conventional machines. AI is simply human intelligence that helps human intellect accomplish better results in a specific effort. Robots are self-operating or partially self-operating machines. They employ AI to increase their independence by teaching themselves new skills. Without any actual humans being involved, they are able to simulate human behavior through the use of computerized information processing and control systems

9.2   Industry 5.0 Revolutionizing Industry

The term "Industry 5.0," also known as the "Fifth Industrial Revolution," refers to a new era of industrialization in which humans collaborate with cutting-edge machinery and AI-enabled robots to improve productivity in the workplace.

There are six types of technology identified by the European Commission as potential drivers of Industry 5.0. Current Industry 4.0 models incorporate the first three classes in full.

- **Simulation and Digital Twinning**: Forward-thinking companies in discrete manufacturing and industrial design are already incorporating these technologies into their processes, which is raising the bar for product quality, productivity, and foresight.
- **Information and Communication Technologies**: Technologies for transferring, storing, and analyzing data. Information is the lifeblood of a competitive business. Cloud and edge-based technologies, as well as other recent innovations in connection, storage, and data analytics, are now indispensable to the success of businesses of all sizes and types throughout the world.
- **Machine Learning and Data Science**: Accelerating at a quantum level, machine learning and AI technologies have become an indispensable toolkit for all manufacturers.
- **Human-Computer Interaction (HCI)**: Human–machine interfaces tailored to the individual user. Industry 5.0 sets itself apart primarily by reintroducing human involvement in formerly fully automated and hyperefficient information technology–based processes and systems. By facilitating collaborative production, humans restore manufacturing's spiritual dimension. Their creativity and adaptability are enhanced by the use of collaborative robots (cobots), and the industry as a whole benefits. True "mass personalization" and quicker turnaround times will be possible when humans and technology work together.
- **Biomimetic Engineering and Intelligent Materials**: Nature is the best example we have of how to live sustainably and efficiently. The limitations of completely industrialized processes present difficulties for producers, especially in light of the extractive nature of some industries, the inefficient use of resources, and the harmful impact on the environment. New developments in bio-friendly materials that are also lighter, stronger, and more flexible promise an improved future for businesses, consumers, and the world.
- **Energy-Efficient Technology**: Energy saving, renewable, storage, and independence technologies. Transportation's shift away from fossil fuels and toward electric vehicles is gaining momentum. However, thanks to falling price curves and economies of scale in battery production, the use of wind and solar power is on the rise. Improvements in autonomous technology have the potential to usher in novel transportation business models while also enhancing safety and lessening environmental impact.

9.3 Cobots in Smart Manufacturing and Production

In contrast to Industry 4.0, Industry 5.0 prioritizes the use of highly skilled humans and robots in tandem with manufacture of consumer-specific goods such as smartphones and vehicles. This is in addition to maintaining consistency in quality, flow, and data collection.

Humans and machines collaborated separately in Industry 4.0. Even though the robot and the human worked on the same assembly line, their tasks were clearly separated. With the advent of Industry 5.0, the barrier between human and robotic labor is blurring. Collaborative robots like KUKA and FANUC can do the hard lifting and ensure consistency while a human craftsperson offers the necessary cognitive abilities. In the manufacturing setting, cobots have the potential to alter the dynamic between humans and machines [6, 7].

Collaborative robots and Industry 5.0 have far-reaching ramifications, one of which is a higher priority placed on human input than in previous generations. Cobots facilitate collaborative robot–human efforts. Speaker cabinets, for instance, can be rough-polished by a robot and then refined and quality-checked in real time by a person. Robotics and manufacturing have entered a new era with version 5.0. Figures 9.2 and 9.3 show the KUKA and FANUC industrial cobots which have been widely acclaimed and used for various industrial purposes.

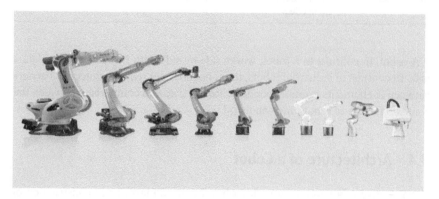

Figure 9.2 KUKA industrial cobots.

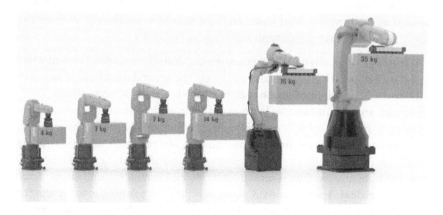

Figure 9.3 FANUC industrial cobots and their size of payload.

Table 9.2 Robots versus cobots.

Aspect	Robots	Cobots
Purpose and design	Designed for independent automation	Designed for collaborative work with humans
Safety features	May require safety barriers or cages	Equipped with advanced safety features for safe human-collaboration
Programming and flexibility	Predefined programming, may require expertise	User-friendly interfaces, easy programming, and reprogramming
Interaction and collaboration	Operate independently, limited interaction with humans	Designed to collaborate and work alongside humans
Cost and deployment	Larger, more expensive, infrastructure modifications may be needed	Smaller footprint, easier integration, more affordable

A cobot, in contrast to a robot, works side by side with humans to complete a task. In contrast to industrial robots, the cobot does not require protective barriers between itself and its human operators. Table 9.2 gives a comparison between the context of usage of robots and cobots [8].

9.4 Architecture of a Cobot

Collaborative robot architecture describes the overall layout and construction of such a machine [9]. It includes the cobot's computer and software, as well as its means of communicating with humans and carrying out their orders. Though cobot architectures may differ based on the maker and model, they all share the following features and components as shown in Figure 9.4:

- **Arm**: Cobots typically have a robotic arm equipped with joints and end-effectors (such as grippers or tools) to carry out tasks. The appendage is built to be versatile and functional, capable of handling a wide variety of responsibilities.
- **Sensors and Actuators**: To learn about their environments and work safely alongside humans, cobots are outfitted with a number of sensors. These sensors can also include vision systems, depth cameras, and proximity sensors. The data collected by these sensors aids the cobot in recognizing and reacting to the presence of humans, avoiding potential obstacles, and facilitating safe interactions.
- **Safety Critical Applications**: Cobots have built-in safeguards to protect the well-being of humans working in close proximity. Included in this category are

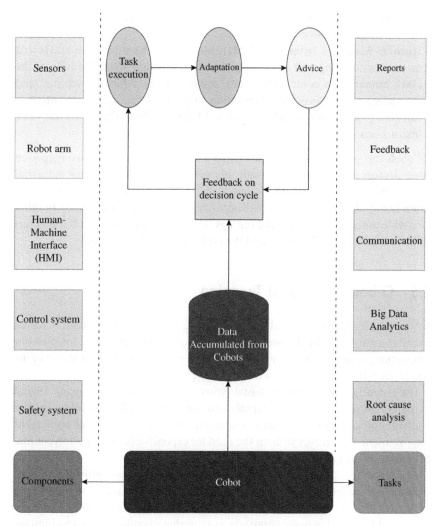

Figure 9.4 Cobot architecture.

methods that allow the cobot to detect and respond to external forces, such as collision detection, emergency stop switches, power and force constraints, and compliance mechanisms. Standard cobot design incorporates risk analysis and safety standards.

- **Control Paradigm**: The hardware and software that make up a cobot's control system direct and coordinate the robot's actions and tasks. Processors, microcontrollers, or programmable logic controllers (PLCs) are the brains of this system;

they take sensor readings, crunch the numbers, and issue commands to the robotic arm and other components.

- **Human–Machine Interface (HMI)**: Most cobots have intuitive HMIs that make programming and controlling the robot a breeze. With the help of the HMI, humans can control the cobot, assign it tasks, adjust its settings, and keep tabs on its performance. Depending on its architecture, a cobot's interface can be a touch screen panel, a Graphical User Interface (GUI), or even voice instructions.

Cobots can share information and expand their capabilities by linking up with other devices and systems. Through this network, cobots and other industrial equipment can work together, share data, and be remotely monitored and controlled. Cobots are designed with security, ease of use, and flexibility in mind for a wide range of manufacturing settings. The specific design of a cobot depends on its function, industry norms, and the preferences of the producer.

9.5 Cobots and Digital Twinning

When the physical and digital worlds interact, digital twins are born. Created simultaneously with the product concept, the digital twin serves as a template for production, is fine-tuned alongside the actual product throughout its lifecycle, and is never treated as an independent entity [10].

The potential advantages of digital twins are enormous for industry. Digital twinning allows developers to rapidly run through countless situations, create and sort through multiple solution strategies, and investigate and implement improvement alternatives without the need for expensive physical prototypes and countless testing iterations.

A producer using this technology can digitally map, test, and certify its products. Optima can model and optimize product flows across the machine's entire life cycle. It is also the foundation for emerging service types like predictive maintenance and its inspiring novel approaches to running a company. It is a technology not only about speed, it is about optimization. The following are the attractive features of cobots.

- **Assistive Cobots**: The Robotics Industry Association gives a vivid picture stating that 5Kg payload Cobots are capturing 47.3% market share and are in huge demand. By 2025 they predict cobot sales will be 33% of overall industrial robot market.
- **Programmable Cobots**: Cobots' simple programmability is one of their most appealing features. Some are even self-learning, picking up new motions after

being touched once or twice by a professional who walks them through the steps. They can be relocated and modified for use in different stages of the manufacturing process, which is even another advantage.

- **From Artificial Intelligence to Machine Learning**: Artificial intelligence (AI) applications like machine learning will also be front and center of cobots. With data and cloud integration, model building is feasible. The outcomes are harnessed to optimize, predict, and classify the informational resources at hand.
- **Widespread Appeal**: The term "machine learning" refers to a branch of AI that encompasses a wide variety of software technologies used in areas as diverse. It is able to harness big data for managing all information that is built around the model. In the near future cobots may optimize on their own.

9.6 Cognitive Digital Twins and Cobots for Collaboration

The cognitive cobots aligned with cyber-physical systems are upcoming in manufacturing industry in highly mission-critical applications [11]. The architecture of the system is shown in Figure 9.5. The components are explained as below.

- Cyber-physical system
- Machine learning layer
- Service layer
- Actual data collection layer

9.6.1 Cyber-physical System

Cyber-physical systems use a physical twin to generate a digital twin, which can then be used for process or product improvement and testing. A digital twin is an identical copy of a physical thing made in a computer. Sensors measuring performance in key areas are attached to the thing being investigated. Information on the object's energy output, temperature, environmental conditions, and more is gathered by these sensors. This information is then sent to a computer where it can be used to modify the digital replica.

Once this data is amassed, the virtual model may be used to run simulations, probe performance issues, and generate ideas for enhancements, all with the aim of getting actionable insights that can be applied to the real thing [12].

Despite the fact that both digital twins and simulations use digital models to mimic a system's numerous activities, simulations are significantly richer

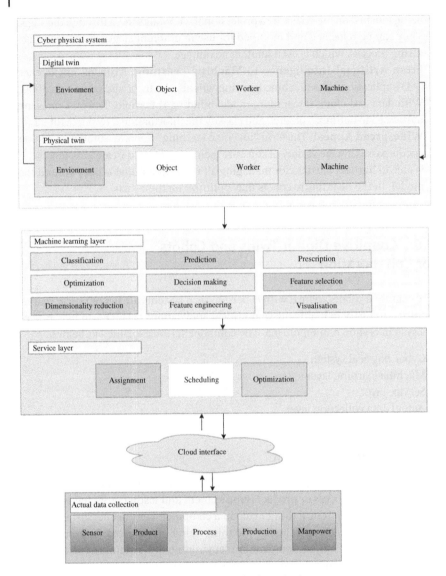

Figure 9.5 Architecture of cognitive digital twin–based cobots.

for research because they are essentially a virtual world. Scope is the primary differentiator between a digital twin and a simulation. Digital twins can explore as many processes as needed, unlike traditional simulations which can only study a single.

That is not all that separates them, either. For instance, real-time data is typically unnecessary for simulations. To counter this, digital twins are built to facilitate two-way communication between the original object and the system processor (which receives data from sensors on the object) and the original object (which receives insights generated by the system processor).

Because digital twins give enhanced and continuously updated data relating to a wide range of sectors, they have a higher potential to improve products and processes than traditional simulations provide [13].

9.6.1.1 Types of Digital Twins

The type of digital twin developed depends on the degree of product enlargement. The primary dividing line between these two cousins is their fields of application. Within a given system or procedure, it is not uncommon for many digital twins to coexist. Let us look over the many kinds of digital twins and their uses to get a better grasp on the field as a whole.

- **Parts Twins**: The simplest operational example is a component twin, the fundamental unit of digital twin. Parts twins are equivalent; however, they refer to slightly secondary parts.
- **Assets Twins**: An asset is the sum of two or more parts that work together successfully. By establishing a "twin" of an asset, you can analyze its components and their interactions in detail, yielding a plethora of performance data that can be analyzed and put to good use.
- **Unit/System Twins**: System or unit twins, the next level of magnification, show how the components depend on each other to form the whole. System twins reveal the interplay between assets and allow for the possibility of performance enhancement recommendations.
- **Process Twins**: It is defined as macro degree of magnification, offered by process twins, that sheds light on the interconnected nature of the systems that comprise an entire factory. How well do these systems work together, and what happens if one of them experiences a delay? Using a process twin can help pinpoint the optimal timing schemes that boost efficiency [14, 15].

9.6.2 Machine Learning Layer

Machine learning's ultimate aim is to develop autonomously analyzing software. When talking about machine learning, deep learning is a subfield. Automatic feature extraction from raw data is made possible by deep learning. It can be used for preprocessing, feature selection and reduction, prediction, and classification. It can also be used for optimizing a process, for making informed decisions, and for drafting prescription from the data to control the process for fewer surprises.

9.6.3 Service Layer

The job is assigned, scheduled, and optimized using the service layer from the input from the machine learning that learns from the cyber-physical systems in turn [16].

The cobot systems are autonomous mobile robots which execute service duties in manufacturing units with a high degree of reliability. Automated services, such as mail delivery, coffee service, and tour guiding, are provided by the robots in response to web-based requests from the building's occupants. The following list provides description of Job scheduling and planning carried out by Cobots.

- **Job Scheduling**: Planning when a robot will carry out each task in accordance with the limits provided by the users is essential for satisfying their requirements. We use a mixed integer programming method to schedule the robots.
- **Job Planning**: Cobot operates as a finite state machine in order to complete tasks. It speaks and displays information on its screen, finds its way to destinations, and makes subplans to get to elevators and get help based on its current location and the locations of its destinations.

9.6.4 Actual Data Collection Layer

The data is collected from automated systems that contain sensors and actuators, product data, process data, production data, and data from employees. It can be data about the product or service also.

Cobots can be used for execution of task, adaptation of a technique, or prescribing an advice. The data can be drafted accordingly. The data sources are planned and incorporated as per requirement.

9.7 Cobots and Artificial Intelligence

With widespread usage of cobots, incorporation of AI is a natural evolutionary idea [17]. Cobots can be used for the following purposes:

- Machine learning with adaptation
- Pattern recognition
- Task planning and optimization
- Natural language processing
- Preventive maintenance
- Efficient data acquisition.

The incorporation of AI helps a cobot to realize efficient processing as explained below:

(a) **Machine Learning with Adaptation**: Model based adaptations can be performed by cobots.
(b) **Pattern Recognition**: Cobots can be deployed for vision-based applications involving image or video. They can automate complicated vision-based applications.
(c) **Task Planning and Optimization**: The digital twinning has helped to plan or optimize the process in efficient way. The process can be good and no surprises are expected from the end process and the process or product is definitely profitable.
(d) **Natural Language Processing**: Speech or unstructured text-based applications can be solved using cobots.
(e) **Preventive Maintenance**: Predictive and preventive maintenances are possible with cobots in manufacturing perspective.
(f) **Efficient Data Acquisition**: The data is efficiently acquired in a timely manner without loss.

9.8 Frontiers Uncovered with Cobots

The cobots are deployed in various areas left to the creativity of the human expert. Few applications of cobots in vertical applications are explained as below.

9.8.1 Cobots in Agriculture

The employment of robots in agriculture is on the rise today. In order to better estimate harvests, savvy farmers employ a variety of analytics and machine learning methods. Agriculture robots are one such instrument, and they are commonly employed in teams (thus the name "cobots"). Because of the robots' artificial arms, harvesting has become more simpler for farmers.

Industrial robots designed to collaborate with people are known as cobots, providing businesses with the best of both worlds. The automobile industry has been employing cobots for some time, notably on the assembly line. However, other sectors are beginning to see their potential as well.

9.8.2 Cobots in Education

Having cobots in the classroom can help in many ways. To begin, cobots have the potential to enhance student engagement and immersion in their studies. Students can better grasp concepts when a robot is used as an interactive teaching aid. They can gain experience in thinking critically and addressing problems.

Cobots can also aid in creating a more level playing field for education. A cobot can help students who do not have the same opportunities as their classmates learn new skills. Those with mobility issues, for instance, can rely on the robot to carry out classroom assignments that would otherwise be impossible [17].

At the end of the day, cobots can assist lighten the load on educators. Cobots can help free up classroom teachers' time for more interesting and innovative projects by taking on some of their routine duties. The classroom experience for both instructors and students could benefit from this.

There are obvious benefits to incorporating cobots into the educational setting. Cobots are becoming increasingly frequent in classrooms as technology improves. It is conceivable that cobots will play a more significant role in the classroom in the not-too-distant future as educators continue to investigate their potential [18].

9.8.3 Cobots in Space Technology

Robots have been used for decades by NASA and other space organizations in a variety of scientific capacities. NASA and other top institutes in the field of space exploration are working to find new applications for cobots as their capabilities improve.

NASA has unveiled its plans to send 12 small shape-shifting cobots to Saturn's moon, Titan. The cobots will be outfitted with a propeller that will allow them to fly or swim in Titan's hazardous environment, which includes freezing temperatures, lakes, caves, oceans, cryovolcanoes, and shower of liquid hydrocarbons. The Jet Propulsion Laboratory (JLA) at NASA is currently putting a 3D-printed prototype through its paces. The purpose of the cobots is to go where no human or simple robot has reached before [19].

9.8.4 Cobots in Quality Control

Quality control inspections and tests are increasingly being carried out by cobots. This leads to more worker output, less time spent testing, and a low error rate. A cobot is ideally suited for testing operations like Printed Circuit Boards (PCB) testing, chip testing, and touch screen device testing. The cobots' inbuilt sensors allow them to behave with great precision and handle different objects with care. The precision and precision vacuum–grasping technique of a cobot allow it to handle even the smallest and most fragile chips. Optical examination is a viable option now that we have vision technology.

9.8.5 Cobots in Warehouse

Cobots in the warehouse reduce the amount of human labor needed to run the facility, which in turn saves money. Cobots do not take the position of employees; rather, they supplement the efforts of human workers to raise output. Cobots like Chuck by six River Systems, for instance, use AI and machine learning to optimize activities and create the most efficient picking routes in real time, all while reducing the need for human workers to physically move around the warehouse. Staff time can be freed for more sophisticated work by streamlining logistics operations.

9.8.6 Cobots in Logistics

With the help of flexible and adaptable robotics, businesses may make use of cobots throughout the year. Because of this, logistics and warehousing firms may make the most of a variety of collaborative robots in operations without incurring any additional costs in the event of a sudden uptick in demand.

9.8.7 Cobots in Healthcare

Cobots are sophisticated robotic arms designed to work side-by-side with people in a risk-free setting. This is a huge improvement over older, bulkier robots that could cause harm to humans [20]. Therefore, cobots are the ideal answer to safely aid doctors and medical practitioners in operating rooms and other applications in the medical industry, where they may improve accuracy and precision without putting patients at risk. In industrialized nations like Canada and the United States, for instance, cobots equipped with cameras aid surgeons in locating malignancies for removal during robotically assisted surgeries.

While robotic surgery may be one of the more visible areas where cobots are being used, their adaptability makes them useful in a wide variety of other areas of the healthcare industry, thanks to Nanyang Technological University (NTU), Singapore. One such charity is Aurolab (a division of Arvind Eye Care Systems) in Madurai, which operates solely on donations. It has put into action eight collaborative robots made by the Danish industry leader Universal Robots, which will aid in Aurolab's mission to end blindness worldwide [21, 22].

9.8.8 Cobots in Hospitality

In the hospitality industry, cobots can replace human concierges by providing automated services. They can answer guests' queries, show them around, and direct them to their rooms or other amenities. These cobots' linguistic flexibility makes them ideal for serving a wide range of customers.

9.8.9 Cobots and Retail Industry

Collaborative robots, often known as "cobots," are smart, adaptable, and easy to use since they are designed to work alongside humans without the fencing that a standard robot demands. Cobots are simple to use and can radically improve the efficiency and quality of an e-commerce company's warehouse and production facility. Cobots can assist with tedious activities, taking them over with pinpoint accuracy and freeing up human workers' time. In addition, cobots may be easily reprogrammed for new purposes, and their simple design means that they can be operated by persons without any experience with robot programming.

There are two big players in the e-commerce business. The first are e-commerce sites that sell only one brand and often produce those products themselves. The second kind of online store serves as a marketplace where different manufacturers can offer their products. Cobots provide optimal solutions for both of these organizational structures [23].

Cobots, when used in a factory setting, can eliminate the need for human labor in the production process [24].

9.8.10 Cobots in Food Production and Delivery

Cobots are capable of nonstop, around-the-clock labor. This is perfect for seasonal surges, as cobots can handle increased output without hiring seasonal or temporary staff. Cobots' adaptability allows factories to make the most of fluctuating production volumes by reassigning them to new product lines or operations. The designed and verified safety mechanism of Universal Robots (UR) cobots makes this possible. When operators are present, they can be set to halt or slow down.

Models of UR cobots accommodate various pick/pack/pal applications thanks to their adaptability in reach and payload. There are built-in palletizing wizards that make deployment simple, especially when catering to unique customer requirements like those found in the food and beverage distribution industry. They guarantee security, high output, and adaptability [25, 26].

9.9 Cobots Safety Standards

When developing an automated system, safety must always come first. ISO/TS 15066, titled "Collaborative Robot Safety," and ANSI/RIA R15.606, titled "Collaborative Robot Safety," are the two primary global safety standards for cobots [27]. Information vital to the safe implementation of a collaborative robot system is provided by these guidelines. The summary of the standards is given in Table 9.3.

Table 9.3 Cobot safety standards.

Safety standard	Description
ISO 10218-1 and ISO 10218-2	International standards providing guidelines for the safe use of industrial robots, including cobots.
ISO/TS 15066	Collaborative robot systems require a technical specification outlining best practices for their development and deployment.
ANSI/RIA R15.06	Standard developed by RIA and ANSI, addressing safety requirements for industrial robots, including cobots.
EN ISO 12100	European standard providing principles for risk assessment and reduction in machinery design, including cobots.
OSHA guidelines	Guidelines provided by OSHA for the safe use of cobots and other machinery in workplaces.

The International Organization for Standardization mandates that collaborative robots implement one of four safety precautions in ISO/TS 15066:2016 "Collaborative Robot Safety" (R2019):

- **Speed and Separation Monitoring**: As soon as the cobot's sensors detect a person it comes to a complete stop. If the robot detects that it is getting too close to a worker, it will either slow down, turn around and move away, or stop altogether.
- **Hand-guided Controls**: Movement of the cobot can only occur when an operator is physically guiding its movements. To load a hefty box onto a vehicle, for instance, an employee can instruct a cobot to do the lifting for them. In this scenario, the cobot will not do anything until the operator presses a "hold to run" button.
- **Power and Force Limiting**: In the case of a collision, this safety feature limits the cobot's ability to exert force. Cobots can come into close proximity to humans if their speed, force, and power are restricted. These constraints also lessen the damage that can be done by an accidental cobot strike. Cobots equipped with force and power limits are often slower and more suited to working with lighter things. These cobots often have softer surfaces and rounder corners to lessen the possibility of injury upon contact [28].
- **Safety Rated Monitored Stop**: Like other industrial robots, the cobot is equipped with a monitored stop that causes it to stop working entirely if it senses the presence of a human worker.

9.10 Conclusion

Growth has become inevitable with the advent of cutting-edge technologies like AI, Big Data analysis, cloud computing, robotics, and automation. Industry 5.0 has come to evolve and the incorporation of Industry 5.0 for welfare of mankind has in turn spawned the collaborative robots called cobots. Cobots aligned with cognition, AI, and cyber-physical systems have become very indispensable in industrial applications. The applications are becoming wide ranging from domestic, commercial, healthcare, educational, and so on. The challenge of cobots is the security threat, but incorporation of safety standards can thwart this issue also. The technology can lobby into many diverse applications in the near future and become part and parcel of day-to-day human life.

References

1 Maddikunta, P.K.R., Pham, Q.-V., Prabadevi, B. et al. (2022). Industry 5.0: a survey on enabling technologies and potential applications. *Journal of Industrial Information Integration* 26: 100257. ISSN 2452-414X, https://doi.org/10.1016/j.jii.2021.100257.

2 Fazal, N., Haleem, A., Bahl, S. et al. (2022). Digital management systems in manufacturing using industry 5.0 technologies. In: *Advancement in Materials, Manufacturing and Energy Engineering*, Lecture Notes in Mechanical Engineering, vol. II (ed. P. Verma, O.D. Samuel, T.N. Verma, and G. Dwivedi). Singapore: Springer https://doi.org/10.1007/978-981-16-8341-1_18.

3 Pizoń, J. and Gola, A. (2023). Human–machine relationship—perspective and future roadmap for industry 5.0 solutions. *Machines* 11 (2): 203. https://doi.org/10.3390/machines11020203.

4 Pizoń, J., Cioch, M., Kański, Ł., and Sánchez-García, E. (2022). Cobots implementation in the era of industry 5.0 using modern business and management solutions. *Advances in Science and Technology Research Journal* 16 (6): 166–178. https://doi.org/10.12913/22998624/156222, http://hdl.handle.net/10045/130076, https://doi.org/10.12913/22998624/156222.

5 Moor, M., Sarkans, M., Kangru, T., Otto, T., and Riives, J. (2024). AI Functionalities in Cobot-Based Manufacturing for Performance Improvement in Quality Control Application. *Journal of Machine Engineering* 24 (2): 44–55. https://doi.org/10.36897/jme/189169.

6 Calitz, A.P., Poisat, P., and Cullen, M. (2017). The future African workplace: the use of collaborative robots in manufacturing. *SA Journal of Human Resource Management* 15 (1): 1–11.

7 Vicentini, F. (2021). Collaborative robotics: a survey. *Journal of Mechanical Design* 143 (4): 040802.

8 Müller-Abdelrazeq, S.L., Schönefeld, K., Haberstroh, M., and Hees, F. (2019). Interacting with collaborative robots—a study on attitudes and acceptance in industrial contexts. In: *Social Robots: Technological, Societal and Ethical Aspects of Human-Robot Interaction*, 101–117.

9 Doyle-Kent, M. and Kopacek, P. (2021). Adoption of collaborative robotics in industry 5.0. An Irish industry case study. *IFAC-PapersOnLine* 54 (13): 413–418.

10 Saenz, J., Elkmann, N., Gibaru, O., and Neto, P. (2018). Survey of methods for design of collaborative robotics applications-why safety is a barrier to more widespread robotics uptake. In: *Proceedings of the 2018 4th International Conference on Mechatronics and Robotics Engineering* (American Society of Mechanical Engineers Digital Collection), 95–101.

11 Quenehen, A., Pocachard, J., and Klement, N. (2019). Process optimisation using collaborative robots-comparative case study. *Ifac-papersonline* 52 (13): 60–65.

12 Aaltonen, I. and Salmi, T. (2019). Experiences and expectations of collaborative robots in industry and academia: barriers and development needs. *Procedia Manufacturing* 38: 1151–1158.

13 Lima, F., De Carvalho, C.N., Acardi, M.B. et al. (2019). Digital manufacturing tools in the simulation of collaborative robots: towards industry 4.0. *Brazilian Journal of Operations & Production Management* 16 (2): 261–280.

14 Bi, Z.M., Luo, C., Miao, Z. et al. (2021). Safety assurance mechanisms of collaborative robotic systems in manufacturing. *Robotics and Computer-Integrated Manufacturing* 67: 102022.

15 Elprama, B.V.S.A., El Makrini, I., and Jacobs, A. (2016). Acceptance of collaborative robots by factory workers: a pilot study on the importance of social cues of anthropomorphic robots. In: *International Symposium on Robot and Human Interactive Communication*, vol. 7. pp. 1–14.

16 Liu, L., Zou, Z., and Greene, R.L. (2023). The effects of type and form of collaborative robots in manufacturing on trustworthiness, risk perceived, and acceptance. *International Journal of Human Computer Interaction* 1–14.

17 Drolshagen, S., Pfingsthorn, M., Gliesche, P., and Hein, A. (2021). Acceptance of industrial collaborative robots by people with disabilities in sheltered workshops. *Frontiers in Robotics and AI* 7: 541741.

18 Ronzoni, M., Accorsi, R., Botti, L., and Manzini, R. (2021). A support-design framework for cooperative robots systems in labor-intensive manufacturing processes. *Journal of Manufacturing Systems* 61: 646–657.

19 Jabrane, K. and Bousmah, M. (2021). A new approach for training cobots from small amount of data in industry 5.0. *International Journal of Advanced Computer Science and Applications* 12 (10): 634–646.

20 Turner, C., Oyekan, J., Garn, W. et al. (2022). Industry 5.0 and the circular economy: utilizing LCA with intelligent products. *Sustainability* 14 (22): 14847.

21 Prassida, G.F. and Asfari, U. (2022). A conceptual model for the acceptance of collaborative robots in industry 5.0. *Procedia Computer Science* 197: 61–67.

22 Saxena, A., Pant, D., Saxena, A., and Patel, C. (2020). Emergence of educators for Industry 5.0: an Indological perspective. *International Journal of Innovative Technology and Exploring Engineering* 9 (12): 359–363.

23 Aheleroff, S., Huang, H., Xu, X., and Zhong, R.Y. (2022). Toward sustainability and resilience with Industry 4.0 and Industry 5.0. *Frontiers in Manufacturing Technology* 2: 951643.

24 Mehdiabadi, A., Shahabi, V., Shamsinejad, S. et al. (2022). Investigating Industry 5.0 and its impact on the banking industry: requirements, approaches and communications. *Applied Sciences* 12 (10): 5126.

25 Doyle-Kent, M. and Kopacek, P. (2021). Adoption of collaborative robotics in industry 5.0. An Irish industry case study. *IFAC-PapersOnLine* 54 (13): 413–418.

26 Pizoń, J. and Gola, A. (2022, June). The meaning and directions of development of personalized production in the era of Industry 4.0 and Industry 5.0. In: *International Conference Innovation in Engineering*, 1–13. Cham: Springer International Publishing.

27 Doyle Kent, M. and Kopacek, P. (2021). Do we need synchronization of the human and robotics to make industry 5.0 a success story? In: *Digital Conversion on the Way to Industry 4.0: Selected Papers from ISPR2020*Online-Turkey, 302–311. Springer International Publishing.

28 Sarıoğlu, C.İ. (2023). Industry 5.0, digital society, and Consumer 5.0. In: *Handbook of Research on Perspectives on Society and Technology Addiction*, 11–33. IGI Global.

10

Edge Computing and Artificial Intelligence-Based Internet of Things for Industry 5.0: Framework, Challenges, Use Cases, and Research Directions

Ramya Ramanathan[1], Sriramulu Ramamoorthy[2], and Simon Grima[3]

[1]*Department of Computing Technologies, SRM Institute of Science and Technology, Katankulathur, Tamil Nadu, India*
[2]*Department of Computing Technologies, SRM Institute of Science and Technology, Kattankulathur Campus, Chennai, India*
[3]*The Department of Insurance and Risk Management Faculty of Economics, Management and Accountancy University of Malta, Msida, Malta*

10.1 Introduction

Cloud computing technology innovation makes people's internet usage level unlimited for different requirements. Basically, cloud computing provides three broad types of services for internet users; they are "Software as a service (SaaS), Infrastructure as a service (IaaS), and Platform as a service (PaaS)" [1]. Through these kinds of different services, any single user or any industry can use cloud computation, storage, and applications using the internet with free or paid services as per requirements. Examples of cloud service providers are as follows: Google provides Google Cloud services, Amazon provides Amazon Web Services, and Microsoft provides Microsoft Azure cloud services. Because of so many innovations, the future world become a smart world through the combination of smart cities, smart homes, smart healthcare, smart manufacturing, and so on, but the kind of increasing number of Internet end users makes it difficult for the cloud service providers in terms of network traffic, transmission cost, network delay, and so on. All cloud service providers are focused to develop a new technology to overcome these difficulties in the future, the new technology named edge computing. By using edge computing technologies, cloud services like computation and storage are brought close to the end user's gadgets.

Digital Twins in Industrial Production and Smart Manufacturing:
An Understanding of Principles, Enhancers, and Obstacles, First Edition. Edited by Rajesh Kumar Dhanaraj, Balamurugan Balusamy, Prithi Samuel, Ali Kashif Bashir, and Seifedine Kadry.

10.1.1 Edge Computing Technology

More Internet-connected application developments for end users require more sensors/tools, which may generate more data from the environment per the application requirements, which is named the IoT applications [2]. All IoT-generated contents are transmitted via the Internet to the cloud where it is stored, processed, and computed based on the needs of the various applications. The overall worldwide Internet user's IoT application data generations and transmissions make network traffic, transmission delay, computation delay. These will increase the overall transmission cost, network energy cost, calculation cost, and so on. To solve these difficulties, all cloud services providers are searching for new technology. To reduce the total application processing cost and time for future end users, edge computing technologies will be deployed.

The edge processor will be positioned between the cloud and end user gadgets since edge computing technology may hinder data transit from end user gadgets to the cloud. Figure 10.1 depicts the edge layer that exists between the cloud's infrastructure and end user IoT gadgets [3]. For instance, in healthcare applications, some sensor devices are used to generate data regarding the human body; those sensors are called IoT devices. Those generated pieces of information are collected by one nearby device, which is named an edge gateway device. Finally, all collected information is stored and processed in the cloud.

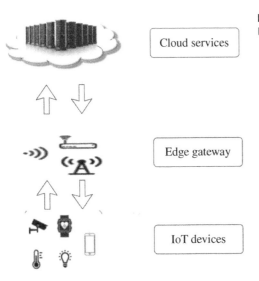

Cloud services

Edge gateway

IoT devices

Figure 10.1 The cloud-edge- IoT layers.

10.1.2 Artificial Intelligence

The technology of AI is widely used, which will make a machine itself solve some problem or task with intelligence. This technology brings more comfort to human life in numerous domains, including cities with smart technology, homes that are smart, intelligent manufacturing, and so on. AI has two broad types named machine learning (ML) and deep learning (DL) [4]. ML is one of the main subparts of AI, which will make a machine learn itself to do some intelligent processes. DL is one of the subdivisions of ML, which must include a few or more neural layers to attempt to work like a human brain [5]. Figure 10.2 represents how AI, ML, and DL are interconnected and subdivided based on their operations [6].

More widespread AI and its subtypes have become solutions for big data manipulation such as IoT sensor data information. The implementation of artificial intelligence (AI) and machine learning (ML) techniques in edge computing technology enables the development of edge intelligence. This allows for storage and computations to be performed in close proximity to the end user application [7–17].

10.1.3 Importance of Edge Computing

Edge computing technology is positioned between a cloud service provider and the end user's IoT devices, and it offers advantages to both parties from several angles. Managing massive amounts of IoT data, maintaining network stability and reaction times at low cost, and ensuring security and confidentiality are the three biggest benefits.

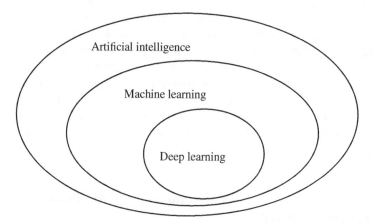

Figure 10.2   AI-ML-DL representation.

- **Enamors IoT Data Handling**:
  With the integration of many daily necessities like intelligent homes, intelligent cities, smart medical facilities, smart the business world, and so on, IoT technology now spans a much larger region. The huge development of Internet-connected devices in various fields such as sensors, online platforms, smartphones, healthcare, and industrial equipment is increasing, which will result in the overall global data increment to 175 zettabytes (ZB) by 2025, according to the prediction of International Data Corporation (IDC) [18–20]. These large data generation processed as big data processing. Data transferring to the cloud for processing will make network traffic and many networks related problems. To prevent these problems, the new concept of edge computing technology is required for processing data generated by IoT devices.
- **Network Stability and Response Time with Minimum Cost**:
  Some IoT application scenarios call for incredibly quick response times. For instance, under the scenario of intelligent driving, autonomous vehicles are equipped with sensor devices like cameras. Throughout the autonomous driving mode, these sensor units can continuously gather data from the environment. According to the cloud computing concept, these data will be sent to the cloud for processing, and the answers will then be sent back to the control chip in the car. This strategy is actually quite time-consuming, and it may even result in the smart vehicle failing to make the proper decision in a timely manner, which might have major repercussions [3] given the complex driving environment of the vehicle.
- **Security and Privacy**:
  There are outsourcing elements in cloud computing. Those who use cloud computing must host local data there. A number of data security and privacy risks result from this [20–23]. Data integrity and accuracy can be harmed by data loss during long-distance transfer between devices and the cloud. Also, excessively centralized processing and storage might develop into significant issues. Other devices in a centralized system will suffer damage if one device malfunctions due to innocent mistakes or malicious attacks. The difficulty with data privacy involves its theft and illegal use by other people, businesses, or organizations. In reality, data owners no longer have control over the data they upload to the cloud, making data privacy difficult to ensure [24].

10.2 The Motivation for Edge Computing and Artificial Intelligence Combination

Edge computing and AI are two remarkable technologies that can collectively result in a powerful combination. The following are the main factors that motivate the integration of edge computing and AI:

- **Latency Reduction**: The latency problem in data processing may be considerably reduced by combining edge computing with AI. By bringing the processing and interpretation of data nearer to the consumers, edge computing can cut down on the quantity of information that must be sent over the network's infrastructure. AI, on the other hand, uses complex algorithms and parallel processing to speed up processing, making it perfect for edge computing.
- **Privacy and Security**: By processing data locally rather than sending it to a central place for analysis, edge computing and AI can improve privacy and security. It is feasible to implement AI models inside of hardware or local gateways to process data while protecting the security and privacy of users.
- **Real-time Decision-making**: AI and edge computing can work together to provide real-time decision-making in various applications. For instance, in a self-driving car or an industrial automation system, decisions must be taken immediately based on complex sensor data. AI algorithms in edge devices can analyze and process the data within milliseconds, providing timely and accurate decisions.
- **Cost-effectiveness**: Due to the lack of expensive central processing units, cloud storage, and other resources, edge computing, and AI may become economically advantageous.
- **Scalability**: Edge computing with AI can offer scalable solutions for a variety of applications because edge devices that are closer to the consumers can be used to deploy and run AI algorithms, bypassing the limits of centralized cloud-based AI models.

In conclusion, the performance, security, privacy, and scalability of numerous applications may all be improved by combining edge computing and AI.

10.2.1 Advantages of Edge Computing with Artificial Intelligence

There are several possible benefits of edge computing using AI as a language model:

- **Reduced Latency**: Edge computing reduces the distance between processing and decision-making from the data source, shortening the time it takes to get that data. This can be beneficial in applications like autonomous vehicles, where real-time decisions can be critical.
- **Increased Privacy**: With the increasing concerns around data privacy, edge computing can help keep personal data within a closed system. This is particularly useful when dealing with sensitive data, such as medical records.
- **Enhanced Reliability**: Edge computing can help ensure the reliable operation of devices by allowing them to function without a constant connection to the cloud. By processing data locally, devices can continue to operate even if the internet connection is lost.

- **Efficient Use of Bandwidth**: Only pertinent data needs to be transferred to the cloud for additional processing owing to edge processing. This allows for quicker replies, lower expenses, and bandwidth conservation.
- **Improved Scalability**: Edge computing can help distribute computational resources more efficiently, making it easier to scale systems up or down as needed. This ensures that the right amount of computational power is available exactly where it is needed.

10.2.2 Advantages of Artificial Intelligence with Edge Computing

There are a few possible benefits of AI integrating edge computing as an AI language model:

- **Speed and Efficiency**: Combining AI with edge computing can enable faster processing of data and decision-making. This can be particularly beneficial in time-critical situations such as autonomous driving, where instant decisions need to be made without delay.
- **Greater Privacy and Security**: Sensitive data can be kept on local devices rather than being sent to the cloud for processing by doing AI computations at the edge (closer to where the data is generated). This results in greater privacy and security.
- **Cost Savings**: Edge computing lessens the data needs to be transmitted to the cloud through edge computing, and infrastructure costs and network bandwidth requirements are reduced.
- **Increased Dependability**: Businesses can lessen their dependency on cloud services and eliminate single points of failure in their systems by processing data and running AI models locally.
- **Scalability**: Edge computing can offer the high-performance processing power required for AI applications that require a lot of computation, allowing AI models to be scaled easily across many devices and platforms.
- **Improved User Experience**: AI services become much more responsive and quicker than previously, which improves customer happiness. This is accomplished by reducing communication between edge and cloud.
- **Lessening of Latency**: AI services that rely on the cloud have latency since it takes time for data to be transmitted, processed, and then the results to be delivered back. Edge computing helps to significantly alleviate this issue.

Overall, edge computing and AI can improve decision-making and data processing by enabling the seamless scaling of AI models across various devices and platforms.

10.2.3 Mutual Beneficial Relationship Between AI and EC

The concept of a mutually beneficial interaction between AI and edge computation (EC) is quite valuable.

AI is the area of computer science that focuses on building systems or machines that can behave intelligently like humans, while EC is a branch of AI that draws its inspiration from the ideas of natural selection and evolution. It solves complex problems using genetic algorithms, swarm intelligence, and other evolutionary methods.

Since both fields have the potential to improve one another's skills, the relationship between AI and EC is advantageous to both parties. AI can benefit from EC algorithms, which are useful for dealing with challenging nonlinear optimization issues. By employing strategies like genetic programming, which allows an AI algorithm to evolve its own code to address a specific problem, EC can be used as a tool to enhance the performance of AI algorithms.

Additionally, AI can assist EC academics and practitioners in creating optimization algorithms that are more accurate and efficient by offering the requisite computational capacity, data processing, and analysis skills. Together, AI and EC can develop new optimization algorithms or enhance those that already exist to address issues in the real world more quickly and accurately.

The mutually beneficial interaction between AI and EC is a promising research area that has the potential to unlock new and advanced AI capabilities using evolutionary approaches. Through collaboration, they may broaden the scope of potential and construct a more promising future for everyone.

10.3 Algorithms of AI in Edge Computing

The method by which ML models are applied to edge devices is referred to as AI algorithms in edge computing. Edge computing is a methodology that allows facts processing and evaluation with the units themselves, as adversarial to the cloud or information center, at the network's edge. This is especially beneficial for real-time applications, like driverless vehicles, or for applications that need privacy, like medical equipment.

AI for edge computing uses a variety of algorithms. These consist of:

○ **Convolutional Neural Networks (CNNs)**: This technique is especially helpful for analyzing images and videos. For tasks like object identification, classification, and segmentation, CNNs are employed.

○ **Recurrent Neural Networks (RNNs)**: This technique is particularly beneficial for speech recognition and natural language processing (NLP). Sentiment analysis, chatbots, and voice assistants are a few applications for RNNs.

○ Decision trees are a very helpful algorithm for decision-making procedures. Decision trees are employed in the detection of fraud, the prediction of customer turnover, and risk analysis.

○ Support vector machines (SVMs) are a very helpful approach for classification issues. SVMs are employed in processes including anomaly detection, text classification, and image classification.

○ **Random Forests**: This approach is especially beneficial for regression issues. They are employed for purposes including forecasting demand, sales, and stock prices.

○ Deep belief networks (DBNs) are a very helpful algorithm for unsupervised learning issues. DBNs are employed in processes like data compression, clustering, and recommendation engines.

Overall, edge computing requires the usage of AI algorithms to process and analyze data quickly along with efficiency. Edge computing is anticipated to play a big role in allowing faster and highly accurate decision-making processes results the increased need for AI across a variety of industries.

10.3.1 Basics of Machine Learning

A subfield of AI called "machine learning" allows computers to learn from data without having to be explicitly programmed. The goal is to teach computers to spot patterns in data so they can make predictions or take appropriate action. Supervised, unsupervised, and reinforcement learning are the three categories of ML.

○ **Supervised Learning**: This kind of ML uses labeled data to learn from. The aim is to predict the outcome for new input data after providing the computer with a collection of input data and related output data. For works like speech monitoring, image processing, and NLP, this kind of learning is employed.

○ **Unsupervised Learning**: In this method, the computer is provided with a collection of input data without any associated output labels. The computer must examine the data and look for structures or patterns. Tasks like clustering, anomaly detection, and dimensionality reduction require this form of learning.

○ **Reinforcement Learning**: This kind of ML relies on trial-and-error learning. The environment is used to teach the computer new skills, and it also receives feedback in the form of incentives or punishments. Uses for this kind of learning include robots, autonomous cars, and game playing.

Several well-liked ML techniques are decision trees, random forests, support vector machines, neural networks, and linear and logistic regression. Applications for these algorithms include sentiment analysis, recommendation systems, fraud detection, and predictive maintenance. In general, ML is revolutionizing industries and allowing computers to carry out complicated tasks more precisely and effectively.

10.3.2 Federated Learning

A decentralized approach to ML called federated learning enables several devices to collaborate to create a single machine learning model without sharing their data with one another or a central server. This method keeps the data on the user's device while occasionally updating the model with data from all connected devices. By allowing consumers to continue participating in the training of a model while retaining their data on their devices, this strategy helps to protect their privacy. Federated learning has a wide range of practical uses, such as enhancing photo or text prediction software.

10.3.2.1 Edge Intelligence Collaborative Learning

Through the use of federated learning, numerous edge devices can work together to collectively train a model without sharing their original data with a centralized server. In order to conduct complicated calculations and analyses locally without depending on a centralized cloud server, edge devices such as cell phones, IoT sensors, and other embedded devices are said to possess edge intelligence.

Edge devices can learn and perform better with the help of federated learning while still maintaining the privacy of the user's data. Edge devices can use federated learning to jointly learn from the data they produce without sending it to a central computer. When confidential data is sent to a central server, privacy worries can be alleviated in part thanks to this strategy.

Federated learning can be used in edge intelligence to enhance a variety of uses, including automated vehicles, image identification, and NLP. For instance, a collection of autonomous cars can work together to learn from their mistakes and enhance their driving abilities. Similar to this, personal digital assistants can benefit from federated learning by being able to learn from their users' behavior and tastes without jeopardizing their privacy.

Overall, federated learning in edge intelligence is a fascinating field of study that could lead to the development of smart, privacy-preserving edge apps.

10.3.3 Evolutionary Algorithms

- Natural selection and genetics are two classes of principles that evolutionary algorithms use to discover the best answer. By surviving and reproducing, the fittest individuals pass on their advantageous qualities to the following generation, simulating the process of natural selection.

- Evolutionary algorithms can be employed in the context of edge intelligence to optimize and hone the performance of the edge devices to the particular task at hand. These algorithms can be used to determine the best configuration options for edge devices so that they are working as effectively as possible.
- The adaptation of DNNs for edge devices is one example of how evolutionary algorithms can be employed in edge intelligence. Since DNNs require a lot of computation, running them on edge devices with limited resources might be difficult. By reducing the model size, increasing the model accuracy, and decreasing the inference time while running on an edge device, evolutionary methods can be utilized to optimize DNNs. Other characteristics of edge intelligence, including power usage, latency, and data processing, can also be optimized using evolutionary algorithms. Edge intelligence can benefit from faster and more effective computing at the edge by utilizing these methods, which will cut down on the price and time needed for cloud computing.

In summary, evolutionary algorithms can be extremely useful for lowering the reliance on cloud computing, increasing the overall effectiveness of edge intelligence, and optimizing edge devices for particular tasks.

10.3.4 Edge Computing for Optimization Solutions with AI

The optimization environment is changing as a result of edge computing and AI. Edge computing enables data to be processed at the network's edge, nearer the data source, reducing latency and enhancing reaction times. On the other side, AI technologies enhance data analysis and decision-making. The optimization solutions can be significantly improved by combining these two technologies.

Real-time optimization is made possible by edge computing since it processes data as it is generated. By analyzing gathered data and making modifications in real time, edge computing, for instance, can assist in optimizing the production process in a smart manufacturing setting. Detecting and responding to quality issues, modifying machinery performance to prevent equipment breakdowns and decreasing delays in the production line.

Reduced data communication to a central server lowers the amount of data that must be stored and transmitted, which is another benefit of edge computing. It is feasible to transfer only the crucial data to a central server for additional processing by performing data processing and analysis at the edge.

Edge computing devices gain intelligence from AI solutions. AI algorithms are able to examine the information gathered by edge devices, spot trends, and then base choices on those patterns. AI is also capable of learning from previous data and making predictions based on it.

For instance, AI-based optimization solutions can assist in lowering operational costs in the logistics and transportation sector by forecasting the most effective routes for shipments. At the network's edge, data processing from GPS devices and traffic sensors can be used to do this.

To sum up, edge computing and AI are crucial technologies for optimization. Combining the two enables real-time optimization and enhances decision-making, resulting in cost reductions, increased productivity, and better customer service.

10.3.4.1 Optimization Solutions Through AI

The optimization environment is changing as a result of edge computing and AI. Edge computing enables data to be processed at the network's edge, nearer the data source, reducing latency and enhancing reaction times. On the other side, AI technologies enhance data analysis and decision-making. The optimization solutions can be significantly improved by combining these two technologies.

Real-time optimization is made possible by edge computing since it processes data as it is generated. By analyzing gathered data and making modifications in real time, edge computing, for instance, can assist in optimizing the production process in a smart manufacturing setting. Detecting and responding to quality issues, modifying machinery performance to prevent equipment breakdowns and decreasing delays in the production line.

Reduced data communication to a central server lowers the amount of data that must be stored and transmitted, which is another benefit of edge computing. It is feasible to transfer only the crucial data to a central server for additional processing by performing data processing and analysis at the edge.

Edge computing devices gain intelligence from AI solutions. AI algorithms are able to examine the information gathered by edge devices, spot trends, and then base choices on those patterns. AI is also capable of learning from previous data and making predictions based on it.

For instance, AI-based optimization solutions can assist in lowering operational costs in the logistics and transportation sector by forecasting the most effective routes for shipments. At the network's edge, data processing from GPS devices and traffic sensors can be used to do this.

Edge computing and AI are crucial technologies for optimization fixes, to sum up. Combining the two enables real-time optimization and enhances decision-making, resulting in cost reductions, increased productivity, and better customer service.

10.3.4.2 Optimization Solutions Through Computing Offloading

Offloading computational duties from a device to an edge computing node or cloud server is a process known as edge computing. This strategy can lessen the burden on the network while also enhancing the device's performance and energy efficiency.

Edge computing offloading can be used to accomplish a number of optimization solutions:

○ **Reduced Latency**: By shifting some of the processing duties to a server or edge device, the system's overall latency can be decreased, leading to quicker reaction times.
○ **Energy Efficiency**: By offloading some of the computational duties to a more potent and energy-efficient edge device or server, edge computing offloading can help to reduce the energy consumption of the device.
○ **Bandwidth Optimization**: Edge computing offloading can assist in reducing the amount of data that needs to be transferred over the network, resulting in lower bandwidth requirements and lessened network congestion.
○ **Security and Privacy**: By keeping sensitive data and computations near the source and lowering the likelihood of data breaches or unauthorized access, edge computing offloading can help to improve the security and privacy of the system.

In general, edge computing offloading can offer a number of optimization options that can help to raise a system's speed, energy efficiency, and security. The trade-offs between the advantages and disadvantages of edge computing offloading, such as the extra latency brought on by the offloading process and the heightened complexity of the system design, must be carefully considered.

10.3.4.3 Other Solutions Without Computing Offloading

While offloading computation to an edge device or computer is one way to use edge computing, there are other approaches that can be used instead. These remedies consist of:

○ **Caching**: Edge devices can store frequently accessed data locally rather than having to obtain it from a cloud or data center, which lowers latency and boosts performance.
○ **Edge Analytics**: Without the need for the data to be sent to the cloud or data center, edge devices can carry out local data analytics to glean insights and useful information from sensing data.
○ **Decentralized Control**: Edge devices can be used to create decentralized control systems in which choices are made locally as opposed to being imposed by a centralized server or cloud.
○ **Distributed Data Storage**: By storing and managing data locally, edge devices can eliminate the need to send data to a data center or the cloud.

○ **Hybrid Approaches**: Depending on the needs of the application, many edge computing solutions combine offloading and local processing, dividing computation between the periphery and the cloud or data center.

Overall, there are numerous uses for edge computing that go beyond simply offloading work to an edge computer or device. The best option will rely on the application's particular needs, including those related to latency, energy efficiency, and security.

10.3.4.4 Optimization for Resource Allocation

In many edge computing applications, resource allocation is a crucial problem because it affects how effectively computing power, memory, and network bandwidth are managed to satisfy the system's performance needs. There are several methods that edge intelligence can be applied to optimize resource allocation, including:

○ **Predictive Resource Allocation**: Based on past usage trends, user behavior, and other variables, edge intelligence algorithms can forecast resource demand. As a result, there is less chance of resource overload and performance bottlenecks because resources can be allocated in preparation.

○ **Dynamic Resource Allocation**: In reaction to shifting workloads or network conditions, edge intelligence can be used to allocate resources dynamically. This enables real-time resource optimization, guaranteeing that the system functions effectively at all times.

○ **Quality of Service Optimization**: By prioritizing resource allocation based on the importance of the application or job, edge intelligence algorithms can be used to improve the Quality of Service (QoS) of the system. This makes sure that the most crucial tasks get the resources they require, while less crucial tasks only get resources when they actually need them.

○ **Energy Efficiency**: By dynamically adjusting the system's power consumption in response to workload and network circumstances, edge intelligence can be used to optimize resource allocation for energy efficiency.

○ **Fault Tolerance**: By dynamically reallocating resources in the event of a failure or other problem, edge intelligence can be used to optimize resource allocation for fault tolerance.

In conclusion, edge intelligence has the potential to be an effective instrument for maximizing resource allocation in edge computing systems. Edge intelligence can aid in ensuring that resources are allocated effectively, efficiency is maximized, and system reliability is increased by utilizing cutting-edge algorithms and data analytics.

10.4 An Overview of the Symbolic Connection Between Industry Capabilities and Edge Computing

In the business world, edge computing can support industrial automation, allow real-time data analysis and decision-making, and increase operational effectiveness. Edge computing, for instance, can be used in manufacturing to watch equipment and foresee maintenance requirements to minimize downtime.

In general, edge computing expansion is viewed as a key enabler for corporate digital transformation and improved productivity.

10.4.1 Existing Industrial Internet of Things (IIoT)

The important parameters of existing IIoT are listed below:

○ **Predictive Maintenance**: IIoT systems are used to track equipment performance data in real time and plan maintenance before breakdowns happen.
○ **Quality Control**: IIoT systems are much faster and more accurate than people at spotting product flaws. Businesses can reduce waste and increase efficiency by identifying quality problems early in the production process.
○ **Supply Chain Management**: IIoT offers real-time surveillance and monitoring of goods and shipments, which can help supply chain management be optimized. This allows businesses to decide with confidence on inventory, logistics, and delivery.
○ **Energy Management**: By automating procedures and maximizing energy use, IIoT can assist businesses in reducing energy waste. This can result in substantial financial savings and advantages for the environment.
○ **Remote Monitoring**: Even in remote areas, IIoT systems can provide real-time monitoring of machines and processes. This allows businesses to address problems quickly and decide on maintenance and repair with knowledge.

IIoT has enormous potential to boost productivity, cut expenses, and enable better decision-making in industrial settings.

10.4.2 Challenges of Edge Intelligence-based IIoT

The integration of edge intelligence with the IIoT has many challenges; some important challenges are listed below:

○ **Scalability and Integration Issues**: IIoT edge intelligence solutions encompass a significant number of sensors, devices, and networks, which creates significant scaling and integration issues. Data must flow freely throughout the complete network and all components must be properly connected and configured.

○ **Security and Privacy**: Since edge intelligence includes the storage, processing, and transmission of sensitive data to the cloud, it is essential to make sure that the data is safe from cyberattacks.

○ **Price**: Edge computing comes at a high expense in terms of infrastructure, infrastructure investment, maintenance, security, and management. It is a challenging job for small and mid-level industries due to the cost and scalability of edge resources.

○ **Data Management**: Edge intelligence systems produce large amounts of data that need to be instantly cleaned, processed, and analyzed. Data management becomes more challenging as the quantity of data elements rises, necessitating the implementation of a strong and efficient data management strategy.

○ **Interoperability**: To achieve a higher level of integration and operational efficacy, systems must be interoperable because devices and sensors must function with various vendors, protocols, and platforms.

○ **Skill Set**: Professionals with the necessary skills to design, develop, and implement these solutions are in short supply because edge intelligence technologies call for specialized knowledge and abilities.

○ **Reliability**: Edge intelligence systems depend on an ongoing, secure link between the cloud and the devices. Any lapse or failure in connectivity can jeopardize the smooth operation of the complete network, making it more difficult to use and less effective.

○ **Regulation**: As the IIoT and edge intelligence work to address complicated and varied machinery and automation issues in various sectors, industrial regulations across various industries present a challenge for integration. These tools must be used in accordance with laws, standards, and ethical principles.

10.4.3 Difficulties in Developing an Edge Intelligence-based IIoT

An IIoT system with edge intelligence technology has some difficulties:

○ Data handling and gathering managing the enormous amounts of data produced by machines and devices on the factory floor is one of the main difficulties of IIoT. Although edge intelligence technology aids in local data processing, managing the ever-growing amount of data presents difficulties.

○ **Security Issues**: There is always a chance for cyberattacks and data breaches with IIoT devices. To protect sensitive data, edge intelligence technology needs added security measures, which adds costs and personnel.

○ **Compatibility Problems**: In IIoT systems that use peripheral intelligence technology, integrating various devices and platforms can be very difficult.

Legacy hardware and systems may have compatibility problems that can impede efficient data transmission.

○ **Lack of Necessary Skills**: To deploy and manage edge intelligence technology, specialized knowledge is needed. It can be difficult for businesses to meet the talent requirements for deploying and maintaining the technology.

○ **Constant Updating**: To stay competitive, businesses must keep their systems up to speed as edge intelligence technology develops. Technology updates can be expensive and demand extensive preparation and resources.

In conclusion, careful thought and planning are necessary for IIoT systems based on edge intelligence technology to surmount the difficulties presented by data gathering and management, security concerns, compatibility issues, skillset shortages, and continuous upgrading.

The application of IoT technology in the industrial sphere is known as the IIoT. In order to watch, gather, and analyze data in the industrial, transportation, energy, and other related sectors, it is necessary to use interconnected machines and devices that are embedded with sensors, software, and network connectivity.

10.4.3.1 General Industry End-point Components with Intelligence

Depending on the particular business and application, there are a number of significant industry end-point components that have intelligence. Here are a few illustrations:

○ **Sensors**: A crucial end-point component, sensors can gather information on a range of bodily factors, including temperature, pressure, humidity, and motion. They are employed in a number of sectors, including industry, gardening, and healthcare.

○ **Smart Cameras**: In sectors like retail, production, and transit, smart cameras are becoming more and more common. They are used in applications like security, quality control, and traffic tracking and are capable of a variety of jobs like object identification, face recognition, and license plate recognition.

○ Actuators are tools that can transform electrical impulses into actual bodily movements. They are a crucial part of control systems and are employed in a number of sectors, including robotics, aircraft, and the automobile.

○ **Microcontrollers**: These tiny computers that are built into gadgets are capable of a variety of jobs, including data gathering, processing, and transmission. They are employed in a number of sectors, including industrial automation, transportation, and consumer devices.

○ **ML Algorithms**: To add intelligence and decision-making skills, ML algorithms are increasingly being incorporated into end-point components. They can be used for purposes like predictive maintenance, fraud detection, and individualized suggestions. They are used in a variety of sectors, including healthcare, finance, and marketing.

Overall, because it can help to increase productivity, cut costs, and allow new apps and services, the incorporation of intelligence into end-point components is becoming more and more significant in many sectors.

The IIoT is made up of several important process units, the following are the some the units:

○ **Sensors and Actuators**: These are the gadgets that record information about the outside world and can operate machines and equipment.
○ **Cloud Computing**: To process the enormous amounts of data produced by IIoT devices, cloud platforms offer storage, processing capacity, and analytics capabilities.
○ **Edge Computing**: This is the real-time processing of data at the network's edge, close to where the data is produced, as opposed to sending all of the data to the cloud for processing.
○ **Connectivity**: For IIoT applications, safe and dependable connectivity is crucial. Wi-Fi, Bluetooth, and cellular networks are just a few of the communication protocols used for commercial IoT.
○ **Analytics**: The IIoT produces enormous amounts of data, which must be analyzed to yield useful insights. Processes can be improved, upkeep requirements can be predicted, and inefficiencies can be found by analyzing this data.

The components of work from sensors and actuators of IoT devices to cloud computing through the process/factory monitoring and control would be classified into three major divisions, named analysis layer, control layer, and access layer, which are represented in Figure 10.3. In this, the sensors and actuators are communicated and controlled by the middle process/factory monitoring and control devices. So, the sensors and actuators are named as access layer and process/factory monitoring and control devices are named as control layer. The overall computation and storage are provided by cloud computing so it is named as the analysis layer.

The overall view of processes from the cloud to the endpoint devices is shown in Figure 10.4. The endpoint devices are the operational technology devices of IoT; those are large numbers of small devices for different operations. So, the endpoint devices are placed at the bottom of Figure 10.4. All the endpoint devices are connected and controlled by the middle-edge AI devices; these are information technology-based learning devices. The cloud computing technology, depicted at the apex of Figure 10.4, offers enduring data storage and comprehensive computational capabilities.

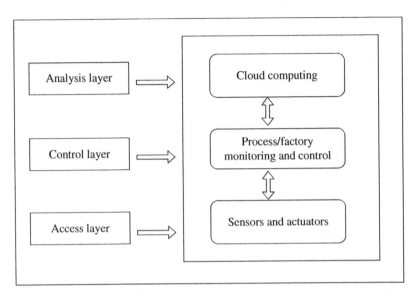

Figure 10.3 Major layers classifications of cloud computing, mid processes, and IoT sensors and actuators.

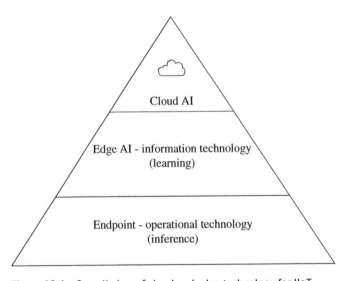

Figure 10.4 Overall view of cloud and edge technology for IIoT.

10.4.4   How Edge Intelligence Supports Industrial Devices?

Edge intelligence is the capacity to handle data, analyze it, and make decisions on devices at the edge of a network as opposed to in a centralized place like a cloud server. Here are some ways that business equipment can benefit from edge intelligence:

1. **Real-time Decision-making**: Without sending data to a centralized server, edge intelligence allows devices to make choices in real time based on local data. This can be especially helpful in sectors where quick action is necessary, like manufacturing, where a broken equipment could result in downtime and lost output.
2. **Reduced Delay**: Edge intelligence can reduce latency, or the amount of time it takes for data to be transmitted from a device to a central computer and back, by processing it nearby. In fields like autonomous vehicles, where choices must be made rapidly to prevent mishaps, this can be especially significant.
3. **Lessening the Amount of Data that Must be Communicated Over the Network**: Edge intelligence can also lessen the amount of data that must be transmitted over the network because only the most crucial data must be sent to a central computer for additional analysis. This can be especially helpful in sectors like oil and gas where devices are frequently found in isolated locations with poor network access.
4. **Increased Security**: Edge intelligence can increase security by lowering the quantity of data that needs to be shared over the network by processing data locally. In fields like healthcare, where private data must be secured, this can be especially crucial.
5. **Increased Scaling**: By enabling autonomous device operation without the need for a centralized computer, edge intelligence can also increase scalability. This can be especially helpful in fields like agriculture where equipment may be dispersed over vast regions and required to function autonomously.

Overall, edge intelligence can help industry devices in a number of ways, including by allowing real-time decision-making, lowering delay and bandwidth needs, enhancing security, and boosting scaling.

The following are some advantages of IIoT:

○ **Enhanced Efficiency**: IoT devices and sensors can watch and regulate industrial processes to optimize operations, leading to heightened productivity and cost savings.
○ **Increased Safety**: Real-time detection of possible safety risks by IoT sensors enables prompt remedial action.
○ **Predictive Maintenance**: IIoT can offer insights into preventative maintenance, resulting in lower maintenance expenses and less downtime.

○ **Improved Product Quality**: By offering real-time tracking and feedback throughout the manufacturing process, IIoT can enhance product quality.

The following are some of the major commercial IIoT applications:

○ **Smart Manufacturing**: IIoT can boost safety, lower expenses, and increase productivity in the manufacturing process.
○ **Connected Transportation**: IIoT can enhance fleet administration and supply chain management while lowering costs and raising safety.
○ **Smart Energy**: IIoT can help energy networks run more efficiently, use less energy, and spend less money.
○ **Predictive Maintenance**: IIoT can aid in the prediction of machine failures, enabling proactive maintenance and minimizing downtime.

10.5 Use Cases of IIoT with Edge Intelligence Technologies

The IIoT with edge intelligence technologies has multidimensional use cases, some important uses cases are mentioned here:

○ **Predictive Maintenance**: Predictive maintenance is one of the most popular IIoT use cases. IIoT sensors can be used to gather information on equipment health, temperature, vibration, and other variables. This data can then be used to forecast when maintenance is necessary and alert the appropriate staff in preparation.
○ **Quality Control**: IIoT can be used in manufacturing for quality control reasons. Manufacturers can spot flaws or anomalies in products and make the required corrections to keep quality and accuracy by using sensors to monitor the production process.
○ **Inventory Management**: By tracking the whereabouts and motion of goods in real time, IIoT can be used for inventory management. This can aid inventory management efforts, cut down on stockouts, and prevent overstocking.
○ **Supply Network Optimization**: IIoT can be applied to improve the supply network. Manufacturers, distributors, and retailers can optimize their logistics and supply networks to lower costs and boost efficiency by tracking goods and shipments in real time.
○ **Energy Management**: IIoT can be used for energy management tasks like observing energy consumption in plants and locating opportunities to increase energy efficiency. The overall expense of energy can be decreased and energy waste reduced as a result.
○ **Safety and Security**: IIoT can be used to monitor worker health and safety, find equipment failures, and spot possible security threats. This could lessen workplace mishaps and boost general security and safety.

10.5.1 Real Time Applications of Industrial Internet of Things with Edge Intelligence

The IIoT with edge intelligence has numerous real-time applications across various industries. Here are a few examples:

1. **Predictive Maintenance**: IIoT with edge intelligence enables real-time monitoring of equipment and machinery, collecting data on performance, temperature, vibrations, and other parameters. By analyzing this data locally at the edge, predictive maintenance algorithms can detect anomalies and predict failures before they occur. This helps optimize maintenance schedules, reduce downtime, and increase overall equipment effectiveness.
2. **Asset Tracking and Management**: IIoT devices with edge intelligence can be used to track and manage assets in real time. For example, in logistics and supply chain management, sensors attached to goods and vehicles can provide location, temperature, and humidity data. This enables better inventory management, improves asset utilization, and ensures the quality and safety of goods throughout the supply chain.
3. **Energy Management**: IIoT combined with edge intelligence can be used for real-time monitoring and optimization of energy consumption in industrial settings. By collecting and analyzing data on energy usage from various sensors and devices, energy management systems can identify energy wastage, optimize energy distribution, and implement demand-response strategies, ultimately reducing energy costs and improving sustainability.
4. **Quality Control and Process Optimization**: IIoT devices equipped with edge intelligence can continuously monitor and analyze data from production lines, ensuring quality control and process optimization. Real-time data analysis can identify deviations, defects, or inefficiencies, enabling immediate corrective actions. This improves product quality, reduces waste, and increases overall productivity.
5. **Safety and Security**: IIoT with edge intelligence can enhance safety and security in industrial environments. Sensors and cameras can monitor critical areas for abnormal activities, unauthorized access, or hazardous conditions. Real-time analysis at the edge can trigger immediate alerts, initiate emergency response procedures, or autonomously take preventive actions to mitigate risks and ensure worker safety.
6. **Supply Chain Optimization**: IIoT with edge intelligence can optimize supply chain operations in real time. By collecting data from sensors placed throughout the supply chain, such as warehouses, distribution centers, and transportation vehicles, intelligent algorithms can optimize routes, minimize delays, and enhance overall efficiency. This results in faster deliveries, reduced costs, and improved customer satisfaction.

These are just a few examples of the many real-time applications of IIoT with edge intelligence. The combination of IoT devices, edge computing, and advanced analytics enables faster decision-making, reduced latency, improved operational efficiency, and enhanced capabilities in various industrial sectors.

10.6 Research Directions of IIoT

1. **Real-time Metrics**: Real-time analytics are a key component of the IIoT. Real-time analytics allows for the collection, analysis, and utilization of large volumes of data generated by the many sensors and systems utilized by an organization. This might enhance product quality, decrease downtime, and increase organizational effectiveness.
2. **Edge Computing**: As the assortment of devices that are connected keeps increasing, edge computing is becoming increasingly important for the IIoT. Edge computing involves conducting data processing and analytics at or close to the edge of a network, as opposed to transmitting enormous volumes of data back to a centralized cloud or data center. In addition to reducing latency and bandwidth requirements, edge computing can help to improve data security.
3. **Cybersecurity**: As there are more connected devices and networks, there is a greater risk of cyberattacks. The security of IIoT systems and data must be ensured, and innovative research is needed to develop efficient cybersecurity solutions.
4. **Lack of Standardization**: One of the biggest problems confronting IIoT is the lack of standardization. It can be challenging to integrate various systems and devices because different manufacturers use a variety of different standards and protocols. Research might concentrate on creating uniform standards and protocols to facilitate easier system and gadget integration and interoperability.
5. **AI and ML**: These technologies can assist businesses in gleaning insightful information from the massive amounts of data produced by IIoT systems. These technologies can support pattern and trend recognition, equipment breakdown prediction, and process optimization. To develop and improve AI and ML algorithms, especially for IIoT applications, more research is required.

10.6.1 Research Directions of Industrial Internet of Things with Edge Intelligence

Edge intelligence and the IIoT offer a variety of research avenues that can improve different elements of industrial operations. The following are some hot spots for this field's research:

1. **Edge Computing Architectures**: To facilitate the deployment and operation of intelligent applications at the edge of the IIoT network, various

edge computing architectures are being investigated. This include investigating approaches for fault tolerance, scalability, and effective resource management.

2. **Edge Analytics and ML**: Creating sophisticated analytics and ML algorithms that can be run locally at edge devices to process and analyze data. This improves real-time decision-making skills in industrial contexts while lowering latency and bandwidth consumption.

3. **Edge-based Security and Privacy**: Looking at techniques that keep edge devices in IIoT networks secure and private. To protect crucial industrial data, this entails investigating authentication procedures, secure communication routes, data encryption, and privacy-enhancing solutions.

4. **Edge-based Data Fusion and Integration**: The IIoT networks are developing methods to integrate and fuse data from various edge devices and sensors. To achieve accurate and thorough insights, this entails resolving issues including data heterogeneity, data quality assurance, and real-time synchronization.

5. **Edge-enabled Predictive Maintenance**: Examining methods for predictive maintenance that make use of edge intelligence to track and assess machine and equipment status in real time. This involves creating algorithms to identify abnormalities, forecast breakdowns, and plan proactive maintenance tasks, which decrease downtime and boost operational effectiveness.

6. **Edge-enabled Energy Optimization**: Investigating methods for monitoring and enhancing energy use in industrial settings by utilizing edge intelligence. Creating algorithms to optimize energy use, reduce waste, and encourage sustainable practices is required.

7. **Edge-based Autonomous Systems**: Investigating autonomous systems that use edge intelligence to take judgements and take action in the present in commercial settings. This entails researching methods like autonomous cars, edge-based robots, and collaborative systems that can function well and adjust to changing circumstances.

8. **Edge-based Human–Machine Interaction**: Researching cutting-edge paradigms for human-machine interaction that make use of edge intelligence to allow for seamless and organic interactions between people and IIoT devices. To improve user experience and productivity, this involves research on augmented reality, virtual reality, and gesture-based interfaces.

9. Developing effective data management methods for edge devices in IIoT networks. In order to enable real-time analytics and decision-making, this entails investigating data storage, caching, and retrieval techniques that can manage substantial amounts of data created at the edge.

10. **Edge-enabled Supply Chain Optimization**: Examining how edge intelligence might improve supply chain management by making it possible to track, monitor, and optimize commodities, inventory, and logistical

procedures in real time. To increase efficiency and traceability, this includes researching strategies like edge-based RFID, blockchain, and predictive analytics.

By using the potential of edge intelligence in IIoT systems, these research initiatives hope to spur innovation, boost operational effectiveness, and allow seismic shifts in industrial processes across several industries.

10.7 Summary

This chapter begins with an overview of edge computing, AI, and its significance. There are justifications for why edge computing technology and AI methodology should be combined, including the benefits of edge computing with AI, the benefits offered by AI with edge computing, and the relationship between AI and edge computing that results in mutual benefits. Then there are the explanations about the different types of artificial algorithms, ML algorithms, federated algorithms, and evolutionary algorithms. Then, a thorough analysis of AI-based optimization techniques for edge computing is presented. The IIoT and edge computing are finally introduced, and the challenges of edge intelligence-based IIoT, the challenges of developing edge technology-based IIoT frameworks, the advantages of edge intelligence in IIoT, applications of edge-based IIoT, use cases, and future IIoT research directions are covered in detail.

References

1 Hua, H., Li, Y., Wang, T. et al. (2022). Edge computing with artificial intelligence: a machine learning perspective. *ACM Computing* 0360–0300. https://doi .org/10.1145/3555802.

2 Ghosh, A.M. and Grolinger, K. (2021). Edge-Cloud computing for internet of things data analytics: embedding intelligence in the edge with deep learning. *IEEE Transactions on Industrial Informatics* 17 (3): 2191–2200.

3 Wang, F., Zhang, M., Wang, X. et al. (2020). Deep learning for edge computing applications: a state-of-the-art survey. *IEEE Access* 8: 58322–58336.

4 Wang, X., Han, Y., Leung, V.C.M. et al. (2020). Convergence of edge computing and deep learning: a comprehensive survey. *IEEE Communication Surveys and Tutorials* 22 (2): 869–904.

5 Shi, Y., Yang, K., Jiang, T. et al. (2020). Communication-efficient edge AI: algorithms and systems. *IEEE Communication Surveys and Tutorials* 22 (4): 2167–2191.

6 Ali, Z., Jiao, L., Baker, T. et al. (2019). A deep learning approach for energy-efficient computational offloading in mobile edge computing. *IEEE Access* 7: 149623–149633.

7 Yahuza, M., Idris, M., Wahid, A. et al. (2020). Systematic review on security and privacy requirements in edge computing: state of the art and future research opportunities. *IEEE Access* 8: 76541–76567.

8 Liu, D., Yan, Z., Ding, W., and Atiquzzaman, M. (2019). A survey on secure data analytics in edge computing. *IEEE Internet Things* 6 (3): 4946–4967.

9 Wang, G., Yang, X., Cai, W., and Zhang, Y. (2021). Event-triggered online energy low control strategy for regionally integrated energy system using Lyapunov optimization. *International Journal of Electrical Power & Energy Systems* 125: 3, 106451.

10 Chen, L., Zhou, S., and Xu, J. (2018). Computation peer loading for energy-constrained mobile edge computing in small-cell networks. *IEEE/ACM Transactions on Networking* 26 (4): 1619–1632.

11 Liu, C., Bennis, M., Debbah, M., and Poor, H.V. (2019). Dynamic task offloading and resource allocation for ultra-reliable low-latency edge computing. *IEEE Transactions on Communications* 67 (6): 4132–4150.

12 Chiang, Y., Zhang, T., and Ji, Y. (2019). Joint task-aware loading and scheduling in mobile edge computing systems. *IEEE Access* 7: 105008–105018.

13 Chen, M. and Hao, Y. (2018). Task offloading for mobile edge computing in software-defined ultra-dense network. *IEEE Journal on Selected Areas in Communications* 36 (3): 587–597.

14 Ning, Z., Dong, P., Kong, X., and Xia, F. (2019). A cooperative partial computation offloading scheme for mobile edge computing enabled the internet of things. *IEEE Internet of Things Journal* 6 (3): 4804–4814.

15 Du, J., Zhao, L., Feng, J., and Chu, X. (2018). Computation offloading and resource allocation in mixed fog/cloud computing systems with min-max fairness guarantee. *IEEE Transactions on Communications* 66 (4): 1594–1608.

16 Wang, Y., Tao, X., Zhang, X. et al. (2019). Cooperative task offloading in three-tier mobile computing networks: an ADMM framework. *IEEE Transactions on Vehicular Technology* 68 (3): 2763–2776.

17 Zheng, Z., Song, L., Han, Z. et al. (2018). A Stackelberg game approach to proactive caching in large-scale mobile edge networks. *IEEE Transactions on Wireless Communications* 17 (8): 5198–5211.

18 Jing, W., Miao, Q., Song, H., and Chen, X. (2019). Data loss and reconstruction of location differential privacy protection based on edge computing. *IEEE Access* 7: 75890–75900.

19 Kang, J., Yu, R., Huang, X. et al. (2019). Blockchain for secure and efficient data sharing in vehicular edge computing and networks. *IEEE Internet of Things Journal* 3 (6): 4660–4670.

20 Wang, Q., Chen, D., Zhang, N. et al. (2017). PCP: a privacy-preserving content-based publish-subscribe scheme with differential privacy in fog computing. *IEEE Access* 5: 17962–17974.

21 Qiao, Y., Liu, Z., Lv, H. et al. (2019). An effective data privacy protection algorithm based on differential privacy in edge computing. *IEEE Access* 7: 136203–136213.

22 Hossain, M.S., Muhammad, G., and Amin, S.U. (2018). Improving consumer satisfaction in smart cities using edge computing and caching: a case study of date fruits classification. *Future Generation Computer Systems* 88: 333–341.

23 Samie, F., Bauer, L., and Henkel, J. (2019). From cloud down to things: an overview of machine learning in internet of things. *IEEE Internet of Things Journal* 6 (3): 4921–4934.

24 Liu, C., Cao, Y., Yan, L. et al. (2019). A new deep learning-based food recognition system for dietary assessment on an edge computing service infrastructure. *IEEE Transactions on Services Computing* 11: 249–261.

11

Smart Manufacturing with a Digital Twin–Driven Cyber-physical System: Case Study and Application Scenario

P. Abinaya¹, S. Prem Kumar¹, P. Sivaprakash², K. Arun Kumar³, B. Shuriya⁴, and Surbhi Bhatia Khan⁵

¹*Department of Information Technology, Rathinam College of Arts and Science, Bharathiyar University, Coimbatore, India*
²*Department of Computer Science and Business Systems, Dr. N.G.P. Institute of Technology, Anna University, Chennai, India*
³*School of Computer Science, Rathinam Technical Campus, Anna University, Chennai, India*
⁴*Department of Computer Science and Engineering, PPG Institute of Technology, Anna University, Chennai, India*
⁵*Department of Data Science, School of Science, Engineering and Environment, University of Salford, Manchester, United Kingdom*

11.1 Introduction

Technology had a significant impact in real world and it is evolving. Revolution, a word to make the mighty tremble, lights a flame of hope and inspires way too many protest songs. The industrial revolutions change the way civilization works, and now another revolution is waiting. New technologies surround us. They change the way we communicate, travel, and see the world. Most fundamentally, they change the way we make stuff cheaper, better, smarter, and hopefully, safer too. The past industrial revolution was brought about by engineering innovation, and the current revolution is being brought about by smart manufacturing. So, in a smart manufacturing methodology, one can connect these things together, have maps to do things, and also be able to solve complicated problems. Smart manufacturing has great impact in the real world. The first Industrial Revolution was powered by steam machines. Let us use an example to better grasp what is meant by the term "digital twin," which is defined as an up-to-date representation of a real physical asset that is now in use. Future environments can communicate with robots and networked computers thanks to technological advancements called "cyber-physical systems" [1]. These systems might be operating in our industries by 2050, traveling beside us on our daily commutes, and driving on

Digital Twins in Industrial Production and Smart Manufacturing:
An Understanding of Principles, Enhancers, and Obstacles, First Edition. Edited by Rajesh Kumar Dhanaraj, Balamurugan Balusamy, Prithi Samuel, Ali Kashif Bashir, and Seifedine Kadry.
© 2024 The Institute of Electrical and Electronics Engineers, Inc. Published 2024 by John Wiley & Sons, Inc.

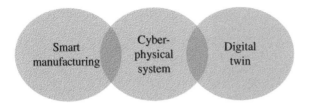

Figure 11.1 Integrating smart manufacturing with cyber-physical systems and digital twin technology for enhanced efficiency and insight.

our highways, By incorporating cyber-physical systems into society, numerous promises are made, including the creation of new markets and economic growth as well as an increase in the efficiency and sustainability of many of our current practices (Figure 11.1).

11.2 Smart Manufacturing

Through the integration of information islands, smart manufacturing enables manufacturing organizations to optimize their production and supply networks. Modern IT infrastructure is really what smart manufacturing is all about. Support manufacturing is absolutely necessary for firms to be globally cost competitive. This one is motivated by unheard-of data availability. These recent innovative developments combine the real and virtual worlds, opening up whole new fields of invention and improving the cost-effectiveness, efficiency, sustainability, and safety of manufacturing. The need for a new IT infrastructure that all manufacturers and suppliers can use and contribute to while keeping their own security and proprietary systems means that smart manufacturing also faces considerable obstacles [2]. Smart manufacturing has such enticing potential to create new business ventures and employment prospects that decision-makers from numerous organizations are collaborating to make it a reality. The creation of new tools that enable manufacturers to use data more effectively is the driving force behind smart manufacturing. The creation of new technologies that enable manufacturers to use data more effectively is known as smart manufacturing. Smart is using that info to create something. In order to control our processes and predict what will happen and manage it, we must first understand it and seek for correlations between various data points. By doing this, we can use that information to make wise decisions that will enable us to manage and optimize our processes.

The ability of manufacturers to develop cyber-physical systems is another way that sensor technology advancements play a part. Smart manufacturing is essential in bringing together people, data, and objects in collaborative cyber-physical systems. And the new Internet of Things (IoT) technologies have made it possible. Cyber-physical systems monitor and gather data from manufacturing processes using inexpensive sensors. The physical and virtual worlds can converge thanks to these systems allowing for creative business solutions. Smart manufacturing allows companies to synthesize information in ways that have never been done before creating a connected enterprise [3]. The goal of connected enterprise is to optimize production and the total supply network by combining information from business systems and plan floor data with information from the supply chain.

The challenge is each system independently has its own methodology of doing things. The second one is assembly lines and electricity. The third is partial automation and computers and now Industry 4.0, the digital robotic age, all the way from steampunk to cyberpunk. How can a normal factory turn into smart factory like industrial sensors, connected systems, robots, software solutions and all the other trimmings? A major role is played by smartphone technology; the sensors become cheaper, they become smaller, and they become more effective. That has the benefit of not just checking the products as they pass through the queue. Furthermore, robots cannot just get more agile. As already known self-driving cars are becoming more and more common. So Industry 4.0 is the advent of utilizing to collect and analyze data to make better decisions. However, there is no denying that technology is advancing and getting better all the time thanks to 5G and other innovations. One of the important components of automation in the form of collaborative robots is the robot that works right alongside people so that one can profit from their cognitive abilities while also delegating the labor-intensive tasks to the robot. Since it is hoped that in the future they will be able to proficiently handle a variety of tasks and begin making decisions even in areas where they were not originally needed. The robots can sense if there is a human being around them, sometimes with vision and sometimes with motion control, and then they can continue to operate around the human at a safer speed. And then, there is 3D printing for potentially everything from machine parts and food to human organs and other 3D printers. The key is telling them when to stop [4]. Prototyping and planning were the uses of technologies such as 3D printing and simulation; by production, however, we mean the ability to actually build things that need to function in the real world.

3D printed homes and buildings are beginning to appear in the construction industry. But the most important thing that today is that no matter if someone is a small player they can have capabilities in smart manufacturing using 3D printers. In the thread tying this together, the industrial Internet of Things (IIoT) just ignores the oil cooler gossip [5]. One can get the insights into where the bottlenecks are. Also, one can improve efficiency and productivity, and the way to do that is you have to have sensors everywhere. To truly engage and convert one's firm to adopt smart practices, one must possess the ability to analyse that data and creating virtual models to track and progress technology is one of the most useful applications of digital twin technology. Everyone may be living in a simulation that replicates some characteristics and dynamics of how an IoT device operates and reacts over the course of its existence. Not only can one simulate and plan in advance how one might lay out a factory or a supply chain warehouse or something like that but one can also then manage more effectively supported by AI and software design to support assist at every step from manufacturing to logistics to retail. Anyway, it is amazing the amount of different points of data that can be gathered now with that amount of data that it becomes really difficult to digest it and understand what is happening, and that is where AI comes into play.

Predictive analytics is a concept that is used in machine learning (ML) and artificial intelligence (AI) to help estimate near-term consumer demand. The ability to make robots and all the production systems more agile, nimbler, and more responsive to your needs so the company can then select the appropriate solutions for their business and then connect them together via an API (Application Programming Interface). Around 81% of manufacturing facilities rely on data analytics or real-time data to improve productivity of manufacturing facilities. All these innovations are affecting everything from automotive to aviation and from pharmaceuticals to oil, gas, and energy for workers everywhere, but it does not mean being obsolete. It can mean better, safer jobs, opportunities to work at home or alongside our future robot overlords, that is, robot partners. There are going to be lots and lots of jobs that are automated, hundreds of millions of jobs that will not be needed. Because robots will do them but there is no shortage of work that only humans can do for the consumer. It is a new world of personalized quality goods made cheaper and delivered faster than ever before. Because of direct consumer production and the emergence of e-commerce platforms like Shopify and WooCommerce, manufacturers are now able to sell directly to customers, which enable them to get goods to customers typically at a cheaper cost. And the side effect of all this efficiency is greener, more planet-friendly manufacturing.

By 2035, increased investments in smart manufacturing could cut the US manufacturing sector's annual energy costs by US$7–25 billion (Ethan A. Rogers, American Council for an Energy-Efficient Economy). Even yet, waste is greatly decreased by intelligent manufacturing, which enables one to appropriately optimise their facility or output. Government initiatives for smart manufacturing are also ballooning across the world with an estimated US market size of US$200 billion by 2026. The industrial robotics sector is not the main driver of the rapidly growing investment in robotics technology; rather, it is the automotive sector, major IT firms, government financing, and other sectors. All governments realize that to stay competitive, they need smart manufacturing. They need to say that they are faster, more efficient, and more agile. In reality, smart manufacturing is the amalgamation of several technologies that facilitate the improvement, speed, and efficiency of businesses in very small amounts of time. Now it depends on the industry and the technologies that one needs, but anything that can help in this process, anything that can help in making the entire process leaner and more efficient, can come under the gamut of smart manufacturing; a smarter way to work, to shop, to build, to create a new world. Intelligent manufacturing frees up a lot of time for people to engage in more creative pursuits. It enables consumers to get products that are made quicker, more efficiently, and better than before and then customize those products to exactly what the market is looking for.

11.2.1 Smart Manufacturing in Medical Field

Example: To ensure a steady supply of goods around the world, Pfizer is employing smart manufacturing to create remotely controlled processes and virtual laboratories. Using comparable ideas, Pfizer has also developed portable manufacturing systems that enable them to effectively produce vaccines for kids in the nations where they are most required. Pfizer is concentrating on creating small-scale, adaptable, modular versions of its larger manufacturing technology. In order to properly implement these manufacturing technologies around the world and build the control strategies for them, Pfizer relies on smart manufacturing.

11.2.2 Steps Taken by Government

Smart manufacturing is going to be the future world. So, every country's government has started to take steps to improve the field's technology using smart manufacturing. In Figure 11.2, the government's efforts to advance smart manufacturing are shown for various nations.

11.3 Cyber-physical System

Cyber-physical system is all about the future. Cyber-physical system is about convergence of technologies, communication, computing, decision-making, perception, and autonomous action. So, examples are manufacturing in Industry 4.0 where there are intelligent machines that do self-diagnosis. They also do the decision-making and fault diagnosis. Next, we have self-governing vehicles venturing far beneath the ocean's surface, travelling to outer space, and even smaller endeavours such as self-driving cars. Take the example of the smart car interacting while parking or smart devices at home sensing the presence and providing information like the room temperature as one enters the home. Cyber-physical

Figure 11.2 Steps taken by government for smart manufacturing.

systems, a novel type of built systems that combine computation and physical processes, power these. Cyber-physical systems encompass a wide range of technological disciplines, including mechatronics, design, embedded systems, IoT, AI, and many more. It is still unclear why and how useful cyber-physical systems are, despite their usefulness and availability. A cyber-physical system is all about the intersection, not the merging, of the physical and the digital worlds. The hypothetical setting in which computer network communication takes place is known as cyberspace. It combines engineering models and techniques from mechanical, environmental, civil, electrical, biomedical, chemical, aeronautical, and industrial engineering with those from computer science. The cyber-physical systems that are connected to the Internet monitor physical components using computer-based algorithms. This indicates that individuals are able to function on their own based on their immediate environment. Because they incorporate the behaviors and actions of all users linked to them, cyber-physical systems have the potential to revolutionize the way electricity is distributed. Perfect intelligent systems have the ability to automatically operate and seamlessly adjust the power supply to meet the varying energy needs of customers in various areas integrating renewable power sources reduction of losses and enhanced safety.

These guarantees include mass-customized products that exactly match consumer preferences and reduce waste in production, as well as automated vehicles that reduce pollution, enhance traffic flow, and allow drivers to work or unwind while driving. Cyber-physical systems and telecare alarm systems are caregiving aids that enable the elderly and sick to live more independently. Intelligent technology tools allow persons with disabilities to participate more fully in society. An enhanced food production rate using the least amount of water, energy, pesticides, and food waste possible is achieved through a cyber-physical agricultural system. Drones and robots used by the military that reduce risk to soldiers and safeguard society. Our prior experiences have shown us that technology never operates exactly as predicted, which might have unintended consequences and policy repercussions. There are always unanticipated outcomes, some of which are favorable, some of which are detrimental, and some of which are never fully understood. One unexpected consequence of 3D printing could be that it alters consumer behavior by making production so simple to obtain that people start producing more and lose attachment to the products. As a result, individuals might be more likely to throw things away and produce more waste. Old jobs will be lost as more duties are given to cyber-physical systems, but new ones will be developed as well, such as fixing robots and acting as a mediator between people and robots. It is desirable to delegate meaningful tasks to robots. Safety should be one of our primary concerns while figuring out how to make human and robot collaboration safe. This is important when robots operate close to humans. These systems are substantial, sophisticated, intelligent, and self-improving. Finding the root cause and assigning blame will also be challenging. In order for cyber-physical systems to function properly, enormous volumes of data are needed, which raises various privacy concerns. Smart home systems, for instance, would wish to keep track of when occupants are away to ensure efficient energy use. This information is also useful to burglars. Should the code of conduct on medical professional secrecy be reconsidered in light of the health information kept on linked medical technology system component? A person's health can be improved by gathering information about their physical characteristics and way of life. Cyber-physical systems will have an impact on how we interact with machines and may possibly cause controversy. Cyber-physical systems can be used to create intelligent prosthetics for the disabled. Research into the possible consequences of cyber-physical systems has shown that they have already had a substantial impact on a variety of elements of our personal and professional lives. A number of legal concerns, such as those pertaining to accountability, liability, data ownership,

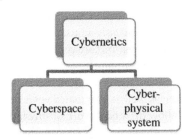

Figure 11.3 Types of cybernetics.

and privacy, are brought up by the usage of networked autonomous working machines in complicated data contexts. When building cyber-physical systems to work close to people, current safety laws must be changed to ensure that no one is wounded and that the benefits outweigh any potential drawbacks (Figure 11.3).

Some other areas where cyber infra systems are likely to have a major role are smart water networks to reduce leakage and better ground water management. Although the word "cyberspace" is frequently used in conjunction with it, its origins are older and deeper. It would be more correct to view the terms "cyberspace" and "cyber-physical systems" as deriving from the same root, cybernetics, rather than as being related to one another. Robotics, the IoT, and smart manufacturing are all examples of cyber-physical systems. This is required in India to improve the quality of life in villages as well as in urban places. A doctor in a city can virtually consult with a patient using telerobotics and telemedicine, and in an emergency, the patient can be brought in right away. Perhaps people need assistance moving from place to place, in which case robots can also aid with tasks. Therefore, villages and other isolated locations can also export the high standard of living found in cities. The next paradigm change in technology, dubbed "cyber-physical systems," is hailed as having the potential to exponentially accelerate growth and development. Cyber-physical systems are intelligent, autonomous, and effective systems that are intended to spur innovation in a wide range of industries, including industry, agriculture, water, energy, transportation, and infrastructure. IoT and cyber-physical systems are playing a bigger and bigger role in daily life, governance, and essential infrastructure. For instance, automobiles, medical devices, building controls, and the smart grid are all examples of cyber-physical systems. The development of software that can help secure this cyberspace is essential. Cyber-physical system technologies provide a competitive advantage to the nation's scientific, engineering, and technical advancements. It aids government initiatives like "Make in India." Although cyber-physical systems have been applied in many fields in the larger context, they are still in the embryonic stage.

11.3.1 Requirements of Cyber-physical Systems in Production Systems

o **Design Methodology:** The design process facilitates the specification, modeling, and analysis of interoperability, networking, clock synchronization, and hybrid and heterogeneous models, or models of computing that are both continuous and discrete. Through synthesis, interaction with older systems, and modularity and composability, it also offers scalability and complexity management. Additionally, validation and verification are supported by design approach through assurance, certification, simulation, and stochastic models.

o **Cyber Security:** Require security from resilience, privacy, malicious attacks, and intrusion detection.

o **Safety:** Need safety measures like hazard analysis which identifies unsafe control actions, safety constraints, and losses.

Figure 11.4 shows the requirements of the cyber-physical system.

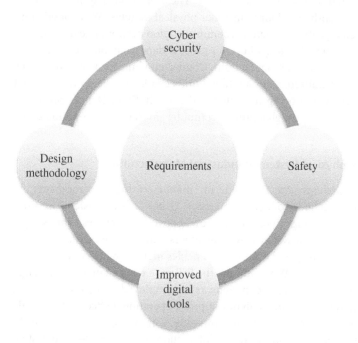

Figure 11.4 Requirements of cyber-physical systems.

11.3.2 Benefits of Cyber-physical System

There are several advantages of cyber-physical systems:

1. Integration
2. Interaction between human and cyber-physical systems
3. Managing uncertainty
4. Better system performance
5. Scalability
6. Agility/Flexibility
7. Faster response time.

11.3.3 Components of Cyber-physical System

The foundation of every cyber-physical system is integrated information processing computer system, Similar to a car, a jet, and other vehicles. These computers are utilized to carry out the particular work. So, for instance, if the ABS system that controls the breaks in an automobile malfunction. These computer systems communicate with wire sensors, wire actuators, and the outside world. These are collectively referred to as embedded systems. These embedded systems will not function independently any longer in cyber-physical systems. With cloud computing, they exchange the data across a communication network like the internet. And that makes it possible to gather and process data from various embedded systems. A system of systems can be built. Connected embedded systems can be managed and decentralized. Computational units can be executed on connected embedded devices while being regulated and decentralized. Data is processed either automatically or through a human computer interface in computing units.

11.3.4 Cyber-physical System in Industrial Manufacturing

Cyber-physical systems are utilized in manufacturing environments to self-monitor and regulate production processes. Cyber-physical systems improve production by improving information flow between machines, the supply chain, suppliers, business systems, and customers. Smart manufacturing, which improves product security and traceability, enables high visibility controls on the supply chain. Cyber-physical systems and the IoT are having an increasingly large impact on the manufacturing sector. Sensors are employed to identify problems and forecast equipment wear. Analytics improve operational effectiveness while lowering maintenance costs. Through computing, communication, and controls, the cyber-physical system can engage in interaction with the real-world system (Figure 11.5).

11.3.5 Applications of Cyber-physical System in Industrial Manufacturing

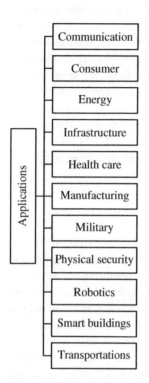

Figure 11.5 Applications of cyber-physical system in industrial manufacturing.

11.4 Overview of Cyber-Physical System

Cyber-physical systems consist of autonomous and cooperative parts and parts that interact with one another depending on the situation at all levels of production, from processes to machines to production and logistics networks. Cyber-physical systems are increasingly being used to promote sustainability. Cyber-physical systems are enhancing agricultural practices, assisting firefighters in detecting and putting out fires, and enabling scientists to reduce underwater oil spills. The cyber-physical manufacturing systems are operated by means of networked systems [6]. The DT, which is a digital representation of a product in use, opens up new opportunities for collaboration between manufacturers, customers, and suppliers beyond pure product development, particularly in collaborative product development and data exchange in mechanical engineering and other technical industries. Products today can be tracked throughout their entire life cycle, from the initial idea to the finished product, including maintenance and service all the way to its modernization. Interoperability is a fundamental necessity of Industry 4.0 since it allows for the continuous transparency of all processes throughout the whole value chain, allowing for the increasingly effective fulfilment of specific client requirements not only during product creation but also throughout the manufacturing of products. However, as production and planning become more complex, they also require close coordination across all value chain actors and continuous management of the product life cycle. This is where the so-called collaborative assembly comes into play. This is made possible by cyber-physical systems' advantages in end-to-end data transmission and information technology assistance even during ongoing manufacturing.

11.5 Digital Twin

A DT is a virtual or digital counterpart of a real-world commodity or object. NASA was one of the first institutions to use this technology for space exploration missions. DTs connect the physical and digital worlds by obtaining real-time data from the placed sensors. The gathered information is either decentralized and saved locally or centralized in the cloud. A virtual clone of the assets is used to assess and simulate the data. Once the information from the simulation has been obtained, the parameters are applied to actual assets. The performance of real assets is improved by this data integration between real and virtual representations. Manufacturing, automotive, construction, utilities, and healthcare are just a few sectors where DTs can be deployed. The Fourth Industrial Revolution's next great item for creating new goods and procedures is DTs. Future innovation will be based on DTs, which will provide virtual replicas of real-world systems, cities, and products. In a recent Siemens study for gas mixing systems, insights from the simulation of form and flow behavior were combined with generative algorithms. The end result is a channel shape and configuration that is truly unique, one that is substantially more efficient than anything previously devised and one that allows for the simulation and testing of large factories down to individual machines. For instance, robots find it challenging to do milling jobs because the strong forces involved in manufacturing cause them to move incorrectly. The DT, however, enables real-time estimation and correction of the forces that pull the robot away from the milling path, keeping it on its planned path. When it comes to operations, DTs enable real-time comparison of sensor data from a real point with a simulation of that site. Sudden interruptions are no longer an issue because it is possible to forecast with certainty if the point parallel to operation will be available. But Siemens is already guiding the future, and this is only the beginning. AI and DTs can be combined to enable computers to autonomously design cutting-edge products. With the Californian start-up Hack Rod, which seeks to create customized sports automobiles, Siemens is currently realizing this promise. The DT disrupts conventional paradigms and creates remarkable opportunities for creation, production, and operation. DTs are the foundation of innovation going forward because of this. Predictive maintenance is one of the key uses for DTs. A world record-setting electric aircraft motor that weighs only 50 kilograms and is five times more powerful than comparable electric motors was created by Siemens Utilities, the DT. But it goes further than that. The power of 3D printing is also unleashed by DTs.

Assume there are locations where several pumps can be used to extract underground oil and gas. A component of a system, like the pump's valve, a system, like the pump itself, or a system of systems, like the well site with several pumps, might all be considered assets under this definition of the DT. Let the pump here be the asset. The creation of a model that will be updated with the incoming data from the pump to represent the current condition will result in an up-to-date depiction of the pump. These pumps have pricey components such as plungers, seals, and valves. As a result, failures must be avoided by foreseeing them, which will help us cut down on downtime. It is essential to locate system flaws and learn which components might require repair or replacement. Additionally, it will make it easier to manage the inventory. But the effectiveness of these pumps' operation will vary depending on the operating environment. To improve the fleet's overall efficiency, it must be able to monitor the entire fleet, model potential future events, and conduct comparisons. This will be beneficial for our operations planning. The modeling technique we must employ truly relies on how we plan to use the DT. For example, to anticipate the pump's remaining useful life and improve maintenance schedules, a data-driven model can be employed. A new degradation model can then be used to determine the pump's remaining usable life. Using information from the pump monitored by many metrics, including pressure, flow, and vibration, this deterioration model is continuously updated. A physical model made by combining mechanical and hydraulic components might serve as an illustration. To stay current, this model is fed data from the pump and its parameters are estimated and modified using the incoming data. With the help of this model, it is possible to mimic the behavior of the pump under various malfunction scenarios. A Kalman filter, which can describe the degradation of the pump as a state and routinely update this state to represent the pump's current status, can also be utilized as a DT in a similar manner. These are some instances of creating a DT. The DT may alternatively be a hybrid of these concepts, depending on its intended function. For the fleet, there is a need to create. A distinct DT must be made for each individual object. As a result, it is necessary to establish a special DT that has been initialized with the parameters relevant to the pump for each of the pumps at the various well locations. A pump could have several DTs depending on its intended application. For example, it is necessary to develop various models that serve the various functions in order to find failure prediction and fault categorization. Through the IoT, these DTs are all linked and exchange information. The history of a real asset's digital counterpart is one of its key characteristics.

The real asset's DT is frequently updated to reflect the state of the real asset. Depending on how the DT is utilized and what is recorded in the current state, the sort of information included in the history may change. The history recorded by each DT can be the operational data from the particular pump and its healthy and defective states if the DT is utilized for fault categorization. The operating information from a single pump can eventually be compared to these DT histories. to comprehend how other pumps responded to failures of a similar nature and how it influenced the effectiveness of the fleet. Utilizing DTs to track the entire fleet has further benefits for organizing operational activities and enhancing repair plans. Think of a scenario in which one of the pumps is predicted to break soon. It is possible to estimate the cost and how this would affect the fleet's efficiency using DTs. In addition to helping with future planning, DTs assist owners understand the history of their assets. To evaluate how various elements like weather, fleet size, or other operating conditions affect the performance, DTs can be utilized to model hundreds of potential future scenarios. By alerting maintenance personnel about impending failures so they can prepare for necessary repairs and replacements, this will help you manage your assets and optimize operations. A DT is an actual representation of a working item. The DT, which is being updated with this data and transformed to its real asset, is periodically sent the data collected from the asset and the environs. Every asset has a distinct DT that likewise records the genuine object's history and modeling technique. Depending on how it will be utilized, the DT's model will need to be chosen. DTs can be used to better manage spare component inventories, monitor and manage the fleet, detect breakdowns in advance, and minimize downtime. DT is one of the outcomes of the technology assessment. This world is about to be destroyed by digital twins [7]. A combination of datasets combined to accurately imitate a real object's behaviour in a platform-style simulation is called a Digital Twin (DT). But its significance and effectiveness much transcend that. First used in "Mirror Worlds," a book by Davin Gellinger published in 1991, the term "digital twin" was later used in relation to manufacturing by Michael Greaves of the Polida Institute of Technology, or more specifically by NASA in their 2010 Roadmap study.

11.5.1 To Create Digital Twin

The first step is to create a design and test it by trial and error methods. Some computational and simulation-based approaches are used. Need some time to check one possible combination of digital twin. Finding the ideal combo strategy is necessary. The time required to verify a single potential combination is considerable when employing the simulation-based method. As a result, a limited number of combinations may be verified quickly. A DT is made up of dataset and it is capable of checking thousands of possible.

11.5.2 Digital Twin in Smart Manufacturing

A virtual representation of a real-world system, procedure, or item is called a "digital twin." It can be used to simulate and study a number of manufacturing-related factors, such as design, testing, and performance. Without the need for expensive and time-consuming physical testing, businesses may acquire important insights into how their goods and processes will operate in the real world by building a DT. There are many advantages to using a DT in smart manufacturing: efficiency and output gains, greater teamwork and communication, enhanced design and testing, and many other improvements.

11.5.3 Cyber-physical System Versus Digital Twin

DT stems from a complex engineering background, but the cyber-physical system originates from a scientific complex computational background. However, it is claimed that they have a number of similarities, including both focusing on the bidirectional flow of data between them and being composed of physical assets and cyber-physical replicas, as well as using applications and technology to produce greater insights and to improve decision-making. They also both have hierarchical structures that are the same and both incorporate bottom-up approaches as necessary beginnings. However, the emphasis is different. The DT, which can simulate this specific spatiotemporal situation, is a highly consistent representation of the physical in terms of geometry and structure. The DT can comprehend and foresee events in the physical world thanks to its behaviors and functionalities that are highly accurate replicas of the real thing. At its core, cyber-physical systems aim to expand the physical system's capabilities by heavily integrating computation and communication into the physical process.

The network of sensors and actuators is thought to be at the center of cyber-physical systems, which conform to the ideas of cooperative computing, communication, and control. Data transfer between physical and cyber systems is also given more consideration. The perceived similarities and differences between DT and cyber-physical systems are now seen as being further apart. The DT is a crucial part of a cyber-physical system. Despite the fact that they have many characteristics in common and employ a variety of applications and technologies to achieve their intended goals, the two concepts are said to differ significantly in their development emphasis, which in turn causes a disparity in the aspects captured model within the respective concepts. Understanding the connection between DTs and cyber-physical systems could significantly change future applications as well as the vocabulary and terminology used when addressing a DT's parts.

11.5.4 Digital and Cyber-physical System Relation

One element of a cyber-physical system is a DT. A technology made possible by the development of cyber-physical systems is the DT. This can then be used as a tool for asset simulation and management. Cyber-physical systems, which include the IoT, AI, big data and data analytics, and cloud computing, are considered to be at the center of the digitalization trend. The DT is supposed to have been developed through this process, the fusion of several technologies, and the gathering of big data. The concept of a cyber-physical system includes both the DT, which is an asset's digital replica, and the IoT, which is a network of sensors and controls deployed inside an asset and serves as a link between digital and physical entities. They are said to differ in their origins.

11.5.5 Digital Framework

A framework for DTs to manage equipment data depicts the production environment digitally and incorporates equipment-related analytical models (Figure 11.6).

11.5.6 Digital Twin in Health Care

Healthcare uses digital twins, which can aid in understanding how AI digital training is used and implemented through real use cases the future of health scale will

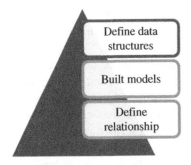

Figure 11.6 Digital twin in health care.

be like that the doctors could visualize health indicators in real time and even simulates the impacts of surroundings on health picture [8]. The physician as well as the hospital only needs to pick up the display next to them in order to monitor and simulate any potential problems or viruses. They can then view the visualisation of the body from sales to organs. AI digital training enables medical professionals to make quick decisions in order to deliver the finest care. While all of this sounds fantastic, these days are not so distant from us. It is not selected by presenting exceptional cases of identity in healthcare; rather, it aids in the advancement of medicine in the real world. First, there is the Dial Beats model that Eddie and Schleskin first presented in 2003. They use different equations to build an edit routing model that has replicated the diabetes at a high level of biological and clinical details. In order to address commodities and treatment with multiple outcomes, this model can be used to a variety of clinical and management problems. It can even be validated for use in diabetes and its complications. This is the kind of innovation that it makes possible. Another fantastic example is the artificial pancreas created by cover chef and the team in 2019. By simulating an individual's metabolic system's operations and automating nutrient administration through the use of a control mechanism, the model makes advantage of readily available real-time inputs, such as continuous glucose monitoring. The image can simulate diabetes testing and treatment in various patients. It makes innovations like this one possible. Please allow me to the patient's surroundings in which their blood glucose is optimally controlled.

The usage of the artificial pancreas in clinical practice has begun occurring similarly working with devices that patients use in the digital therapeutic inconsistent. Finally, but just as importantly, let's discuss Gutierrez and Sean's digital heart trinkets and how each may alter in 2019. They built a digital three model of hearts to detect and predict causes of the strong biases and simulate the flow of the blood through the heart. Imagine if doctors could visualise the state of their patients' hearts and make precise health predictions. They have seen some critical examples of this in action, but because the human body is so complex, this example tends to focus on a single aspect of our health, which is where the thing comes in. They create a digital dream for patients by modeling the whole human body and providing a panoramic view; they can offer personalized systematic and precise treatment plans. The real cool aspect is they can buy graph representation with digital twin to overcome the limitations of the traditional digital tree. They are able to scale various body signals at different levels providing a more accurate and detailed view of the patient's health. The project represents the first proof of concept for a new class of machinery support tools that may be expanded to include runtime monitoring and verification, deployment of healthcare devices, and developing concepts combining scientific computing and machine learning in systems medicine. Babio and his team are the only ones advancing IDT, as seen by the four-layered photographs of patient models.

The open layer, which deals with the network of tissues and cooperating organs that all have related functions, is, in fact, considerably more complicated. It is incredible how everything works together to keep their body functioning properly. And last, but not least, they have the exposure group layer. The layer represents the total amount of exposure experience they are talking, like a toxic substance, treatments, physical activities, posture, and even lifestyle habits. All these factors influence the individual health or disease status and their independence with human genetics health status and physiology. They also show two case studies in the papers. In the first case, they use digital patients' model over time to simulate appropriate treatment plan to treat patients with hypertension and in the second case continuous real-time monitoring and accurate prediction use entity to prevent life-threatening complications of inclusion. The clinical state of the heart in two-dimensional projected phase to fall to different case studies in both cases on GM based model was used to simulate the effects of treatment on blood pressure. The first graph on the left shows the effects of different treatments for high blood pressure. The grid density shows a predicted outcome of the treatment plan that includes increased physical activity, then the reduced calorie in text compared to medication. With these treatment strategies, blood pressure and heart problem fluctuation are reduced overall, which reduces the chance of having a condition known as divorce disease.

When a patient has a vital infection that affects their heart, the simulation displays the following information: the orange density indicates the effects of the current treatment; the red density shows the long-term effects of the unstressed infection on body pressure; and the green density shows the predicted blood pressure trajectory in the heart canvas in real time if the patient needs to be in a healthy state. Agate has a lot of potential applications in healthcare. There are also obstructions they need to overcome. One of the biggest challenges is data security and privacy digital training which needs access to vast amounts of patient data to be effective, but they also need to ensure that patient data is protected only accessed by authorized personnel. Another consideration is the ethics they need to make sure that their agency is fair and unbiased and do not discriminate against certain patients. This is particularly important because an acetone can sometimes cause bile to be present in the data. They are trained on these challenges. The potential of training in healthcare is huge. Technology is changing the way they think about Input/Output Scale and Intelligent Decision Technologies (IDT) is at the forefront of this revolution by integrating data and using ML to critics outcomes. They can provide the structure with a better decision support system and reduce the cost of evaluating optimized zeros and reduce errors. This technology is even extending to medical device deployment and runtime monitoring and validation. It combines ideas from systems medicine, scientific computing and machine learning to give us a more holistic view of humanity and improve the quality of healthcare.

Using all this data, the transcriptome, sales organs, and exposure layers can forecast an individual's potential response to certain treatments or calamities. This is not a static model; rather, it is a dynamic one that will adapt and evolve over time to reflect the individual. Body can change the way it looks using something called graphical neural networks (GNNs). That's a fancy way of saying it can predict important things like blood pressure and it's not just the prison and IDT and use generative adversarial networks against it to combine different types of data like genes and/or proteins to get a more complete picture of what's going on, but what's really exciting about the entity. With CT, we can see various therapies that may have an impact on us and test novel medical interventions. Let us have a brief introduction to what a GNN is. In the past, complex systems like social networks and biological networks were effectively represented by GNNs, which are nodes and edges in a data structure. A graph neural network is a particular kind of neural network that is designed for work with the kind of data GNN, like a tool that can learn, understand, and manipulate. These networks can look at each node and edge in the graph and assign them into the future vector, which is a way to represent the nodes or edges as the number or vectors is flexible and customizable. They are capable of working with a variety of graph architectures and handling a wide range of data types, such as additional attributes that describe the relationships between nodes or global properties that apply to the entire network. In other words, they are powerful tools to understand and analyze complex systems that are represented as the graphs so they need a graph network for this project. There are actually several reasons that make her a perfectionist for the job. First of all, GNNs and Neural Networks are excellent at identifying nonlinear patterns in data; as most systems are nonlinear, this is significant, much as the graph fundamental model is simpler to understand than other neural network types. Utilise the graph since it is a special kind of non-Euclidean data structure and the model is predicting specific terms, which is important in domains like clinical practice.

It sounds complicated, but it just means that they are great for modeling complex biologic systems at different scales, like the tissues and organs in the body. Modernity is another key property of GNN. They are taking advantage of this, which means they can learn independent mechanisms that can be reduced in different parts of the graph making it easier to scale up and model the dynamic properties of the system. Finally, both Graph Networks (GN) and Generative Networks (GN) are really good at combining different types of data sources like structured and unstructured data. They are integrating signals at different biological scale levels, so Graph Neural Networks (GNNs) are the perfect choice for the healthcare project. We will provide plenty of reasons to support this, aiming to break down the complexities of the human body into separate subsystems,

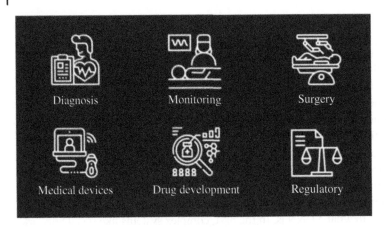

Figure 11.7 Applications of digital twin in healthcare.

each represented by different nodes or a network of nodes in the digital patient model. There are four biologic layers. The first is the transcriptomic layer which is all about the RNA transcript produced by the genome at the specific time, and they can measure the entire genome and use the resulting genie expression data to reveal its mechanisms. Moving on to the cellular layer, they are dealing with biological processes that affect individual cells, such as metabolism, replication, and motility. This translation of all these processes shows how they work together to keep our bodies running smoothly (Figure 11.7).

DTs can model physically accurate hospital environments, operating rooms, medical instruments, and even a patient's individual anatomy. These DTs can be used by medical device developers to build AI-enabled solutions. Neurosurgery is one application. A surgeon must manually move a robotic microscope throughout an operation in order to maintain focus on the area of interest and align it with their tools. As a result, the procedure takes longer and the patient spends more time under anesthetics. A DT–based AI technology can be used to improve this. It starts with Industrial-grade Xavier (IGX) computer running the Clara Holoscan SDK, which includes a pretrained tool segmentation model. Real-time inferencing performance is validated with virtual scenes from Omniverse. The IGX platform has been configured for high-bandwidth data streams to be processed and visualized by AI. For example, 4K 60 fbs video is processed in less than 15 ms. Without AI tracking, the microscope will not maintain alignment with the surgeon's tools. With tracking, the IGX platform automatically tracks the position of the surgical instruments and controls the real-time movements of the microscope. This can shorten the patient's stay in the operating room and increase surgical effectiveness. Even on high bandwidth data streams, the IGX platform is optimized for AI processing and visualization. With IGX and Clara Holoscan, it can build, test, and deploy—all on a single platform.

AI is the most important technology force of our time. They have now entered the wave of transformer AI, now more commonly referred to as large language models, and this week, we will see healthcare keeping pace by leveraging these technologies to deliver cutting-edge solutions. A few remarkable tales in general and neurosurgery in particular use AI digital twins that can observe and fix. Visualization is crucial to their work. If they are able to interact with and visualize the patient's images, they can better plan the right procedure and make sure they are doing the right procedure. Medical imaging is an indispensable tool for healthcare and it can fix and see beautifully. Medical imaging gives us the ability to see inside the body with 2D for screening and early detection, but the human body is volumetric and constantly in motion. So, 3D imaging provides spatial understanding, quantitative measurement, and segmentation, while 4D introduces temporal information that elucidates anatomy function essential to diagnosis and planning treatments using imaging combined with real-time deep learning and computer vision. They enter into the fifth dimension to perceive surroundings and navigate inside the body to plan actions and track objects to understand the dynamics and perform surgical tasks. This is imaging's next frontier and has contributed to innovations in minimally invasive and robotic assisted surgery. Each evolution of imaging applications was enabled by combining breakthroughs in sensor technology and powerful sensor processing technology. They have been working with the medical sensor ecosystem for over 15 years and are now to enable this new world of real time AI sensing. They are announcing the NVIDIA IGX edge AI platform, an all-in-one platform that combines NVIDIA or Ampere Tensor Core GPU for robotics Connect X7 streaming IOS processor and a functional safety island and safety microcontroller unit. Now, all future medical devices can take advantage of the breakthroughs that enabled autonomous vehicles to be powered by AI and develop software definitions.

CT acquisition with intelligent zooming, reducing radiation dose and improving image quality Intuitive ion is a robotic lung biopsy platform that uses real-time fiber optic shape sensors that measure hundreds of times per second, providing precise information about the shape and orientation of the site to reach otherwise impossible locations in the lung. Clarion Hollow Scan is being used by Olympus, the biggest endoscope manufacturer in the world, which holds more than 70% of the global market, to create their next-generation AI endoscopy platform. They are delighted to announce the next generation surgery systems have chosen Clara Holoscan on IGX active surgical active edge platform is powered by Clara Holoscan on IGX to process their unique hyperspectral sensor information and deliver real-time perfusion imaging for blood flow, and providing surgical teams with a visual representation of physical structures. Clara is being adopted by

Moon Surgical to deliver it as the next generation robotic-assisted maestro system, a single platform surgical robot. It offers improved sensory-aided scope control to surgeons and operating room perception proximity in an operating room telepresence. Clara Holoscan delivery real-time remote surgeon collaboration that is both secure and safe to use in a clinical environment. Robotic systems require not only a computing platform to run the robotics pipeline in the physical world that is Clara Holoscan and IGX but also a computing platform to deliver the robotics application in the digital world. They are working with the team to realize the incredible potential of digital twins for surgery [9]. They focus on the power of digital twins as a tool for medical device makers to validate their AI applications with Clara Holoscan and IGX hardware in the loop. An NVidia Omnivores model with realistically accurate hospital settings and operation rooms was constructed using DTs. These DTs can be used by medical device manufacturers to produce AI-enabled solutions for medical equipment and even patient anatomy. One application is in neurosurgery, where a surgeon must manually adjust a robotic microscope throughout an operation to keep it focused on the area of interest and line it up with their tools. This extends the procedure and lengthens the patient's stay with anesthesia. It can be improved with an AI-based solution using a DT. They start with an NVidia IGX computer running the Clara Holoscan SDK, which includes a pre-trained tool segmentation model. The IGX platform has been enhanced for AI processing and visualization of high-speed data streams through the use of Omniverse virtual scenes and validated real-time inference performance. For example, 4K 60 fps video is processed in less than 15 ms without AI tracking. The microscope will not maintain alignment with the surgeon's tools with tracking; the IGX platform automatically tracks the position of the surgical instruments and controls the real-time movements of the microscope. This can shorten the patient's stay in the operating room and increase surgical effectiveness. With IGX and Clara Holoscan, they can design tests and deploy them all on a single platform. The IGX platform is optimized for AI processing and visualization even on high bandwidth data streams. AI is a continuous cycle of generate, build, deploy, and simulate, and they call the AI factory where AI development and application testing can first be done virtually before being deployed in the real world. DTs will be an essential pillar of the medical device company's future AI factories. The world of AI drug discovery genomics has ignited the digital biology revolution with the cost of sequencing outpacing Moore's law.

The production of sequencing costs has kicked off genomics programs across the world from sequencing newborns to population programs like the US's All of Us programs including one million Americans. Sequencing is only the first step, though. With the end of Moore's law, new computing techniques must be created in order to reduce the cost of analysis, as sequence analysis requires computers, data science, and AI to read and interpret the genome and unlock the full potential of the genome sequence analysis, which is complicated and computationally intensive. NVidia Clara Parabricks is a suite of software tools and composable pipelines delivering GPU acceleration and deep learning at every step. Deep learning is now becoming the standard for base calling due to the speed and accuracy, improvements in neural network interpret image and collecting instrument signal data, as well as estimating the three billion nucleotide sequences of the human genome. In their latest generation Hopper architecture, they are able to optimize the inference by five times from current ampere architectures. The next computationally challenging step is alignment, to find the best match between a pair of sequence reads - a step that has been repeated hundreds of millions of times on Hopper. They have a dynamic programming API that dramatically accelerates alignment algorithms such as Smith–Waterman. At a crucial stage of sequence analysis when they discover small insertions, deletions, or complicated rearrangements that differ in a patient sample from a reference genome, they provide a 7× speedup over ampere. The new variant callers use deep learning to increase both speed and accuracy and assist physicians identify potential genetic diseases that a serious patient may have, or they can search the entire population for novel medication targets. NVidia Clara Parabricks offers the highest cost performance on variant calling pipelines reducing CPU turnaround times from days to hours and reducing the cost of the analysis by over half. The next generation of genomics is being made possible by Clara Parabricks, which powers platforms for both short and long read sequencing with faster acceleration and AI based calling and variant calling. Oxford Nanopore Technologies is the world leader in long-read sequencing uniquely enabling real-time sequencing. Just last week, Seattle Children's Hospital used PromethION to rule out a genetic disorder in the first three hours of a newborn life. Ultimate genomics is targeting whole genome sequences at as little as US$100. Singular genomics is the fastest bench top sequencer. An MGI G400 has very high accuracy with adaptive sequencing, and to make Clara Parabricks easy to access it is also available in public cloud services around the world from Alibaba, AWS, Baidu, Google, and Oracle to genomics cloud platform providers like Agilent, DNAnexus, UK Biobank, and Lifebit . In order to make NVidia Clara available on the Terra cloud platform, they are announcing their partnership with the Broad Institute.

We live in a moment in time when the most important problems in biology and medicine require a convergence between life scientists; the Broad–NVidia collaboration is an exemplar. They are compatible with NVidia, and we plan to investigate other applications. One of the most significant pieces of information gleaned from healthcare records is that big language models enable us to analyse these notes and comprehend the course of disease with a level of resolution never before possible. Similar to how big language models are trained to interpret human literature, biology also pays close attention to the languages of DNA,

RNA, and proteins. These same models can be trained to study the language of life. Tarot is a platform for managing, distributing and analyzing biological data that originated in genomics. The platform now supports a wide range of different data formats, and by bringing together different kinds of data it can facilitate considerably more intriguing scientific findings. It was previously impossible for anything as complicated as the human body to be understood using a single modality, thus this partnership with NVidia will increase access to these types of multimodal data analyses and make them available to a wider audience. They do not just use chest x rays or EKGs or DNA sequencing and it needs to have a variety of different tools to interrogate physiology and health and disease. One of the great opportunities of modern machine learning is to develop multimodal representations of the human body. The partnership between Broad and NVidia represents an exciting step forward in which the two organizations will work closely together to integrate diverse genomic and clinical datasets into the representation of disease incidence and can better address large language models as they help us understand the language of biology alpha full demonstrated. Research and the predictive power of protein language models indicate that larger is better. The capacity to predict 3D structure proteins with experimental accuracy has replaced costly and complex experimentation, and models trained on unsupervised learning objectives can learn protein structure from sequence. However, these models only scratch the surface of protein structure. NeMo Megatron, a supercharged framework created by NVidia, employs models with the new Hopper design; data parallelism can scale Large Language Models (LLMs) to hundreds of billions and trillion parameter sizes. They can reduce training time from months to days with bimolecular datasets exploding from sequencing and laboratory automation. Fortunately, digital biology data has its own language that is suitable for unsupervised learning methodologies and massive language models for DNA, proteins, and chemicals. Many petabytes of unlabeled datasets are being produced. For academics and developers to create new, pretrained massive language models at any scale and with any form of biological sequence, they are announcing the domain-specific framework BioNeMo. The chemical space has 10^{60} possible drugs like molecules and the protein space is even larger at 10^{130}. Drug discovery researchers are searching the chemical space for molecules with specific properties, synthetic biologists are looking for proteins with specific functionality, and developers of biologics are looking for antibodies with specific binding characteristics. The numerical methods limit the number of interactions they can compute in the future. These limits can be run in an end-to-end AI drug discovery pipeline generative models like mole fiow and mega-generate thousands of optimized molecules in a second and infer thousands of representations of molecules in seconds than using the likes of OpenFold and AlphaFold predict.

The 3D structure creating the two necessary ingredients to provide an AI model that predicts the drugs interact with proteins with a nearly infinite drug discovery space given the possible combination of proteins and large language models give us a new instrument to explore the infinite world of biomolecules despite the fact that they only approve approximately 50 medications annually at an average cost of US$1 billion and chemistry but limits take weeks to months to train and setting up environments is really hard, so they are announcing BioNeMo limits service providing researchers access to pretrained chemistry and biology language models. ESM1 will be used first to infer representations of amino acid sequences, and OpenFold will soon be used to predict protein structure from sequence mega Malabar. The results can be applied to further activities like creating novel proteins and compounds or forecasting reaction qualities based on structure or function. The world of AI applications will be a mix of large-scale AI frameworks and optimized inference services. AI workloads will require a hybrid infrastructure to optimize scale and cost. AI is the most important technological breakthrough of our time for making decisions in healthcare. Moon Surgical has chosen Holoscan on IGX to power the operating system with AI and AI that understands the language of DNA.

11.5.7 Developing a Digital Twin of the Heart

Heart failure is one of the most significant burdens in terms of healthcare. During the process of the disease, as the heart becomes bigger and sicker, these two chambers actually beat synchronously. So, the idea of cardiac resynchronization therapy is to actually overwrite the electrical asynchrony using an advanced pacemaker. The idea is to put electrodes on the right and on the left, and then time it correctly o that the heart is re-synchronized. And if it is placed well and the patient responds, the quality of life of the patient increases significantly. The problem is that there are patients who still do not respond, about 30–50% depending on the statistics. So, the question is how the therapy can be adjusted to increase the chance that the patient is going to respond. And, eventually, the concept of building a DT emerged and various computer setups were tested. Several recommendations for what might or might not work, and then utilise that knowledge and information to direct the action. An ambitious vision is to create a computer duplicate of the human body, specifically a DT of the human heart. And a great deal of knowledge from various domains is needed for this.

11.5.8 Human Digital Twin

The human DT as modern medicine is not our grandparents' medicine. It will not be their grandchildren's medicine either [10]. They have fast and complete genome sequencing, supercomputers to analyze all types of data, new tools and techniques that yield knowledge unavailable before. As an illustration, this particular set of cells exhibits typical behaviour, but rather of degenerating as expected, they develop into a tumour. This is the computer simulation of the DT of a patient to understand they need to travel to a smaller level. The differences between their genomes define physical features like the color of eyes or blood type but also the chances of developing a disease like diabetes or cancer. They find correlations between genotype and diseases. They analyze thousands of positions in the genome for thousands of patients. This is only possible thanks to the ever-growing amount of genomic data available today and the computational power of supercomputers. They can also look for correlations between disease and gene expression, the process that converts DNA instructions into protein and other molecules. Gene expression is defined as a type of tissue, even if all tissues have the exact same genome by analyzing expression data from different patients.

They can find differences in every cell and connect abnormal expression to disease according to body tissue. For example, the gene involved in glucose transport is under-expressed in the tissues in diabetes patients. Understanding the cause of diseases at this level can help us develop more personalized treatments. With the help of computer simulations, we can design better drugs when proteins are involved in a disease. There are other ways of simulating treatments that require supercomputers. They can modify the concentration, frequency, and composition around the cell of a drug that prevents uncontrolled growth. They can make more models realistic by adding more and better data metabolism, genetic information, or signaling mechanisms. They have learned more to travel again but in the opposite direction. When scientists begin to link data from several patients and discover patterns that were not apparent in the research conducted separately. Diseases that tend to happen together based on similar gene expressions or drugs that are usually taken together can interact and provoke undesired side effects. Despite the fact that they still need to overcome significant technical obstacles to obtain their privacy, storage, and data management. The future of medicine for their grandchildren is likely to depend largely on vast global databases of biological and genetic data being linked together, harnessing the power of supercomputers to create digital twins and using the data to improve the network for others.

11.6 Case Studies

11.6.1 Case Study of a Mine Digital Twin

Sentient better than being there utilizing Indie software technology, Indie software technology solves the problem of access. Our customers have complex assets in remote and dangerous locations and Indie brings those locations to their fingertips. They have made an application that uses game technology and it is very easy for someone who is familiar with 3D application games; they can jump in, navigate around, and it is all intuitive. A DT of their mining sites enables their people to access verified documents and work more effectively both in the office and from home. They were supplied with two items from their client: a Navisworks CAD file, which is an engineering drawing of the installation, and APIs to several data sources, the properties of individual pieces of equipment. For several years, we have been developing pipelines and processes that enable us to edit large, complex 3D models and import them into Unity. Accessing information about individual components in complex systems used to be a frustrating time-consuming task. By building a DT within the app, the user can search for a component or simply click on it to access all of the relevant information they require. Since all the information is accessible digitally, there is only one reliable source of information. In components, real time and historical data can be visibly displayed. By integrating training into our DT, one can work as closely as they would in real life with real data in real space, view components in the context of our environment and see how our condition may relate to other components in our area, and physically view and navigate the assets through the powerful search system. The DT gives us the ability to be able to look at something when they are not there. Traveling to remote sites is dangerous and expensive. The changes of 2022 have forced everyone to look at new ways of working. A DT can transform the way they work with better research, access to data training, and the ability to make better decisions.

11.6.2 Case Study of a Hollis Offshore Installation

With a DT of our offshore installation, we have 24/7 access to a dangerous, remote and expensive place to visit. Indie is an app that can be installed locally on a normal computer. They built a DT using their proprietary software named Indie. They were supplied with two items from their client Hollis, a Navisworks CAD file, which is an engineering drawing, and an Excel CSV file containing all the properties related to the individual pieces of equipment. They used their proprietary tool Indie to bring this information into Unity in an optimized and meaningful way. Accessing information about individual components on complex systems has historically been a time-consuming task. By building a DT that contains this information within the app, the user can simply click on a component and access all the relevant information they require. They can filter out structures or other equipment that are not relevant in order to quickly find objects. The status of equipment can be shown and easily understood through the use of color-coding notifications and other properties can be added and viewed by others. They can even measure distances between objects easily and the user experience is customizable for different user profiles. A DT provides a safe and convenient way to research access data and conduct training. They have just delivered it so it is pretty fresh, but they had really positive feedback so far. The future is really exciting for Indie and really only limited by their imaginations. They can start to integrate VR training and procedures to the environment. They can add augmented reality support and also add data analytics that come from the customer or experts like Hollis.

11.6.3 Case Study of Digital Twin in Offshore Industry

Offshore ventures costing billions of dollars include platforms and wind turbines. Offshore constructions with true DTs, like platforms and wind farms, cost billions of dollars to build, but the investments do not end there. In fact, now is the time when greater sums are at stake in the operation and maintenance phase. The buildings are put to the test in an offshore environment by wind, waves, the soil, and other environmental elements. Operators need to ensure that their structures have sufficient strength to carry these loads in a safe manner to avoid critical failure and reduced lifetime of the structures. The traditional way of doing this is to do on-site inspection in this high-risk environment at intervals specified by norms and best practice. Another, somewhat more affordable method, is to continuously keep track of the structure using a digital model that is updated with information about the loads it is subject to. They call this a "True Digital Twin" [11]. The offshore industry has worked with DTs, a digital copy of a structure, for many years. Ramboll's true DT technology adds an extra layer to the DT by bringing in the real-life behavior of the structure into the model. This is achieved by using an innovative structural health monitoring system combined with digitalization. The comprehensive set of data is then analyzed and the model updated to reflect the actual behavior of the structure. Based on this knowledge, operators can make informed decisions on when to reinforce, upgrade, or end the life of the structure. In this way, the true DT may increase the lifetime of a structure to gain many more years of operation than the original design would suggest. It also reduces time and money spent on unnecessary inspections.

The true DT technology has been developed over the last 25 years and was first used by Ramboll in 2006. Its validity has been verified on projects for many operators in the Northern Sea. Currently, it is being demonstrated for the first time on large-scale offshore wind farms. These tests are part of ROMEO, an industry-driven, EU-backed program to lower offshore wind power's operating and maintenance costs. When it comes to asset and integrity management tools, the DT created for ROMEO will surpass Ramboll and Iberdrola in innovation. The digital twin, which was originally developed for the oil and gas industry, is used in offshore wind farms, including our offshore wind farm Wikinger. Dealing with the vast volumes of data from the operational assets is the key problem for offshore wind operators. The DT is used to close the gap between the installed assets' actual state and their design. The advantages that a DT can offer are extremely obvious. First, they should gain a better understanding of offshore wind assets to cut down on maintenance and operating expenses, and then they should take advantage of lifetime extension prospects. The technology can be transferred to other parts of the energy sector as well as to assets in other sectors where they can monitor their behavior, such as in high rises and bridges.

11.6.4   Case Study of Integrating a Digital Twin in DCMS Framework

In order to meet the demands of High Performance Computing (HPC) AI, the data centre employs 6Sigma Digital Twin from the beginning to the end. This is done in order to validate the design of the data centre using Computational Fluid Dynamics (CFD) simulations up to 30 kw/rack with non-homogeneous density and failure scenarios. To predict the arrival of the first customer and provide a cutting-edge home-away-from-home experience, measure and manage the data center's sustainability and environmental impact, optimise data centre performance, and maximise room space and capacity. Data Center Management System (DCMS) accessible via a web portal for service in SAAS or on-premise multisite platform and hypervisor performance indicators to help in decision-making process catalos of service like security. The DT is seamlessly integrated within the DCMS to easily locate IT equipment and manage the assets to define their power connections and manage the capacity planning. Checking that all the resources are available before deployment here, alerting to insufficient cooling capacity, a risk is identified by background CFD simulations that are run on the 6Sigma Digital Twin to update current performance levels and predict the effects of future changes, so it is seamlessly providing accurate available cooling and power capacity for each rack. No other solution on the market is able to provide this information. The simulation predicts and enables risk mitigation. The DCMS platform manages the workflow's installation of a new HPE PROLiant DL380 Gen10. The DT submits the change request to the DCMS, and the service ticket, CR-35, is automatically created and opened; the marketplace plans to buy a new rack to allocate the necessary power and submit the purchase request after the quote is accepted and the purchase order is confirmed; human resources are allocated, work schedules are updated, calendars are managed, and access to the site is scheduled—scheduled for January 18. The client is assigned access rights to this rack in the DT and the financial module reflects this new purchase. The DCMS has a modular architecture with plug-ins with more facility engineering modeling the power infrastructure to guarantee the power resilience and monitor the power consumption like power containers, transformer HV/LV, UPS, and generators. Power network in the DT from the power supply sources to the server power supply, provided by the blue and green power rails for the two feeds 2 intelligent PDUS in the rack that monitor power and breakers. The DCMS is integrated with the building management system and feeds live power data to the DT and is tightly integrated with the 6Sigma Digital Twin, a unique solution providing operational intelligence for day-to-day operations.

References

1 Baheti, R. and Gill, H. (2011). Cyber-physical systems. *The Impact of Control Technology* 12 (1): 161–166.

2 Tao, F., Qi, Q., Wang, L., and Nee, A.Y.C. (2019). Digital twins and cyber–physical systems toward smart manufacturing and industry 4.0: correlation and comparison. *Engineering* 5 (4): 653–661.

3 Zhang, G., Huo, C., Zheng, L., and Li, X. (2020). An architecture based on digital twins for smart power distribution system. In: *2020 3rd International Conference on Artificial Intelligence and Big Data (ICAIBD)*, 29–33. IEEE.

4 Batty, M. (2018). Digital twins. *Environment and Planning B: Urban Analytics and City Science* 45 (5): 817–820.

5 Jones, D., Snider, C., Nassehi, A. et al. (2020). Characterising the digital twin: a systematic literature review. *CIRP Journal of Manufacturing Science and Technology* 29: 36–52.

6 Tao, F., Cheng, J., Qi, Q. et al. (2018). Digital twin-driven product design, manufacturing and service with big data. *The International Journal of Advanced Manufacturing Technology* 94: 3563–3576.

7 Ambra, T. and Macharis, C. (2020). Agent-based digital twins (ABM-DT) in synchromodal transport and logistics: the fusion of virtual and pysical spaces. In: *2020 Winter Simulation Conference (WSC)*, 159–169. IEEE.

8 Croatti, A., Gabellini, M., Montagna, S., and Ricci, A. (2020). On the integration of agents and digital twins in healthcare. *Journal of Medical Systems* 44: 1–8.

9 Verboven, P., Defraeye, T., Datta, A.K., and Nicolai, B. (2020). Digital twins of food process operations: the next step for food process models? *Current Opinion in Food Science* 35: 79–87.

10 Barricelli, B.R., Casiraghi, E., Gliozzo, J. et al. (2020). Human digital twin for fitness management. *IEEE Access* 8: 26637–26664.

11 Tygesen, U.T., Jepsen, M.S., Vestermark, J. et al. (2018, June). The true digital twin concept for fatigue re-assessment of marine structures. In: *International Conference on Offshore Mechanics and Arctic Engineering*, vol. 51203, V001T01A021. American Society of Mechanical Engineers.

12

Industrial Excellence Through Virtual Factory Simulation and Digital Modeling Innovations for Next-Generation Manufacturing System Design

Pravin Satyanarayan Metkewar

School of Computer Science and Engineering, Dr Vishwanath Karad MIT World Peace University, Kothrud, Pune, Maharashtra, India

12.1 Introduction to the Chapter

Manufacturing is a clear cooperation; the normal substances or part parts are bought and a short time later changed into a finished object. In any case, to succeed, the creator ought to have the choice to deal with the cost of making the thing, fulfilling a need, and making a thing that is alluring to the market.

Section 12.2 covers the analogy of virtual factory simulation [1, 2] including its benefits. Digital modeling innovations [3] provide insight about current trends and technologies in the digital era covered in Section 12.3. Sections 12.4 and 12.5 have put more emphasis on industrial excellence in virtual factory simulation and in digital modeling innovations, respectively, with respect to industry [4] related aspects in order to save time, resource optimization, and cost savings. A virtual factory data model is being discussed in Section 12.6. In fact, importance of manufacturing plan is covered in Section 12.7. Why the viability of product design plays an important role in the industry as far as product specifications are concerned is covered in Section 12.8. In Section 12.9, digital twin–based factory simulation aspects have been covered in detail which helps end users to set up the system accordingly. How to optimize the product design and procedures is being covered with suitable examples. Finally, challenges and future research directions in manufacturing system design are thoroughly explained in order to find more areas for research and development.

Digital Twins in Industrial Production and Smart Manufacturing:
An Understanding of Principles, Enhancers, and Obstacles, First Edition. Edited by Rajesh Kumar Dhanaraj, Balamurugan Balusamy, Prithi Samuel, Ali Kashif Bashir, and Seifedine Kadry.
© 2024 The Institute of Electrical and Electronics Engineers, Inc. Published 2024 by John Wiley & Sons, Inc.

12.2 Virtual Factory Simulation

A virtual manufacturing plant is a coordinated recreation model of significant subsystems in a production line that thinks about the manufacturing plant overall and gives a high-level choice help capacity. It permits mirroring the genuine tasks of the plant. The idea of virtual assembling [5, 6], or virtual plant, is to display and reproduce fabricating frameworks including configuration, machining process, sequential construction systems, and robots notwithstanding items on PCs to make, break down, further develop them by utilizing the innovation of augmented reality.

Let us look at the fundamental advantages of computer-generated in assembling:

- Remote encounters in three aspects.
- Financially savvy site gatherings.
- Expanded quality control.
- Virtual on location preparing.
- Process surveys.

Run-of-the-mill utilizations of virtual assembling are as per the following: VM can be utilized to evaluate the feasibility of an item configuration, approve an assembling plan, and improve item plan and systems. These lower the expense of the item's life cycle.

Genuine individuals utilize reenacted gear in a recreated world (pretend, cockpit reproduction, item reenactment). Programmatic experiences are "virtual" on the grounds that they incorporate reenacted individuals as well as hardware in a PC mimicked setting.

As per the medical services reenactment word reference, computer experience is "the entertainment of reality portrayed on a PC screen." While taking part in programmatic experiences, students experience recreated clinical situations from a PC screen while frequently utilizing a mouse or console to collaborate with the climate.

Virtual assembling is a PC-based innovation for characterizing, recreating, and picturing the assembling system right off the bat in the plan stage, when some, while possibly not all, fabricating related issues can be distinguished and tended to.

What is an illustration of virtual?

The modifier virtual is utilized to depict something that exists generally yet not in reality. You might have made a virtual companion on an internet gaming website, yet do not anticipate that that individual should meet you for espresso.

It includes the utilization of state-of-the-art innovations, including augmented reality (AR), expanded reality (XR), movement catch, and high-level PC designs to establish vivid and reasonable virtual conditions.

The principal interaction behind virtual creation is the reconciliation of various innovations. These incorporate Computer-generated Imagery (CGI), movement catch, facial acknowledgment, laser examining, advanced mechanics, increased reality and more into one stage. Most regularly, this is accomplished through the incredible motor and programming like Industrial Light & Magic (ILM's) showmanship.

Virtual creation consolidates physical and virtual filmmaking procedures to make state-of-the-art media. How it functions: groups utilize constant 3D motors (game motors) to make photorealistic sets, then show them on huge drove walls behind actual sets utilizing the continuous delivering abilities of the game motors.

Benefits of a Virtual Machine

- Different working frameworks are running at the same time.
- Separate memory.
- Security.
- Asset sharing.
- Movability.
- Similarity.

Advantages of Virtual Business

- Financially savvy.
- Zero overheads.
- Amazing skill.
- No drive time.
- Efficiency.
- Increment adaptability of work.
- Admittance to an enormous pool of ability.
- Proficient location.

A recreation mimics the activity of genuine cycles or frameworks with the utilization of models [7]. The model addresses the vital ways of behaving and attributes of the chosen interaction or framework while the reproduction addresses how the model develops under various circumstances over the long run.

Reenactment that is worked with mimicked frameworks and worked by at least one genuine individual is called programmatic experience. For instance, plane reproduction, fire control reenactment, war recreation, and so on.

Fundamental Stages and Choices for Reproduction *Issue Definition.* The underlying step includes characterizing the objectives of the review and figuring out what should be addressed.

- Project arranging.
- Framework definition.
- Model definition.
- Input information assortment and examination.
- Model interpretation.
- Confirmation and approval.
- Trial and error and examination.

In ceaseless microfluidic [8] processes, for example, production line of chips, most significant factors to be detected for executing an ideal control are typically not open. This is, for example, the instance of the shape and position following the stream interfaces in a stream centering microdevice. In this specific circumstance, virtual sensors are promising devices to further develop the ongoing control frameworks focusing on a zero-imperfection system. This hypothesis presents a procedure for building a precise virtual sensor, in light of PC-helped designing recreations, both scientific and mathematical, and model request decrease methods. The system is applied to a given stream centering unit. The result is a continuous model (virtual sensor) ready to foresee the shape and area of the multiphase liquid connection points from the volumetric stream rate estimated in the framework. Results are effectively approved against the exploratory information. The fundamental test of this approach is to limit vulnerabilities related with the microfluidics arrangement.

12.3 Digital Modeling Innovations

Proceeding with vulnerability in the business world has sped up the requirement for advancement. A combination of ongoing occasions—the pandemic, worldwide inventory network disappointments, military contentions, and the approach of enormous language models (like ChatGPT)—have uncovered distinct weaknesses, holes, and remarkable difficulties for organizations. Previously, endurance has relied on how rapidly an organization could adjust, yet the expense of developing is critical concerning cash, time, and chance. To accomplish supportable outcomes, organizations should send a methodical system for making benefit-creating thoughts that influence existing items, markets, and framework.

Safer and less exorbitant than different types of development, retooling a current plan of action can likewise yield strong upper hands. Achievement, however, requires an organized interaction that takes out mystery and subjectivity and can be copied across all regions of the association. Plan of action advancement in the computerized age empowers business pioneers to learn and embrace this cycle—and reveals potential developments that reach from straightforward yet profoundly successful acclimations to industry-upsetting changes that are troublesome or even difficult to duplicate.

Features and Key Results

In plan of action advancement in the computerized age, you will:

- Reveal inside your association presently unidentified approaches to further develop benefit and efficiency emphatically;
- Work on your capacity to get authoritative purchase in and execute new plans of action;
- Figure out how to direct a review of your ongoing plan of action and produce thoughts for developments;
- Become familiar with a methodical system for assessing, exploring different avenues regarding, and prototyping your thoughts;
- Foster guard components past licensed innovation security;
- Ace a system for driving and overseeing imaginative change in your firm;
- Find how to transform problematic innovations into productive organizations inside huge firms.

Most associations methodically and thoroughly keep up with and screen key measurements, for example, quarterly budget summaries, efficiency levels, and market position. Yet, international, mechanical, and financial changes overturn the same old thing, and the subsequent monetary aftermath has offered organizations a chance to completely reevaluate and try and change their plans of action. To make the way forward clearer, plan of action development in the computerized age assists members with opening this possibility by fostering a cycle for advancement that can be overseen and moved toward a driver of development.

Not at all like development that includes the formation of new items, administrations, or advances, this brand of plan of action development can emerge out of any place in the organization. It requires fewer assets and fewer chances to execute, and can bring about developments that are more beneficial and extraordinary than item or innovation driven ones.

Plan of action development in the computerized age gives a precise structure that depends on functional administration as opposed to startup nimbleness or karma. You can start applying this structure to your own business during the program, directing a review of your ongoing models and recognizing regions for expected developments. You may likewise become familiar with a cycle for trial and error that tends to shortcomings in the plans of action and examination with those models to evaluate their worth and practicality.

Since the basic last move toward the cycle is execution, you will be drenched in a half-day reenactment on change authority, reinforcing your capacity to get purchase in and drive change in your association. You and your group will go about as experts to an organization going through a change exertion, attempting to offer that work to those in the association effectively. This strong opportunity for growth, a feature of the program, permits you to apply new methodologies, gain from continuous input, and level up your abilities.

Driven by teacher Serguei Netessine, a worldwide idea pioneer in plan of action development and an Amazon researcher who applies examination to assist with taking care of specialized system issues, this program likewise uses the skill of other Wharton personnel who have some expertise in enormous information, man-made brainpower, business environments, and computerized disturbance. Also, effective plan of action trailblazers will impart their experience and bits of knowledge to members in two profoundly intelligent meetings [9].

The medical services industry is constantly developing with imaginative disclosures and treatments and, simultaneously, there is a decrease in the innovative work efficiency prompting an inflated expense for payers, suppliers, and patients. Regardless of the advantages that computerized advancements can have on medical care development, such an exceptionally controlled industry frequently depends on demonstrated laid out advances and hierarchical methods that can conflict with the new rationales of computerized advancement. The system is planned and approved through an iterative course of constant transformation with nearby practices in a biopharmaceutical organization. Faint gives pragmatic direction to drive computerized developments that involve various rationales contrasted with conventional advancements, by working on the perceivability of the advanced advancement process and expanding authoritative trust in chasing after computerized developments and improving dynamic adequacy.

12.4 Industrial Excellence in Virtual Factory Simulation

The idea of a virtual, computerized comparable to an actual item or the advanced twin was presented in 2003 at the College of Michigan Chief Seminar on Project Lifecycle Management (PLM). At the time this idea was presented, computerized portrayals of genuine actual items were generally new and juvenile. Moreover, the data being gathered about the actual item as it was being created was restricted, physically gathered, and for the most part paper-based. In the ten years that have followed, the data innovation supporting both the turn of events and upkeep of the virtual item and the plan and assembling of the actual item has detonated. Virtual items are rich portrayals of items that are, for all intents and purposes, undefined from their actual partners. The ascent of assembling execution frameworks on the plant floor has brought about an abundance of information gathered and kept up with on the creation and type of actual items. Likewise, this assortment has advanced from being physically gathered and paper based to being computerized and being gathered by a wide assortment of actual non-disastrous detecting innovations, including sensors and checks, coordinate estimating machines, lasers, vision frameworks, and white light filtering. They have presented the expression "Advanced Twin" in Essentially Awesome: Driving Creative and Lean Items through Item Lifecycle and Credited in account of John Vickers of NASA whom work with him.

Advanced twin idea model contains three principal parts: (i) actual items in genuine space, (ii) virtual items in virtual space, and (iii) the associations of information and data that ties the virtual and genuine items together. In the 10 years since this model was presented [10], there have been colossal expansions in the sum, lavishness, and loyalty of data of both the physical and virtual items. On the virtual side, they have further developed how much data are accessible. They have added various conduct qualities with the goal that can imagine the item, can test it for execution abilities, and can make lightweight adaptations of the virtual model. This implies Michael W. Laments (Scientist), LLC 2014 that can choose the calculation, qualities, and traits that expect without hefting around pointless subtleties [3, 11]. This decisively lessens the size of the models and takes into consideration quicker handling. These lightweight models permit the present reenactment items to envision and recreate complex frameworks and frameworks of frameworks, including their actual ways of behaving, progressively and with satisfactory register costs. These lightweight models additionally imply that the time and cost of conveying them electronically is considerably less. They currently can be shared with the association as well as all through the provider organization. This upgrades joint efforts in both diminishing opportunities to comprehend and improving both quality and profundity of comprehension of item data and changes. As significantly, it can reenact the assembling climate that makes the item, including most activities, both robotized and manual, that comprise the assembling system. These activities incorporate, mechanical welding, framing, processing, and other assembling floor tasks.

12.5 Industrial Excellence in Digital Modeling Innovations

Virtualization makes a virtual variant of innovation [12] that has an actual presence, like a work area or server. Instead of having a devoted server for every business capability, you can make virtual servers that all live inside a solitary spot. Other striking instances of VR headsets incorporate Oculus Crack, Samsung Stuff

VR, HTC VIVE, Google Daydream View, or Google Cardboard. These headsets eliminate the vision of this present reality and give a video to each eye considering profundity of vision. VR can give a profoundly vivid encounter, causing you to feel as though you are truly present in a reproduced climate. It gives you the feeling that it is genuine.

12.5.1 Involving VR for Preparing and Treatment Purposes

The preparation interaction can be made simple with computer-generated reality. Augmented reality is most ordinarily utilized in diversion applications, for example, computer games, 3D movies, event congregation rides including dull rides, and social virtual universes. Computer-generated reality is basically utilized in gaming. However, as of late, computer-generated experience has likewise been a part of medical care, instruction, diversion, and car businesses.

Lately, a couple of investigation works have focused in on chipping away at prevalently visual and sound proliferations, which furnish the clients with the impression of being faced with a respectable entertainment of this present reality. Accordingly, most Virtualization Technology (VT) applications rely essentially upon general media overhauls and do not exploit other material signs. In an unexpected way, the VT adds hand joint effort and olfactory lifts. In an unexpected way, the VT adds hand joint effort and olfactory lifts. These efforts are extensively embraced in gaming and social heritage applications, yet they are not totally exploited in promoting and the movement business. Besides, the development of olfaction to VR experience is not totally maintained today, so it tends to be an oddity in VTs. Olfaction has been viewed as outstandingly basic in influencing the human experiences in like manner in VR conditions associated with food examination and data , and some dedicated hardware contraptions are emerging to add scents to virtual scenes, but olfactory features are still missing at whatever point appeared differently in relation to the next substantial VR interfaces, and there is no ready-to-use gear. The usage of fragrances to further develop an overall media is most certainly not a weighty idea, there is a shortfall of contraptions openly accessible, but just relatively few prototypal applications with a couple of limitations for veritable applications. The basic issues rely upon the synchronization of the sensation of sight with that of smell, explicitly as indicated by the perceptual point of view and the age of the aromas. Even more lately, a couple of captivating prototypal game plans have been made in research: a limited, negligible cost olfactory show fitted to the hand controller of the HTC VIVE VR structure to engage control of scent degree. It allows the blend of fragrances with virtual articles, sensible for donning, educational, legitimate, or helpful capacities. However, a skin-associated olfactory analysis system with from a distance, programmable limits considering assortments of versatile and downsized fragrance generators for olfactory VR applications. Anyway, they are prototypal game plans not available in that frame of mind as yet.

Toward this way, the inspected VT gives a prompt examination of the virtual spot by multisensory feeling, by combining VR general media amusement with the olfactory energy given by veritable food scents.

Lately, an arising stream of exploration proposes that organizations ought to utilize mechanical advancement and plan of action development agreeably to amplify firm execution. To enhance the academic discussion on the fit between mechanical advancement and plan of action development, drawing experiences from the powerful capacities point of view, this hypothesis analyzes the fit between offer development and innovative advancement (shifty versus explorative) for the presentation of new companies in the computerized climate. Considering on-location overview information of 285 advanced new companies in one of the world's biggest computerized economies, we find that explorative development fortifies the positive effect of incentive advancement on the exhibition of new businesses, while manipulative development debilitates this constructive outcome. Besides, these directing impacts are amplified in an exceptionally questionable interest climate. These discoveries infuse new experiences into existing insightful discussion on the fit between mechanical development and plan of action advancement by offering a powerful capacities' point of view and by extending it to an inexorably computerized business climate.

The writing contends that a genuinely computerized change of firms requires all-encompassing changes of the plan of action. Regardless of information about this aggressive objective, comprehension of how advanced plan of action change can be accomplished is still exceptionally restricted. They investigate how firms accomplish computerized plan of action change. A creator can apply a contextual analysis plan to examine how officeholders have changed their individual plan of action aspects during computerized change. Results underscore the significance of a preliminary stage in which the essential course is set. Besides, discoveries exhibit that changing an organization's plan of action is best when a solitary individual, specifically the Boss Computerized Official, is capable. Discoveries add to the plan of action writing, by giving a more comprehensive view on how plan of action development can be used during computerized change.

The logical scene of assembling frameworks keeps on changing connected at the hip with the advancement of the actual field. It is guessed that new logical experiences will affect further development around here. The number of staff with significant specialization in complex frameworks, those that get recruited by the scholarly world and industry, stays small; yet, the expert necessities are high and remember exhaustive training for at least one region.

12.6 A Virtual Factory Data Model

Let us assume the storytelling of milk preparation in the milk boilers' virtual data model. After the investigation of the on-location experience, the consideration zeroed in on the plan of the computerized VT ready to make comparable feelings by a far-off advanced insight. Besides, beginning from the material gathered at the chosen dairy, the examination pointed toward making a conventional virtual encounter to advance freely from the creation plant.

12.7 Manufacturing Plan

Fabricating arranging alludes to the week by week or day-to-day creation and machine plans across various plants or lines to fulfil requests and estimate need. Some assembling arranging modules additionally integrate materials arranging.

How would I make an assembling arrangement? Five phases of creation arranging and booking:

- Request estimating and scope organization.
- Material preparation and acquisition.
- Booking and asset assignment.
- Creation control and checking.
- Ace creation plans.

For an assembling business, the assembling and tasks plan requirements to remember strategies for stock control, buying, creation control, and what parts of the item will be bought and which tasks will be performed by your labor force (called settle on or purchase choices).

A creation plan characterizes the creation targets, required assets, and by-and-large timetable, along with every one of the means associated with creation and their conditions. A very much planned creation plan assists organizations with conveying items on time, diminish costs, and answer issues.

Let us look into preparation and control?

- Arranging. Arranging figures out what will be created, by whom, and how.
- Steering. Steering decides the way natural substances stream inside the plant.
- Booking. Booking underscores "when" the activity will be finished.
- Stacking.
- Dispatching.
- Follow-up.

An assembling marketable strategy is a conventional record that frames the objectives and targets of your business. It incorporates nitty-gritty data about items or administrations and target market.

In the present worldwide and unstable market, fabricating endeavors are exposed to extraordinary worldwide contest, progressively abbreviated item life-cycles, and expanded item customization and fitting, while at the same time being compelled to keep a serious level of cost-proficiency. As an outcome, creation associations are expected to present all the newer item models and variations into existing creation arrangements, prompting more incessant increase and slope down situations while changing from an active item to another one. To adapt to, for example, challenge, the arrangement of the creation frameworks necessities to move toward reconfigurable assembling frameworks (RAM) [13, 14], making creation equipped for changing its capability and limit as indicated by the item and client interest. Subsequently, this study presents a reenactment-based multi-objective improvement approach for framework redesign of multipart stream lines exposed to versatile limits, which tends to the task of the errands to workstations and cushion distribution, all the while boosting throughput and limiting all-out support ability to adapt to fluctuating creation volumes. To this degree, the outcomes from the review show the advantages that leaders could acquire, especially when they face compromising choices inborn in the present assembling industry by embracing a reproduction based multi-objective optimization approach.

The physical and control design of assembling frameworks is portrayed, and a various-leveled model for supporting adaptation to noncritical failure exists. The primary benefit of the model is that it permits different adaptation to noncritical failure and recuperation methods to be utilized in various portions of the organization. To help in the decision of explicit procedures, specifically network portions, a bunch of measurements to assess adaptation to internal failure strategies is introduced alongside starter thoughts regarding coordinating the various methods.

12.8 Viability of Product Design

Suitability lets you know whether your item checks out. Regardless of whether you have the best item on the planet, on the off chance that it is excessively costly or is not beneficial, then, at that point, it is anything but a decent plan of action. A genuinely feasible item thought checks out temporarily and into what is to come. Possibility alludes to whether making the product is in fact conceivable. Practicality alludes to whether the item is monetarily feasible. The Diane Von Furstenberg (DVF) structure can be utilized in mix with different apparatuses and processes, for example, client research and prototyping, to make fruitful items.

Feasibility, the plan of action–Is your item monetarily legitimate and commendable interest in accordance with business system? This aspect characterizes incentive of an item according to a business viewpoint.

Suitability—Does our answer add to long haul development? The last test for your development centers around the worth chain of your answer for guarantee that it is feasible now and later on. Testing for reasonability asks, does our plan of action fit with the manner in which our clients need to utilize and pay for our answer? The thing practicality implies the nature of having the option to occur or having a sensible likelihood of coming out on top. The practicality of holding your party at an eatery could rely heavily on the number of visitors that they can situate.

The fundamental motivation behind the test system is to give criticism as engine cost and creation booking during engine plan. It focuses on demonstrating and mimicking a processing plant, displaying creation costs, and giving instances of how the plan of electric engines can be further developed by considering producing data through programming. The test system utilizes standards of framework elements and specialized cost displaying to demonstrate the progressions of materials and the advancement of the item through assembling stages across the processing plant floor. The contributions to the test system are outline information for engine plan and the situation with the industrial facility. The result from the test system is cost and timetable data. The test system likewise gives a component by which plan factors can be tried for varieties that produce cost enhancements.

The utilization of the Design for Manufacturing (DFM) methodology in the design of printed circuit boards is widely recognized and established. An arrangement of configuration rules is utilized to guarantee that plans meet assembling related needs. The guidelines catch the necessities of plant processes, so the requirements of assembling are viewed as consequently in the plan cycle. Further, they are adaptable, and can be handily adjusted to the necessities of various items, plants, and assembling processes. Exposed board testing and in-circuit test reviews are exposed to guarantee that the board will be testable after manufacture and gathering.

12.9 Digital Twin-based Factory Simulation

A computerized twin works by carefully reproducing an actual resource in the virtual climate, including its usefulness, highlights, and conduct. A continuous computerized portrayal of the resource is made utilizing shrewd sensors that gather information from the item.

Digital twins enable intelligent producers to obtain real-time information crucial for making rapid production decisions by producing a virtual counterpart of the physical asset. This eradicates the time-consuming hindrances linked to tangible objects, such as machinery, workforce, and resources.

Advanced twin innovation is like reproductions, yet at the same it is undeniably more precise. Dissimilar to recreations where designers need to set all boundaries physically, computerized twin innovation [15] utilizes genuine information to duplicate cycles in an advanced climate. A computerized twin is a virtual imitation of an actual item, interaction, or framework. It tends to be utilized to reproduce and break down different parts of the assembling system, including configuration, testing, and execution.

What are the three sorts of computerized twins? By and large, there are three kinds of advanced twins—item, creation, and execution. The blend and combination of the three advanced twins as they develop together is known as the computerized string.

This innovation empowers organizations to test and approve an item before it even exists. By making a copy of the arranged creation process, a computerized twin empowers designers to distinguish any cycle disappointments before the item goes into creation.

Computerized twin innovation is utilized to make virtual models of transportation frameworks, like streets, extensions, and trains. These advanced twins are utilized to reproduce the way of behaving of these frameworks in various situations, like gridlock, atmospheric conditions, and support occasions. Producers will generally utilize computerized twin innovation to further develop tasks, for example, plant cycles and machine execution to improve supply chains. Gideon Gartner predicts that by 2021 a portion of all enormous modern organizations will utilize advanced twins.

Computerized twins utilize reproduction related to virtual conditions and various information coming from various plant hardware and actual frameworks to ceaselessly refresh the advanced models of the world in a criticism plan to work with the dynamic cycles. The heterogeneity of existing equipment and programming requires the advancement of programming structures ready to manage the data trade because of the joining and association of a few framework parts and independent dynamic frameworks. In this work, they have examined the plan and development of a product design that coordinates an assembling cycle test system with the notable Robot Working Framework (RWF) to effectively exchange data with an independent dynamic framework. The proposition is tried with the test system Tecnomatix and the free conveyance ROS Melodic. They have presented an occasion of programming design for a normal complex contextual investigation of assembling plants and show its simple joining with an independent dynamic framework in light of the support learning worldview.

12.10 Optimize Product Design and Procedures

The course of presentation, development, and decline of an item is named as product lifecycle management (PLM). The item configuration process regularly comprises of the accompanying advances: Idea meaning of the item or item prototyping.

One of the principal contemplations for a plan streamlined for creation are the materials, parts, and innovation that are integrated into the item. For instance, plastic infusion shaped part configuration requires consideration in a few critical regions as well as a comprehension of what the maker can do.

The reason for streamlining is to accomplish the "best" plan comparative with a bunch of focused on measures or limitations. These incorporate boosting elements like efficiency, strength, unwavering quality, life span, productivity, and use.

The meaning of item configuration depicts the most common way of envisioning, making, and repeating items that take care of clients' concerns or address explicit requirements in each market. The way to effective item configuration is understanding the end client, the individual for whom the item is being made. Item streamlining is the most common way of refining and working on an item to make it more important to current clients and more appealing to new ones. As an item chief, you ought to streamline your item pre-send-off during the underlying turn of events. The item advancement lifecycle, then again, has seven key stages: ideation, approval, prototyping, advertising, improvement, send-off, and improvement.

With advanced innovation turning out to be progressively full grown, the circumstances and techniques for development subjects to partake in information advancement in big business computerized development environments are being contemplated. In view of the developmental game hypothesis of limited sanity, this builds a transformative game model with a center endeavor, a scholar and examination establishment and a data go-between as the game subjects, taking the information development relationship of the center undertaking as the speculation. Joined with reproduction examination, the impact of information advancement sponsorships, information retention, and information osmosis on the planned development methodology of information advancement bunches in the game framework are talked about.

The ends are as per the following: (i) As the manual for information development in advanced advancement environments, the center endeavor has the most noteworthy information potential, and its advancement methodologies straightforwardly influence the inner motor energy of information advancement in the framework; (ii) information digestion limit and information retention limit are decidedly corresponded with the information possible energy of development subjects and with the information development dynamic energy of the framework; and (iii) the appropriations given by the center venture altogether upgrade the excitement of the scholar and examination establishment and data mediator to take part in information development; however, unnecessary development sponsorships might prompt advantage and free-riding and afterward to "pseudo collaboration" inside the framework. Consequently, the center endeavor ought to present a discipline instrument. At long last, a few ideas are given to work on the harmless collaboration and helpful development of information development capacity between computerized advancement biological systems and development subjects.

12.11 Challenges and Future Research Directions in Manufacturing System Design

Nevertheless, enterprises who attempt to create and employ manufacturing platforms still face numerous obstacles. These incorporate the stage improvement process itself, stage documentation, distinguishing proof, demonstrating, and use.

Challenges in Assembling Industry

- Absence of talented specialists.
- Commercialization patterns.
- Selling direct to purchaser.
- Scaling your business.
- Expanding income and deals.
- Support and overheads.
- Robotization.

The assembling business makers face five fundamental difficulties: (i) the significant expense of improvement and joining, (ii) significant expense of creation, (iii) low capacity to change business processes rapidly, (iv) low capacity to produce client devotion, and (v) expanded client imperfection rates. Future exploration bearings are accessible in shrewd assembling which serves to coordinate different advancements connected with assembling, processing, virtualization, availability, taking care of information, and so on. The increasing number of smart manufacturing technologies has broadened as a result of the convergence of different advancements, leading to cost-efficiency, enhanced productivity, seamless implementation, enhanced comprehension, rapid adaptation to market needs, adaptability, and remote surveillance.

12.12 Summary

An assembling framework is any mix of activities and cycles utilized all through the creation of any product. While organizations have created different frameworks and cycles over the long run, they have turned into an undeniably significant component of any creation climate. A cutting-edge Manufacturing Execution System (MES) framework is a SaaS-based task stage that, through steady execution, empowers organizations to focus on and tackle explicit use cases for quicker time to esteem than customary frameworks.

The 2020s will be perceived as the 10 years in which worldwide financial and natural disturbances provoked driving associations to rehash their business thinking and embrace advanced change by carrying out composable systems.

Simply prioritizing reliability is no longer enough, since the impact of global supply chain and labor disruptions, together with changing market expectations, has demonstrated that enterprises must integrate agility into their technological infrastructure in order to stay competitive.

Light-footed execution is conceivable through cutting edge MES arrangements, which utilize a human-driven approach that enables forefront laborers with the devices they need to address the trouble spots regularly connected with conventional MES frameworks, particularly those including innovative restrictions that defer time to esteem.

To additionally show the benefit hole between customary MES and cutting-edge MES, think about the accompanying:

- **Client Experience**

  Customary MES frameworks depend on dissimilar frameworks and siloed information structures that are inclined to human blunders, with innovation that frequently includes a resolute, dated client experience.

  End clients might battle to embrace this innovation, because of extended expectations to absorb information and unpleasant UIs, all of which might add to diminished efficiency.

  A cutting-edge MES framework considers the end client's insight, with no-code applications that make it more straightforward for forefront laborers to construct answers for their issues without depending on IT or sellers, while giving the easy-to-understand interface that a more youthful, well-informed labor force has generally expected.

- **Arrangement Execution and Adaptability:**

  Conventional MES frameworks focus on strength over versatility, responding to change to keep up, not to develop, and depending on huge frameworks that require costly forthright establishment, administrations, and yearly agreements, while requiring a very long time to execute.

  A cutting-edge MES framework is planned with the expectation of progress as a learning experience, not an obscure variable, and supports generally safe execution since it tends to be sent in layers to oblige necessities over the long haul without tearing and supplanting existing frameworks.

- **Information Availability and Network**:
  Customary MES situation normally highlights unbending, hard to-work information structures where the assortment of normalized information is not dependable and group storehouses restrain the trading of data expected for a comprehensive perspective on tasks.

  A cutting-edge MES framework offers adaptable, open, and contextualized information gathered from sensors, machines, cameras, and extra information sources like gadgets and destinations, giving bits of knowledge that permit clients to make a move continuously.

- **IT Dependence/Resident Turn of Events**:
  Customary MES frameworks are restricted by bottlenecks forced by IT or sellers, considering how each change demand is a tedious discussion, obstructing upgrades. A cutting-edge MES framework created on a no-code or low-code stage empowers the nearest laborers to fabricate and design the instruments they need now and later as new circumstances emerge.

- **Cost of Proprietorship**:
  Conventional MES frameworks typically involve a long organization process that prohibits the capacity to scale across the undertaking and includes devoted IT sources and dependence on merchants for customization and updates, bringing about high direct foundation and upkeep costs.

  A cutting-edge MES framework is a SaaS-based task stage that, through gradual execution, empowers organizations to focus on and tackle explicit use cases for quicker time to esteem than conventional frameworks.

  Tasks that depend on conventional MES frameworks as of now battle with the difficulties from utilizing exorbitant, out of date innovation that needs adaptability, disappoints educated forefront laborers, and neglects to respond rapidly to changes in the business and on the lookout.

  The Manufactures are looking to expand return for capital invested in weeks, not months, ought to furnish their bleeding edge labor force with the devices and information they need to execute significant bits of knowledge that main a cutting-edge MES framework can give. Changing to a cutting-edge MES, whether to some degree or completely, is an okay, minimally expensive way for makers to rapidly address the significant impediments thwarting their development and strength.

References

1 Diaz, C.A.B., Aslam, T., and Ng, A.H.C. (2021). Optimizing reconfigurable manufacturing systems for fluctuating production volumes: a simulation-based multi-objective approach. *IEEE Access* 9: 144195–144210. https://doi.org/10.1109/ACCESS.2021.3122239.

2 Tucci, C.L., Lang, J.H., Tabors, R.D., and Kirtley, J.L. (1994). A simulator of the manufacturing of induction motors. *IEEE Transactions on Industry Applications* 30 (3): 578–584. https://doi.org/10.1109/28.293702.

3 Totterdill, P. (2020). Chapter 10 – Workplace innovation and Industry 4.0: creating synergies between human and digital potential. In: *Digital Innovation and the Future of Work*, 197–224. River Publishers.

4 Satwekar, A., Volpentesta, T., Spagnoletti, P., and Rossi, M. (2023). An orchestration framework for digital innovation: lessons from the healthcare industry. *IEEE Transactions on Engineering Management* 70 (7): 2465–2479. https://doi.org/10.1109/TEM.2022.3167259.

5 Turner, C.J., Hutabarat, W., Oyekan, J., and Tiwari, A. (2016). Discrete event simulation and virtual reality use in industry: new opportunities and future trends. *IEEE Transactions on Human-Machine Systems* 46 (6): 882–894. https://doi.org/10.1109/THMS.2016.2596099.

6 Saavedra Sueldo, C., Villar, S.A., De Paula, M., and Acosta, G.G. (2021). Integration of ROS and Tecnomatix for the development of digital twins based decision-making systems for smart factories. *IEEE Latin America Transactions* 19 (9): 1546–1555. https://doi.org/10.1109/TLA.2021.9468608.

7 Guo, H., Yang, J., and Han, J. (2021). The fit between value proposition innovation and technological innovation in the digital environment: implications for the performance of startups. *IEEE Transactions on Engineering Management* 68 (3): 797–809. https://doi.org/10.1109/TEM.2019.2918931.

8 García-Camprubí, M., Bengoechea-Cuadrado, C., and Izquierdo, S. (2020). Virtual sensor development for continuous microfluidic processes. *IEEE Transactions on Industrial Informatics* 16 (12): 7774–7781. https://doi.org/10.1109/TII.2020.2972111.

9 Jiang, S., Xu, C., Gupta, A. et al. (2016). Complex and Intelligent Systems in Manufacturing. *IEEE Potentials* 35 (4): 23–28. https://doi.org/10.1109/MPOT.2016.2540079.

10 Klos, C., Spieth, P., Clauss, T., and Klusmann, C. (2023). Digital transformation of incumbent firms: a business model innovation perspective. *IEEE Transactions on Engineering Management* 70 (6): 2017–2033. https://doi.org/10.1109/TEM.2021.3075502.

11 Li, L. and Zhou, M.C. (2023). Mathematical modeling of manufacturing systems. In: *Sustainable Manufacturing Systems: An Energy Perspective*, 139–180. IEEE https://doi.org/10.1002/9781119578314.ch4.

12 Zou, H., Ji, H., Qin, H. et al. (2022). Research on the strategy evolution of knowledge innovation in an enterprise digital innovation ecosystem: kinetic and potential perspectives. *IEEE Access* 10: 78764–78779. https://doi.org/10.1109/ACCESS.2022.3194071.

13 Pennino, T.P. and Potechin, J. (1993). Design for manufacture. *IEEE Spectrum* 30 (9): 51–53. https://doi.org/10.1109/6.275165.

14 Chintamaneni, P.R., Jalote, P., Shieh, Y.-B., and Tripathi, S.K. (1988). On fault tolerance in manufacturing systems. *IEEE Network* 2 (3): 32–39. https://doi .org/10.1109/65.3271.

15 Schaffers, H., Vartiainen, M. and Bus, J. (Eds.) (2020). *Digital Innovation and the Future of Work*, i–xxiii. River Publishers.

13

Digital Transformation in the Pharmaceutical and Biotech Industry: Challenges and Research Directions

S. Subha[1], M. Shanmugathai[2], A. Prasanth[3], S. Sree Varagi[1], and V. Dhanashree[1]

[1]*Department of Electronics and Instrumentation Engineering, Sri Sairam Engineering College, Chennai, Tamilnadu, India*
[2]*Department of English, Sri Sairam Engineering College, Chennai, Tamilnadu, India*
[3]*Department of Computer Science and Engineering, Vel Tech Rangarajan Dr. Sagunthala R&D Institute of Science and Technology, Chennai, Tamilnadu, India*

13.1 Introduction

The pharmaceutical and biotech business significantly contributes to improving global healthcare by developing advanced medications and treatments. The industry has undergone a significant shift in recent years, propelled by digital technologies. Many aspects of pharmaceutical and biotech operations, including drug research, clinical trials, manufacturing, supply chain management, and patient care, could be completely transformed by this digital transformation.

Big data analytics, AI, the Internet of Things (IoT), blockchain, and cloud computing are just a few examples of digital technologies that are being used to improve the speed, accuracy, and efficiency of pharmaceutical and biotech processes [1]. Numerous prospects for reducing expenses, enhanced productivity, customized therapy, and better patient outcomes are provided by these technologies.

The digital revolution of the pharmaceutical and biotech industry may also provide significant challenges further to the prospects. Organizations must deal with difficulties relating to data privacy and security, legal compliance, interoperability, talent recruiting and retention, and cultural resistance to change as a result of the integration of digital technology. Furthermore, to stay ahead in this dynamic environment, the rapid speed of technical breakthroughs needs ongoing research and development.

To appropriately address these difficulties, it is necessary to investigate the obstacles that the pharmaceutical and biotech industry is facing as it adopts digital

Digital Twins in Industrial Production and Smart Manufacturing:
An Understanding of Principles, Enhancers, and Obstacles, First Edition. Edited by Rajesh Kumar Dhanaraj,
Balamurugan Balusamy, Prithi Samuel, Ali Kashif Bashir, and Seifedine Kadry.

transformation. This research aims to provide insights into the current state of digital transformation in the industry, highlight the obstacles encountered, and propose potential areas of future research to accelerate and optimize the adoption of digital technologies.

In this study, we will examine the difficulties the pharmaceutical and biotech industry encountered while it underwent digital transformation [1, 2]. We shall look at how digital technologies have changed various facets of the market and talk about the particular challenges that businesses face when implementing these technologies successfully. We will also pinpoint and investigate prospective research opportunities that could assist in resolving these issues and opening up opportunities for a successful digital transformation of the pharmaceutical and biotech industry.

Stakeholders in the pharmaceutical and biotech sector can obtain crucial insights to encourage innovation, simplify operations, and ultimately improve patient care by understanding the challenges and research directions in the context of digital transformation.

13.2 Personalized Medicine and Precision Health

The biopharmaceutical industry is currently undergoing a digital transformation that is opening up exciting opportunities for positive change. Advanced technologies like AI, ML, and big data analytics are driving this transformation and enabling the development of more personalized and precision-based healthcare solutions.

One of the significant advancements in this field is the rise of personalized medicine, also known as precision health. By leveraging AI and ML algorithms, healthcare professionals can analyze vast amounts of data to gain insights into how diseases manifest and progress differently in individuals. This knowledge allows for the customization of treatment plans, ensuring that patients receive interventions that are tailored to their unique characteristics. Personalized medicine holds great promise in enhancing treatment outcomes and increasing patient satisfaction. By tailoring medical interventions to individual characteristics, this approach has the potential to bring about substantial improvements in patient well-being.

Moreover, wearable devices and mobile applications are empowering individuals to actively engage in their healthcare journey. These technologies provide continuous monitoring of vital health metrics, such as heart rate, activity levels, and sleep patterns. With this data, individuals can identify potential health risks early on and manage chronic conditions more effectively. Patients become partners in their care, leading to improved health outcomes and a greater sense of control over their well-being [3].

The positive implications of this digital transformation extend beyond individual patients. The use of AI and ML in drug discovery and development processes can accelerate the identification of potential drug targets and enhance the efficiency of clinical trials. These advancements hold the promise of bringing innovative treatments to market faster, benefiting patients worldwide.

While challenges exist, such as ensuring data security and addressing privacy concerns, the overall trajectory of the biopharmaceutical industry's digital transformation is undoubtedly positive. By embracing these technologies responsibly, pharmaceutical companies, healthcare providers, and patients can collaborate to unlock the full potential of personalized medicine, leading to improved healthcare outcomes, reduced healthcare costs, and ultimately, a healthier future for all.

13.3 Automation and AI for Drug Discovery

Automation and AI are transforming drug discovery. AI has the potential to vastly accelerate the process of identifying new drug candidates and speed them to the market [4]. ML algorithms can analyze huge datasets to detect patterns that humans often miss. By applying ML to massive amounts of biomedical data, researchers can uncover new insights into disease mechanisms and identify potential drug targets. Robotic systems equipped with AI are automating laboratory processes like high-throughput screening, in which large libraries of compounds are tested against potential drug targets. AI planning and scheduling algorithms can optimize the sequence of experiments to maximize efficiency. Deep learning models can generate novel molecular structures that may have desired properties. These AI-designed compounds can then be synthesized and tested, accelerating the discovery of new drug candidates. Some biotech startups are already using AI to design new molecules.

13.3.1 The Drug Discovery Process: A Long and Arduous Journey

The process of discovering and developing new drugs is an incredibly long and challenging one. On average, it can take over a decade and cost billions of dollars to bring a new drug to the market.

The first step is identifying a biological target, such as a gene or protein that is involved in a disease pathway. Researchers then work to understand the target's 3D structure and role in the disease to determine how best to affect it. Next, chemists synthesize thousands of small molecules that may bind to the target. These "hits" are then tested in cells and animal models to determine which are most likely to be safe and effective.

Promising candidates move on to clinical trials in humans, beginning with Phase 1 trials to evaluate safety and dosage. Phase 2 trials assess efficacy and side effects in a few dozen to hundreds of patients. Finally, Phase 3 trials involve hundreds to thousands of patients to confirm the drug's effects. The entire clinical trial process can take six to seven years or more.

Once trials are complete, the manufacturer submits a new drug application to regulatory agencies like the FDA, who review the evidence to determine if the benefits outweigh the risks. If approved, the drug can be marketed to the public. However, the job is not done yet; post-market surveillance continues to monitor safety, efficacy, and optimal use.

Advancing technologies like AI can help identify new biological targets, predict how molecules will behave, and analyze huge datasets to uncover connections that humans might miss. Automation allows chemists to synthesize and test far more compounds in less time. These tools will not replace human researchers but augment and enhance their efforts, ushering in a new era of faster, cheaper drug discovery [5].

Advancements in AI and ML are enabling drug discovery at an unprecedented scale and speed. AI systems can analyze massive amounts of data to identify promising drug targets and compounds much faster than humans.

13.3.1.1 AI Accelerates Target Identification

ML algorithms can scan huge molecular databases to detect patterns and identify molecules that are most likely to bind to a target. AI models can also analyze how molecules interact with biological targets or pathways to predict their effects.

- AI screening of molecular libraries allows researchers to evaluate millions of molecules to pinpoint those with the highest potential to affect disease mechanisms or modulate therapeutic targets. This high-throughput screening dramatically accelerates the target identification process.
- AI models can analyze the 3D structures of biological targets like proteins or DNA to detect binding sites for drug compounds. By identifying these druggable pockets, AI enables researchers to focus their search on molecules most likely to bind to these sites.
- AI systems can also analyze vast amounts of biomedical data to uncover new disease mechanisms and potential drug targets. By spotting previously unknown correlations in huge datasets, AI can point researchers to molecular targets they may never have considered.

13.3.2 Automated Screening of Compound Libraries

13.3.2.1 Screening Large Compound Libraries

One of the most significant advances in drug discovery has been the automation of screening processes to evaluate potential drug compounds. Automated screening systems can quickly test thousands of compounds from libraries to determine their effects. This high-throughput screening allows researchers to rapidly identify compounds that could be further developed into drug candidates.

Traditionally, compounds were tested manually using in vitro methods on cells or tissues to evaluate their biological activity and potential as drug leads. This was an extremely slow, labor-intensive process that limited the number of compounds that could be screened. Automated systems use robotics and bioassays to speed up screening. Robotic arms can precisely add minute amounts of compounds to assay plates holding cells or biochemical systems. Instruments can then automatically analyze the effects and determine a compound's potency, efficacy, and selectivity.

13.3.2.2 Bioassays and Reporter Genes

Bioassays that can be automated include binding assays using purified drug targets, as well as cell-based assays. Reporter genes linked to regulatory regions of a gene of interest allow indirect assessment of a compound's effect. For example, luciferase assays use light emission to indicate the activity of a compound. Fluorescent assays are also commonly used. These types of assays can provide high throughput while reducing the amount of compound needed.

13.3.2.3 In Silico Screening

In addition to high-throughput screening of physical compound libraries, virtual compound libraries can be screened using computer modeling. This in silico screening uses software to simulate the interaction between compounds and a drug target to predict which may have biological activity. Compounds selected through in silico screening can then be prioritized for testing in in-vitro assays. By focusing high-throughput screening on compounds predicted to be active, in silico modeling helps maximize the efficiency of the overall drug discovery process.

Automated screening systems have revolutionized drug discovery by enabling the rapid testing of vast libraries of compounds. When combined with advances in chemistry for generating diverse compound libraries, high-throughput screening allows researchers to sift through millions of possibilities to find promising leads that would otherwise have remained undiscovered. Continued progress in bioassays, robotics, software, and parallel processing will only enhance this capability and accelerate the identification of new drug candidates.

13.3.3 Virtual Patient Models and Simulations for Clinical Trials

The use of virtual patient models and simulations is transforming the clinical trial process as shown in Figure 13.1. Advanced computational models can simulate human physiology, diseases, and medical interventions to generate virtual patient cohorts for testing new drugs [6]. These "in silico" methods are accelerating

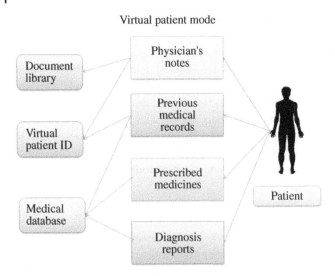

Figure 13.1 Virtual mode in clinical trials.

drug discovery by enabling low-cost, high-throughput testing and improving trial design.

13.3.3.1 Reducing Costs and Speeding Up Trials

Virtual patients reduce the time and money required to screen new drugs. They can simulate thousands of patients to identify candidate compounds, optimal dosages, and likely side effects before moving to human trials. This helps researchers avoid dead-ends and focus resources on the most promising avenues. The simulation also allows for rapid iterations as models are refined based on human testing data.

13.3.3.2 Improving Trial Design

Complex computer models incorporating genomic, biological, and medical data can closely mimic specific diseases and patient populations. This enables researchers to simulate a diverse range of patients, including those that are rare or hard to recruit. Simulations can also help determine optimal trial protocols by modeling how different cohorts may respond to a new treatment. The result is better-designed trials, more representative patient groups, and higher quality data.

13.3.3.3 Personalized Treatments

In the future, virtual patient models may enable personalized medicine by simulating how a drug will interact with a specific patient's unique biological systems and genetic makeup. Doctors could test various treatment plans on a patient's virtual model to determine the optimal approach before administering therapy. This has the potential to improve outcomes, reduce side effects, and avoid unnecessary procedures.

13.3.4 Reducing Costs and Increasing Efficiency with Automation

Automating routine tasks and leveraging AI for drug discovery can significantly reduce costs and improve efficiency, reduce repetitive lab works.

Manual labor makes up a large portion of costs in drug discovery. Automating routine experiments and lab processes with robotics and AI frees researchers to focus on more complex work. Robotic systems can perform repetitive pipetting, sample handling, and liquid transfers much faster and more accurately than humans. AI algorithms can also analyze samples and detect patterns to identify new drug candidates, reducing the time spent on early screening [7].

13.3.4.1 Accelerating the Analysis of Research Data

AI excels at quickly analyzing huge amounts of data to uncover insights and patterns that would be nearly impossible for humans to identify on their own. ML algorithms can sift through massive amounts of chemical, biological, and patient data to predict which compounds might have the highest potential as new drugs. AI can also analyze data from failed drug trials to determine why treatments were ineffective, helping researchers avoid repeating the same mistakes.

13.3.4.2 Optimizing the Drug Discovery Process

Many steps in the drug discovery process remain inefficient, with researchers often repeating work that has already been done or pursuing dead ends. AI and automation can help optimize the process by:

- Analyzing all available data on disease targets and drug candidates to determine the most promising areas of focus.
- Tracking all completed and ongoing experiments to avoid duplication of effort.
- Identifying knowledge gaps and suggesting targeted experiments to address them.
- Predicting possible side effects and drug interactions early on to avoid wasted time on unsuitable candidates.
- Streamlining clinical trial design and patient recruitment to accelerate the testing process.

While AI and automation cannot replace human researchers entirely, they are invaluable tools for reducing costs, improving efficiency, and optimizing the drug discovery process. By taking over time-consuming routine work, these technologies allow researchers to focus their efforts where human intelligence and creativity remain essential. Overall, the combination of human and AI may prove key to developing innovative new treatments and improving human health.

13.3.5 The Future of AI and Automation for Drug Discovery

Looking ahead, AI and automation will continue to transform and accelerate drug discovery. The future is promising for improving health outcomes through advanced technologies.

13.3.5.1 More Sophisticated Algorithms and Computing Power

AI algorithms and computing power will become more sophisticated, enabling systems to analyze massive amounts of data to uncover complex patterns and insights for drug discovery. Quantum computing may radically advance computational power, though it is still in the early stages.

13.3.5.2 Integration of Multiple Data Sources

Integrating multiple data sources—genomic, clinical, real-world evidence, and more—will enhance AI modeling and predictions. As interoperability between data sources improves, AI can gain a more complete understanding of diseases and potential treatments.

13.3.5.3 More Advanced Simulation and Modeling

Simulation and modeling techniques will become more advanced, allowing for in-silico testing of drug candidates to predict effectiveness and safety before clinical trials. Virtual patients, organs-on-chips, and other approaches can simulate human physiology to evaluate how people may respond to new drugs.

13.3.5.4 Personalized and Precision Medicine

AI and automation will enable more personalized and precision medicine. Analyzing a patient's genetic profile, health history, lifestyle, and environment can help determine the best course of treatment for their unique situation. AI can suggest tailored drug regimens, dosages, and combinations for precision care.

13.3.5.5 Continuous Learning and Optimization

AI systems will continuously learn and optimize as they are exposed to more data. Self-learning algorithms can identify new patterns to refine predictions and insights over time without needing to be reprogrammed by humans. Continuous learning will make AI more adaptable and scalable for accelerating drug discovery.

With increased computing power, data integration, simulation capabilities, and continuous learning, AI and automation will transform drug discovery in the coming years and beyond. While human expertise will still be essential, advanced technologies can enhance our understanding of diseases and our ability to develop life-saving treatments. The future of AI and automation for drug discovery looks bright.

13.4 Robotic Process Automation for Clinical Trials

13.4.1 Automating Clinical Trial Processes

Clinical trials are complex processes that involve numerous repetitive and routine tasks, such as collecting and processing participant data, managing documents and forms, and monitoring compliance. Many of these tasks can be automated using robotic process automation (RPA) technologies [8]. RPA bots can perform rule-based, highly structured, and repetitive computer-based tasks on a large scale. They do not replace human researchers but rather take over routine tasks to free up time for more strategic work. Figure 13.2 shows the robotic hand control in the hospital's clinical trials.

13.4.2 Improving Data Quality and Consistency

Software bots can assist with clinical data entry, validation, and query management. They enter data from paper records into electronic systems, double-check that the data meets predefined rules, and generate queries to resolve any inconsistencies. This helps reduce human error and ensures high-quality, standardized

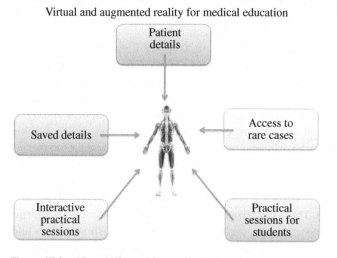

Figure 13.2 AR and VR used for medical education.

data. The bots also log their activities for auditing purposes to comply with regulatory requirements.

13.4.3 Streamlining Document Management

RPA bots excel at systematically controlling and organizing electronic documents and forms required for clinical trials. They can rename, sort, and file documents according to predefined taxonomies. They fill in electronic forms with information from various data sources and store completed forms in the appropriate folder structures. The bots keep an audit trail of all document access and changes. This simplifies document control and change management for researchers.

13.4.4 Monitoring Trial Progress

Software bots can also assist with routine monitoring tasks like tracking patient recruitment numbers, follow-up visit schedules, and protocol deviations [9, 10]. They generate reports and alerts to notify researchers about the progress and potential issues. The constant monitoring and notifications help keep the trial on schedule and budget.

13.4.5 Challenges Faced in RPA

RPA has emerged as a promising technology in the field of clinical trials, offering opportunities to streamline processes, increase efficiency, and reduce manual labor. However, like any transformative technology, RPA implementation in clinical trials comes with its own set of challenges.

One of the significant challenges faced in adopting RPA for clinical trials is the complexity and variability of trial protocols. Clinical trials involve numerous processes and workflows, each with its unique requirements and guidelines. Developing RPA solutions that can effectively navigate and automate these diverse protocols can be a complex task. It requires a deep understanding of the intricacies involved in trial procedures and the ability to design flexible automation workflows that can adapt to different trial designs and protocols.

Another challenge is the integration of RPA with existing systems and technologies used in clinical trial management. Clinical trials often involve multiple stakeholders, including sponsors, contract research organizations (CROs), and regulatory authorities. These organizations may use different software platforms and data management systems. Integrating RPA with these existing systems and ensuring seamless data flow and compatibility can be a technical challenge. It requires effective coordination and collaboration among various stakeholders to align processes and systems for successful RPA implementation.

Data security and compliance are also crucial challenges to address when implementing RPA in clinical trials. Clinical trial data, including patient information, is highly sensitive and subject to stringent privacy and regulatory requirements, such as those outlined by the Health Insurance Portability and Accountability Act (HIPAA) and the General Data Protection Regulation (GDPR). Ensuring that RPA solutions adhere to these regulations and maintaining data security throughout the automation process is paramount. Organizations must establish robust data protection measures, implement encryption and access controls, and regularly audit and monitor RPA processes to safeguard patient privacy and maintain regulatory compliance.

Despite these challenges, the potential benefits of RPA in clinical trials are substantial. For example, RPA can automate data extraction and data entry tasks, reducing human errors and improving data accuracy. It can also automate repetitive administrative tasks, such as document management and report generation, freeing up valuable time for researchers and trial coordinators to focus on more critical activities. Additionally, RPA can enhance process efficiency by enabling real-time data monitoring and ensuring timely data collection, analysis, and reporting.

13.5 Virtual and Augmented Reality for Medical Education

Virtual and augmented reality technologies are poised to transform medical education and training as shown in Figure 13.2. As digital innovations continue to reshape the pharmaceutical and biotech industry, virtual and augmented reality can provide valuable opportunities for improving how medical students and professionals learn and develop skills [11, 12].

13.5.1 Enhanced Learning Through Immersion

Virtual and augmented reality enables an immersive learning experience that provides a safe space for medical students and professionals to develop skills through interactive simulations. Learners can engage with virtual patients, practice procedures repeatedly, and gain exposure to a range of scenarios that would be difficult to experience in real-world settings. This immersive learning environment leads to improved knowledge retention and the development of muscle memory, which translates to enhanced real-world performance.

13.5.2 Access to Rare or Complex Cases

Virtual and augmented reality also provides access to rare medical cases or complex procedures students may not encounter during their practical training. By exposing learners to a wider range of scenarios through simulations, they can develop a broader skill set and learn how to handle high-risk, high-complexity cases, even if they have limited opportunities to experience them in person. This access to a wider range of learning opportunities helps to better prepare the next generation of medical professionals.

13.5.3 Reduced Costs and Improved Safety

Compared to traditional medical training, virtual and augmented reality offer more cost-effective and safer learning experiences. Simulations reduce costs associated with medical equipment, facilities, and additional instructors or standardized patients [13]. They also eliminate risks to live patients during the learning process. Students can make and learn from mistakes without causing harm.

13.5.4 Challenges in Augmented Reality

Several challenges need to be addressed for the successful implementation of AR in medical education.

One of the significant challenges is the limited availability of high-quality AR content tailored specifically for medical education. Developing accurate and detailed 3D models, simulations, and interactive modules requires significant resources and expertise. Medical educators and content creators must ensure the accuracy and relevance of the AR content to align with the curriculum and learning objectives. Additionally, the content needs to be regularly updated to reflect the latest advancements in medical knowledge and technologies.

The technological infrastructure required to support AR in medical education, implementing AR in a classroom or clinical setting requires robust hardware, such as head-mounted displays or mobile devices, capable of rendering high-resolution AR content in real time. Ensuring the availability and affordability of these devices can be a barrier, particularly for educational institutions with limited budgets. Moreover, the seamless integration of AR technology with existing educational platforms and systems is crucial for a smooth learning experience.

Ethical and privacy concerns also arise with the use of AR in medical education. As AR often involves real-time data collection, storage, and sharing, ensuring patient privacy and data security becomes paramount [14]. Educators and institutions must adhere to strict privacy regulations and establish clear guidelines on the responsible use and storage of sensitive patient information. Consent management and anonymization techniques should be employed to protect patient identities and maintain confidentiality.

Furthermore, the adoption and acceptance of AR in medical education may face resistance from educators and learners who are unfamiliar with or hesitant to embrace new technologies. Educators need to be trained and supported in integrating AR into their teaching methodologies, ensuring that they understand the potential benefits and can effectively utilize AR tools in the classroom. Similarly, learners may require time and guidance to adapt to the new learning environment facilitated by AR, addressing any initial challenges or concerns they may have.

While AR holds great promise in transforming medical education, challenges related to content development, technological infrastructure, ethical considerations, and adoption barriers need to be overcome. Addressing these challenges requires collaborative efforts among educators, content creators, technology developers, and regulatory bodies to ensure the effective and responsible integration of AR into medical education. By doing so, AR can enhance the learning experience, improve knowledge retention, and ultimately contribute to developing highly skilled healthcare professionals.

13.6   3D Printing of Drugs and Medical Devices

13.6.1   Cost Reduction and Improved Accessibility

3D printing technology has the potential to significantly reduce costs and improve access to pharmaceuticals and medical devices. By printing drugs and devices on-demand and on-site, 3D printing can eliminate the need for large production facilities and complex supply chains. This could lower manufacturing costs by up to 50–90% for some products. Lower costs and simplified logistics will make healthcare more accessible in developing countries and rural areas.

13.6.2   Personalized and Precision Medicine

3D printing enables customized production of drugs and medical devices tailored to individual patients. Doctors can use 3D printing to create implants, prosthetics, and surgical tools matched to a patient's precise anatomy. 3D-printed drugs can be tailored to a patient's specific dosage needs and biochemistry. This will allow for optimized treatments and improved patient outcomes. 3D printing is a key technology enabling the shift to precision and personalized medicine.

13.6.3   Reduced Side Effects and Improved Compliance

The advent of 3D printing in the pharmaceutical industry brings forth a multitude of positive and advantageous possibilities. One remarkable advantage is the ability to develop precisely customized 3D-printed drugs, tailored to individual patients' specific requirements. By achieving optimal release profiles and doses, these customized drugs can mitigate the risks associated with both under- and overdosing, thereby reducing side effects and potential complications. Furthermore, 3D-printed drugs offer opportunities for customization in terms of flavors, shapes, and textures, enhancing the patient experience and promoting better adherence to medication regimens. For instance, patients who face difficulties swallowing can benefit from 3D-printed pills designed to dissolve slowly, ensuring ease of ingestion.

The scope of 3D printing in the pharmaceutical and medical device industry extends far beyond customized drugs. The technology enables the printing of organs and tissues, presenting groundbreaking possibilities in organ transplantation and regenerative medicine. With personalized prosthetics, individuals can receive tailored devices that perfectly fit their unique anatomical structures, improving comfort, mobility, and overall quality of life. Additionally, the flexibility of 3D printing allows for the creation of medical devices customized to meet individual patient's needs, fostering better treatment outcomes and patient satisfaction.

The transformative potential of 3D printing in healthcare is immense. By revolutionizing the development, manufacturing, and delivery of drugs and medical devices, 3D printing can make high-quality healthcare more accessible, affordable, and effective on a global scale. As technology continues to advance and regulatory frameworks evolve to ensure safety and efficacy, the benefits of 3D printing in healthcare are poised to positively impact patients and healthcare systems worldwide.

13.6.4 Challenges Faced in 3D Printing of Drugs and Devices

3D printing of pharmaceuticals and medical devices is a rapidly evolving field that offers promising opportunities for personalized medicine and customized healthcare solutions [15]. However, like any new technology, some challenges need to be addressed before it can reach its full potential and be safely and effectively implemented.

A major challenge is the regulatory framework for 3D-printed pharmaceuticals and medical devices. As this technology evolves, regulators must set guidelines and standards to ensure the quality, safety, and efficacy of 3D-printed products. This includes addressing issues such as material biocompatibility, manufacturing processes, sterilization methods, and quality control measures. Establishing a clear regulatory pathway for approval and ensuring compliance with existing regulations is critical to enabling the widespread adoption of 3D-printed medicines and medical devices.

Another challenge is the scalability and reproducibility of 3D printing in medicine. Because 3D printing enables the manufacturing of complex custom structures, it is necessary to ensure consistent quality and repeatability across multiple prints. Development of standardized processes and materials, optimization of printing parameters, and implementation of stringent quality control measures are essential to achieving reliable and reproducible results. Moreover, expanding 3D printing capabilities to meet high-volume production needs while remaining cost-effective remains a challenge that requires further research and development [16].

Material selection is another key challenge in the 3D printing of pharmaceuticals and medical devices. Materials used for 3D printing must meet stringent requirements for biocompatibility, mechanical strength, and durability. Ensuring the availability of the right material with the properties required for a particular application can be a challenge. Additionally, the compatibility of printed materials with regulatory requirements and existing manufacturing processes must be carefully evaluated.

Data security and intellectual property protection are also important considerations when 3D printing. The digital nature of 3D printing files and the ease with which designs can be replicated raise concerns about possible unauthorized use and intellectual property infringement. Protecting sensitive patient data and unique design information is critical to maintaining confidentiality and preventing unauthorized use and counterfeiting.

Despite these challenges, the potential benefits of 3D printing for pharmaceuticals and medical devices are significant. This enables personalized medicine, patient-specific therapies, and the fabrication of complex structures that are difficult to achieve with traditional manufacturing methods. By addressing regulatory, scalability, materials, and data security challenges, the healthcare industry is leveraging the full potential of 3D printing to improve patient outcomes, increase treatment efficacy, and added innovation in smart healthcare [17].

13.7 IoT and Connected Health

The Internet of Medical Things (IoMT) is revolutionizing healthcare by bringing together a vast array of connected medical devices and applications that monitor and transmit data wirelessly as shown in Figure 13.3 [18]. Imagine a world where individuals with diabetes can effortlessly track their glucose levels using smart glucose monitors that seamlessly transmit data to their healthcare providers. Furthermore, patients with heart conditions can utilize EKG monitors that continuously monitor their heart rhythms and transmit real-time data for remote analysis. Additionally, blood pressure cuffs equipped with IoMT capabilities enable patients

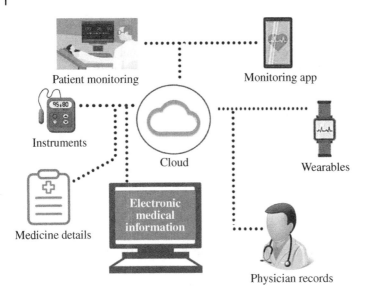

Figure 13.3 Example of IoMT.

to measure their blood pressure at home and instantly share the data with their physicians.

These examples represent just a fraction of the incredible possibilities that IoMT offers. By enabling patients to monitor their health conditions in the comfort of their homes and transmit vital data to healthcare professionals, IoMT empowers individuals to actively participate in their care. Pharmaceutical companies can leverage this wealth of real-world data to gain insights into how patients are responding to medications. By analyzing the data collected from connected devices, they can identify trends, optimize treatment strategies, and ultimately improve health outcomes.

For instance, pharmaceutical companies can collect data from IoMT-enabled inhalers used by asthma patients. This data can provide valuable insights into medication adherence, frequency of inhaler usage, and environmental triggers affecting patients' symptoms. Armed with this information, pharmaceutical companies can develop personalized treatment plans, provide targeted interventions, and enhance patient outcomes [19, 20].

Furthermore, IoMT can enable pharmaceutical companies to conduct remote clinical trials by remotely monitoring patients and collecting real-time data. This approach eliminates geographical limitations, enhances patient recruitment and retention, and accelerates the drug development process. It also allows for the collection of comprehensive, real-world evidence on the safety, effectiveness, and adherence to medications, offering valuable insights for regulatory agencies, healthcare providers, and patients.

13.7.1 Connected Health and Clinical Trials

Pharma companies can leverage IoMT and connected health technologies to enhance clinical trials. Connected devices can collect real-time biometric data from study participants between site visits. Remote monitoring reduces the burden on patients to travel to study sites and increases the frequency of data collection [21]. More data points lead to faster detection of health changes and identification of safety issues. Connected trials also improve patient retention since participants can engage from home. Improved data and higher retention rates result in higher quality, lower cost clinical trials.

13.7.2 Challenges and Concerns

While IoMT and connected health offer benefits, there are also challenges to consider. Connected medical devices must meet strict data privacy, security, and regulatory requirements to protect patients. Pharmaceutical companies need to implement strong cyber security controls and get appropriate patient consent to collect and use health data [22, 23]. There are also interoperability issues to address, as connected health systems and trial data platforms often cannot share data seamlessly. Patient access and digital literacy also remain barriers, as some populations lack access to technology or the skills to use connected health tools.

Overall, digital health and IoMT are shaping the future of healthcare and the pharmaceutical industry. Connected medical devices and health platforms enable new insights into patient care and more efficient clinical research [24, 25]. However, privacy, security, interoperability, and access must be addressed to realize the full potential of digital health. With the rapid pace of technological change, pharmaceutical companies need to pursue digital transformation to gain a competitive advantage, reduce costs, and improve health outcomes in the coming decades.

13.8 Predictive Analytics and Real-World Evidence

13.8.1 The Promise of Predictive Analytics

Predictive analytics is a powerful tool for pharmaceutical companies, leveraging historical data to identify patterns and predict future outcomes and trends. By analyzing vast amounts of data, including clinical trial data, genetic information, and patient records, predictive analytics can help improve R&D productivity in the pharmaceutical industry.

One area where predictive analytics can make a significant impact is in optimizing clinical trial design. By analyzing real-world data from electronic health records, claims databases, and other sources, pharmaceutical companies can gain insights into patient populations, treatment outcomes, and potential safety issues.

This information can be used to design more efficient and effective clinical trials, reducing costs and time-to-market for new drugs.

Furthermore, predictive analytics can help pharmaceutical companies predict treatment outcomes for individual patients, enabling personalized medicine [26]. By combining real-world data with genetic information and patient characteristics, ML algorithms can generate predictive models that help determine the most suitable treatment options for specific patients. This can lead to improved patient outcomes and a more targeted approach to healthcare.

Access to real-world data also enables pharmaceutical companies to monitor the safety and effectiveness of their products in real-world settings. By analyzing data from electronic health records and adverse event reporting systems, predictive analytics can identify potential safety signals and provide early warnings of adverse drug reactions [27]. This allows companies to take timely action and ensure patient safety.

13.8.2 Challenges Around Data Quality and Governance

However, real-world data is often unstructured, inconsistent, and lacks critical details like patient demographics or lifestyle factors. Data governance is crucial to ensure high-quality, standardized data that can produce meaningful insights. Pharma companies must invest in data management platforms and staff to clean, integrate, and structure real-world data from disparate sources.

13.8.3 The Need for Advanced Analytics Skills

Extracting valuable predictions from data also requires highly skilled data scientists, statisticians, and ML engineers. These skills are in high demand, so attracting and retaining top talent is challenging but critical for success in predictive analytics. Some companies may need to retrain existing staff or acquire smaller analytics firms to build up internal capabilities.

13.8.4 Real-world Evidence and AI Improving Clinical Trials

Real-world evidence from electronic health records, claims data, patient registries, and other sources is enabling AI and ML to improve clinical trial design and execution. AI can analyze large datasets to identify eligible patient populations, select study sites, and optimize trial protocols.

13.8.5 Optimizing Patient Recruitment

One of the biggest challenges in clinical research is identifying and recruiting suitable patients for trials. AI is helping solve this problem by analyzing patient data to find candidates that meet study criteria. For example, an AI system can review electronic health records and claims data to identify patients with a specific condition, of a certain age, medication history, and other attributes that match a trial's eligibility requirements [28]. This allows researchers to target potential participants more, improving recruitment rates and speeding up the enrollment process.

13.8.6 Choosing Ideal Trial Locations

Selecting the right locations to conduct a clinical trial is key to its success. AI is enabling researchers to perform advanced analytics on healthcare data to determine optimal sites based on factors like patient demographics, standard of care, healthcare infrastructure, and costs. An AI system can assess data on disease prevalence, patient flow, and availability of resources across different hospitals, clinics, and physician practices to recommend locations likely to yield the highest quality data and most cost-effective results.

13.8.7 Refining Trial Protocols

AI and ML techniques are also helping improve clinical trial design through the analysis of real-world data. By examining how patients respond to treatments in actual practice, researchers can gain insights into protocols that may work better in trials. For example, an AI system could analyze patient outcomes associated with different drug dosages, combinations, or treatment schedules used in real-world settings to help determine an optimal approach for testing in a clinical trial. Real-world evidence may also point to patient subgroups that could benefit more from an experimental therapy, allowing researchers to target a trial to specific populations.

In summary, real-world data and AI are transforming clinical research by enabling more efficient patient recruitment, optimal trial location selection, and refined study protocols. By leveraging insights from how treatments perform in actual practice, researchers can design trials that have a higher likelihood of success and better reflect patient needs. AI will continue to enhance clinical research techniques, accelerating the drug discovery process and speeding up innovative new therapies for patients.

13.9   Digital Medicine

Digital medicine refers to the use of digital technologies, such as apps, sensors, and AI, to improve human health. In the pharmaceutical and biotech industry, digital medicine aims to enhance drug discovery and development, improve clinical trials, and enable personalized patient care.

Digital technologies can accelerate pharmaceutical research and development in several ways:

- AI and ML can help identify new drug candidates by analyzing huge datasets to detect patterns and insights that humans may miss. AI models can also predict the properties and effects of new compounds to focus research on the most promising leads.
- Wearable sensors and mobile apps can provide real-world data on patients' health, symptoms, and responses to treatments. Access to this data can enhance clinical trial design and recruitment. Digital monitoring also enables virtual clinical trials, reducing costs and increasing participation.
- Digital medicine allows for personalized therapies tailored to individuals' unique genetics, lifestyle, and health profiles. By capturing detailed patient data, companies can better understand how people respond to drugs and develop precision treatments targeting specific patient groups [29].

While promising, digital medicine also introduces challenges and uncertainties for pharmaceutical and biotech companies:

- New digital technologies require significant investments in R&D, infrastructure, and talent acquisition. The costs and risks of investing in digital medicine can be high, especially for smaller companies.
- Data privacy and security concerns may limit access to patient data. Regulations like HIPAA establish strict rules about handling personal health information.
- Doctors and patients may be hesitant to adopt new digital tools and therapies without proven benefits. Companies will need to demonstrate the value of digital medicine through evidence from clinical studies and real-world use.
- Business models are gradually advancing. It is unclear how companies can generate revenue from digital medicine and who will pay for new digital therapies and services. Partnerships and collaborations may be needed.
- While digital medicine is still a nascent field, it shows great promise for improving health outcomes and transforming the pharmaceutical and biotech sector. With continued innovation, investment, and real-world evidence, digital technologies are poised to reshape the future of pharmaceutical industry [30].

13.9.1 Challenges Implementing in Digital Transformation

- Data management and privacy challenges: Digital transformation implementation involves a lot of work in the sector of data management and privacy issues related to customer data.

13.9.2   Vast Amounts of Data and Data Integration

The pharmaceutical and biotech industry generates massive amounts of data throughout the drug development lifecycle, including clinical trial data, genomic data, real-world evidence, and patient data. The challenge lies in efficiently capturing, storing, and processing this data for analysis and decision-making. Additionally, integrating data from various sources, such as electronic health records, wearables, and research databases, poses significant challenges due to differences in data formats, standards, and systems. Research directions in this area could focus on developing scalable data management frameworks, data integration strategies, and advanced data analytics techniques to derive meaningful insights from heterogeneous datasets.

13.9.3   Data Privacy Regulations and Compliance

Data privacy and protection are crucial considerations in the digital transformation journey. The pharmaceutical and biotech industry must adhere to stringent data privacy regulations, such as GDPR and HIPAA. Ensuring compliance with these regulations while leveraging data for research and innovation poses a challenge. Research directions may include developing privacy-preserving data analytics methods, exploring differential privacy techniques, and implementing privacy-enhancing technologies to protect patients and sensitive data.

13.9.4   Research Directions for Data Management and Privacy

Developing robust data governance frameworks that outline data ownership, access controls, and data lifecycle management.

- Exploring advanced data integration techniques such as semantic interoperability, ontologies and data harmonization standards.
- Investigating novel approaches for secure and privacy aware data sharing such as federated learning and secure multiparty computation.
- Designing User centric consent management systems that empower patients and research participants to control their data-sharing preferences.
- Conducting studies on the ethical and legal implications of emerging data technologies and their impact on data privacy, regulatory compliance challenges.

13.9.5   Complex Regulatory Landscape

The pharmaceutical and biotech industry operates in a highly regulated environment to ensure patient safety and efficacy of therapies. However, the rapid pace of digital transformation introduces complexities in meeting regulatory requirements. Navigating through the evolving regulatory landscape, which includes regulations specific to clinical trials, pharmaceutical co-vigilance, and quality control, poses challenges for organizations. Research directions may involve developing AI-based tools and natural language processing algorithms to automate regulatory intelligence gathering, monitoring regulatory changes, and assessing compliance risks.

13.9.6 Adapting to Changing Regulations

Regulatory agencies continually update guidelines and regulations to keep pace with technological advancements and address emerging challenges. Pharmaceutical and biotech companies must remain agile and adapt their digital transformation initiatives to align with evolving regulations. Research directions may include developing dynamic regulatory compliance frameworks that can quickly adapt to regulatory changes, leveraging AI for real-time compliance monitoring, and exploring blockchain-based solutions for transparent and auditable regulatory reporting.

13.9.7 Research Directions for Regulatory Compliance

The main research focus for regulatory compliance will be to develop ML algorithms and interpret regulatory text, enabling automated identification of relevant regulations and guidelines to cope up for the same. Leveraging natural language processing and text mining techniques to extract structured information from unstructured regulatory documents, facilitating compliance assessments are few examples of this category.

13.9.8 Talent Acquisition and Upskilling Challenges

The digital transformation of the pharmaceutical and biotech industry requires a workforce equipped with advanced digital skills, including data analytics, AI, bioinformatics, and digital project management. However, there is a significant skills gap in the industry, with a shortage of professionals possessing these specialized skills. The challenge lies in attracting and retaining talent with the right skill set to drive digital initiatives forward. Research directions may include:

- Identifying the specific digital skills required in the industry and conducting workforce gap analysis.
- Collaborating with educational institutions to develop specialized programs and certifications tailored to the industry's digital needs.
- Encouraging cross-disciplinary collaboration and knowledge sharing between traditional pharmaceutical/biotech professionals and digital experts.

- Exploring partnerships with technology companies to provide training and upskilling opportunities for industry professionals.

13.9.9   Cybersecurity Challenges – Threats to Sensitive Data and Intellectual Property

13.9.9.1   Threats to Sensitive Data and Intellectual Property

As the industry becomes more digitally connected, the risk of cyber security threats increases. Protecting sensitive data, including patient information, clinical trial data, and intellectual property, is of paramount importance. The challenge lies in implementing robust cyber security measures to prevent data breaches, unauthorized access, and cyber-attacks. Research directions may include:

- Developing comprehensive cyber security frameworks and policies that address the unique challenges of the pharmaceutical and biotech industry.
- Implementing advanced authentication and encryption techniques to protect data at rest and in transit.
- Leveraging AI and ML algorithms to detect and respond to cyber security threats in real-time.
- Conducting regular vulnerability assessments and penetration testing to identify and address potential security loopholes.
- Collaborating with cyber security experts and organizations to share best practices and stay updated on the latest threats and mitigation strategies.

13.9.10   Research Directions for Regulatory Compliance

Developing AI-powered systems that can continuously monitor and interpret regulatory changes, provide real-time alerts, and help update compliance processes is an innovative approach to managing regulatory requirements. Using AI, pharmaceutical companies can automate the monitoring of regulatory updates such as new policies, guidelines, and legal requirements. AI algorithms have access to databases across various sources – the government's websites, industry publications and government bulletins which could help in identifying relevant changes and notify the compliance team promptly. This proactive approach helps organizations stay current with evolving regulations, update compliance processes on time, and mitigate potential risks and non-compliance issues.

The use of predictive analytics models to assess and mitigate regulatory risk is another valuable application of data-driven technology in the pharmaceutical industry. By analyzing historical data, industry trends, and regulatory enforcement actions, predictive analytics models can identify patterns and factors that contribute to non-compliance. This enables organizations to assess their compliance posture, identify potential gaps, and take proactive steps to address them. Using predictive analytics, pharmaceutical companies can make data-driven decisions, mitigate risk, and effectively allocate resources to ensure compliance with regulatory requirements.

Blockchain technology offers great potential to improve regulatory reporting in the pharmaceutical industry. By leveraging blockchain's decentralization and transparency, businesses can build a secure and immutable system for regulatory reporting. Blockchain can ensure data integrity and prevent falsification or unauthorized alteration of government records. In addition, audit capabilities built into this technology make it easier to track and verify regulatory information, streamlining the compliance process. By using blockchain, pharmaceutical companies can improve transparency, accountability, and credibility in regulatory reporting while simplifying audits and regulatory inspections.

13.9.11 Deep Learning Techniques

- Convolutional neural networks (CNNs) have excelled in the processing of images and signals. By using filters and pooling operations, CNNs excel in learning hierarchical representations from input data, allowing them to recognize both local and global features.
- Recurrent neural networks (RNNs) are designed to process sequential or time-varying inputs, so they can analyze the temporal aspects of target detection. Recurrent connections in RNNs enable them to recognize dependencies between sequence elements. RNNs may analyze time-series data to find patterns and associations that could point to possible objectives such as gene expression profiles or longitudinal patient data.
- The use of biological organisms as nodes and their interactions as edges in a network is used as a tool for network analysis. Based on their connectivity inside the network, graph techniques including centrality measurements, community discovery, and network diffusion can be used to identify important targets. In the setting of intricate biological networks, network-based techniques offer a systems-level perspective and make it possible to recognize hub targets, critical pathways, and future research drug targets.

13.10 Conclusion

The future of the pharmaceutical and biotech industry faces a profound digital transformation. The emergence of new technologies creates many opportunities for innovation and progress. This transformation has the potential to accelerate research and development, optimize operational efficiency, improve the patient experience, and push the boundaries of personalized medicine. However, managing this digital revolution requires investment in partnerships, skills, and a culture of data-driven decision-making. Organizations that actively pioneer and adopt digital technologies will benefit from a more efficient, patient-centric, and impactful healthcare ecosystem. Despite the potential challenges, the industry has a unique opportunity to harness the power of digital technology to shape a healthier future. By actively participating in this digital revolution, pharmaceutical and biotech companies can not only benefit their businesses but also contribute to medical progress and human well-being. The future is still unknown, but with a shared vision and determination to move forward, the opportunities to leverage digital transformation in our industry are limitless.

13.11 Summary

The pharmaceutical and biotech industry is undergoing a profound digital transformation driven by technological advancements and data-driven approaches. This transformation presents numerous opportunities to enhance various aspects of drug discovery, development, manufacturing, and patient care. However, along with these opportunities, several challenges need to be addressed to fully leverage the benefits of digital transformation.

One of the primary challenges lies in effectively integrating and managing the vast amounts of diverse data generated within the pharmaceutical and biotech industry. With data sources ranging from genomic data to electronic health records and real-world evidence, harnessing and integrating this wealth of information can provide valuable insights for target identification, drug repurposing, and personalized medicine. However, challenges such as data privacy, interoperability, and standardization need to be overcome to ensure seamless data sharing and analysis.

Another critical challenge is the need for advanced analytics and AI capabilities. The complex and heterogeneous nature of biological systems requires sophisticated computational models and algorithms to extract meaningful insights. AI techniques, including ML and deep learning, have the potential to analyze vast datasets, predict drug-target interactions, optimize drug formulations, and enable precision medicine. However, successful implementation of AI in the industry necessitates addressing issues related to data quality, model interpretability, and regulatory considerations.

Regulatory and compliance issues also pose significant challenges in the digital transformation of the pharmaceutical and biotech industry. As technology and data-driven approaches advance, regulatory frameworks must adapt to ensure safety, efficacy, and ethical standards are maintained. Ensuring data security, patient privacy, and compliance with regulatory requirements are crucial aspects of leveraging digital tools and technologies effectively. Collaborative efforts among industry stakeholders, regulatory agencies, and policymakers are essential to establish frameworks and guidelines that promote innovation while maintaining necessary safeguards.

Alongside the challenges, research directions in the field of digital transformation in the pharmaceutical and biotech industry focus on several key areas. These include the development of advanced computational models and algorithms for drug discovery, the utilization of real-world evidence and patient-generated data for research and development purposes, the exploration of digital therapeutics and wearable devices for enhanced patient care, and the integration of blockchain and IoT technologies to improve supply chain management.

In conclusion, the digital transformation in the pharmaceutical and biotech industry presents significant opportunities for advancements in drug discovery, development, manufacturing, and patient care. However, addressing challenges related to data integration, analytics, regulatory compliance, and privacy is crucial. By focusing on these challenges and pursuing research in key areas, the industry can effectively harness the power of digital transformation to accelerate innovation, improve patient outcomes, and shape the future of healthcare.

References

1 Sharma, A. and Kumar, R. (2023). Artificial intelligence in health care sector and future scope, *International Conference on Innovative Data Communication Technologies and Application (ICIDCA)*, Uttarakhand, India, pp. 210–214.

2 Hole, G., Hole, A.S., and McFalone-Shaw, I. (2021). Digitalization in the pharmaceutical industry: What to focus on under the digital implementation process. *International Journal of Pharmaceutics: X* 3: 100095.1–100095.2.

3 Landwehr, K. (2015). Digital Transformation in the Pharmaceutical Industry - From Products to Services 'Beyond the Pill'. https://doi.org/10.13140/RG.2.1 .1668.5047.12-19.

4 Blanco-González, A., Cabezón, A., Seco-González, A. et al. (2023). The role of AI in drug discovery: challenges, opportunities, and strategies. *Pharmaceuticals* 16 (6): 891.MDPI,2-5.

5 Schneider, G. (2018). Automating drug discovery. *Nature Reviews Drug Discovery* 17 (2): 97–113.

6 Kononowicz, A.A., Zary, N., Edelbring, S. et al. (2015). Virtual patients - what are we talking about? A framework to classify the meanings of the term in healthcare education. *BMC Medical Education* 15 (11): 1–3.

7 Jiménez-Luna, J., Grisoni, F., Weskamp, N., and Schneider, G. (2021). Artificial intelligence in drug discovery: recent advances and future perspectives. *Expert Opinion on Drug Discovery* 16 (9): 949–959.

8 Anagnoste, S. (2018). Robotic process Automation in Pharma: Three Case Studies, *New trends in Sustainable Business and Consumption*, pp. 779–784.

9 Ni, Y., Wright, J., Perentesis, J. et al. (2015). Increasing the efficiency of trial-patient matching: automated clinical trial eligibility pre-screening for pediatric oncology patients. In: *BMC Medical Informatics and Decision Making*, vol. 15, 1–3. Springer.

10 Maibaum, A., Bischof, A., Hergesell, J. et al. (2022). A critique of robotics in health care. *AI & SOCIETY* 37: 467–477.

11 Kamphuis, C., Barsom, E., Schijven, M. et al. (2014). Augmented reality in medical education? *Perspectives on Medical Education* 3: 300–311.

12 Rezepa, S. Rezepa, S., and Pekar, J. (2021). Strategy and Planning and Implementation and Impact for the Pharmaceutical Industry with and by Industry 4.0.

13 Leal, F., Chis, A.E. et al. (2021). Smart pharmaceutical manufacturing: ensuring end-to-end traceability and data integrity in medicine production. *Big Data Research* 24: 100172.

14 Lahjouji, M., El Alami, J., Hlyal, M., and Lahjouji, O. (2023). A systematic literature review: the power of the block chain technology to improve pharmaceutical supply chain. *Journal of Theoretical and Applied Information Technology* 101: 957–959.

15 Williams, K. and Johnson, M. (2021). Advancing Patient-Centric Healthcare through Digital Transformation: Opportunities and Challenges for the Biopharmaceutical Industry.

16 Elbadawi, M., McCoubrey, L.E., Gavins, F.K.H. et al. (2021). Harnessing artificial intelligence for the next generation of 3D printed medicines. In: *Advanced Drug Delivery Reviews*, vol. 175, 1–3. Elsevier.

17 Sanal, M.G., Paul, K., Kumar, S., and Ganguly, N.K. (2019). Artificial intelligence and deep learning: the future of medicine and medical practice. *The Journal of the Association of Physicians of India* 67 (4): 71–73. Pubmed, Europe PMC.

18 Ghubaish, A., Salman, T., Zolanvari, M. et al. (2021). Recent advances in the Internet-of-Medical-Things (IoMT) systems security. *IEEE Internet of Things Journal* 8 (11): 8707–8718.

19 Qureshi, F. and Krishnan, S. (2018). Wearable hardware design for the Internet of Medical Things (IoMT). *Sensors* 18 (11): 3812.

20 Chen, Y. and Liu, F. (2021). Exploring the role of artificial intelligence in digital transformation of biopharmaceutical manufacturing. *Biotechnology Journal* 1: 76–82.

21 Chen, J. and Wu, C. (2020). Digital transformation in biotechnology: unlocking value through data analytics. *Journal of Industrial Information Integration* 6: 558–564.

22 Desai, C. and Joshi (2020). Digital transformation in the pharmaceutical and biotechnology industries: drivers, trends, and challenges. *Journal of Drug Delivery and Therapeutics* 8: 66–72.

23 Mitchell, J. and Johnson, L. (2020). Leveraging digital technologies for clinical trials in the pharmaceutical industry. *Drug Discovery Today* 1: 7–15.

24 Gupta, R. and Singh, S. (2022). Digital transformation in the pharmaceutical industry: current trends and future directions. *Current Pharmaceutical Biotechnology* 3: 345–352.

25 Saha, E., Rathore, P., Parida, R., and Rana, N.P. (2022). The interplay of emerging technologies in pharmaceutical supply chain performance: an empirical investigation for the rise of Pharma 4.0. *Technological Forecasting and Social Change* 181 (1): 121768.

26 Miglierini, G. (2021). Time for change: The digital transformation of the Pharmaceutical Industry.

27 Braveen, M., Subramanian, N., Seetha, R. et al. (2023). ALBAE feature extraction based lung pneumonia and cancer classification. *Soft Computing* 1–14. https://doi.org/10.1007/s00500-023-08453-w.

28 Sharma, A. (2021). Artificial intelligence in health care. *International Journal of Humanities, Arts, Medicine and Science* 5: 106–109.

29 Yang, J., Li, Y., Liu, Q. et al. (2020). Brief introduction of medical database and data mining technology in big data era. *Journal of Evidence-Based Medicine* 13 (1): 57–69.

30 Erolin, C. (2019). Interactive 3D digital models for anatomy and medical education. In: *Biomedical Visualisation. Advances in Experimental Medicine and Biology*, vol. 1138 (ed. P. Rea). Cham: Springer.

14

Overcoming the Obstacles of Motion Sickness in the Metaverse's Digital Twins

Erik Geslin[1] and Diego Saldivar[2]

[1]*Faculty of Interactive Media – Games, CNAP Lab, Noroff University College, Kristiansand, Norway*
[2]*Faculty of Interactive Media – Games, Noroff University College, Kristiansand, Norway*

14.1 Introduction

Combined with immersive virtual reality (VR) technologies, digital twins (DTs) are emerging as a very promising technology. Unfortunately, despite grandiose promises from corporations, VR is not yet a smooth experience. In fact, for some users, VR technologies may not only fall short of said ambitious promises but can also be actively detrimental to their health. The potential harm in VR simulations stems from a disruption of proprioception between the afferent expected by the central nervous system (CNS) during locomotion and the afferent perceived. This leads to VR sickness, break in presence (BIP) [1, 2], and an inability to be immersed in an extended reality (XR). However, for several decades, and since the emergence of telepresence, many academic studies defined design rules to limit or avoid simulation sickness. These rules often remain unknown to VR designers. In this chapter, we will attempt to make a functional synthesis of them in the form of best practices. To this end, it is first necessary to define the cognitive mechanisms that come into play in the functional maintenance of balance and proprioceptive sensitivity of our CNS in both everyday life and in hostile environments. Thanks to numerous afferents capable of reflecting environmental stimuli, the vertebrate brain constantly calculates and adjusts the body's position in space. This cognitive mechanism invokes conscious and unconscious processes related to kinesthesia, forces, heaviness, and force sense [3, 4].

This chapter will first define the well documented physical side-effects of VR known by many names: VR sickness, motion sickness (MS), simulation sickness, cybersickness, and others. Unlike motion-induced illness (e.g., space sickness, seasickness or road sickness) VR sickness is not induced by real environmental

Digital Twins in Industrial Production and Smart Manufacturing:
An Understanding of Principles, Enhancers, and Obstacles, First Edition. Edited by Rajesh Kumar Dhanaraj, Balamurugan Balusamy, Prithi Samuel, Ali Kashif Bashir, and Seifedine Kadry.
© 2024 The Institute of Electrical and Electronics Engineers, Inc. Published 2024 by John Wiley & Sons, Inc.

motion, but more often by vection [5, 6]; which is the dissociation between simulated movement and non-movement perceived by the vestibular system or other cognitive systems subject to external afferents. This process is related to visually induced motion sickness (VIMS) [7]. Fortunately, the brain can train itself to ignore some of the external stimuli. Thus, resilience to DT environments can be built through incremental exposure, prolonged training, and the construction of an adaptative VR experience from external stimuli. However, VR sickness is a huge obstacle to the adoption of DTs and the Metaverse. As shown in research conducted by Nguyen and Aaron in 2020 on 292 VR users, 57.8% of the users reported motion sickness, 13.7% experienced frequent VR sickness, 19.1% "sometimes" did, and 24.9% did only on rare occasions [8].

In this chapter, we argue that, with the use of several cognitive user experience (UX) engineering rules, DT designers can build virtual environments (VEs) that help users adapt faster and circumvent easily preventable virtual discomfort. Indeed, the central concept of this chapter revolves around proposing a design tool to guide DT designers on how to actively prevent simulation sicknesses in their VEs. This tool represents a synthesis of the scientific research carried out to date on the topic of VIMS. It should, moreover, continue to be updated as more research is carried out in the fields of cognition and VR sickness and as DT technologies and standards continue to advance.

Now then, to start exploring this topic, it is important to consider that some of the causes of simulation sickness are linked to fundamental and uncontrollable parameters of human perception. Others, however, are due to technological limitations, often dependent on plain design aspects in the engineering process. If virtual experiences are not developed with a careful and enlightened approach, they may cause the strongest forms of preventable VR sickness.

14.2 Motion Sickness, Cognitive Principles

Let us start by defining the crux of the problem: motion sickness (MS), a term unfortunately familiar to some air or sea travelers. Dynamic environments challenge not only our cognition but also our physical performance. Our gross perceptual motor skills are challenged, as well as our finer motor skills (e.g., dexterity and smooth ocular pursuit). An imbalance in our perceptual systems or their cognitive analysis induces severe discomfort, malaise, nausea, pallor, cold sweating, and vomiting. If initially the term "motion sickness" was essentially attributed to vomiting induced by certain kinds of motion, the term is now attributed to each of these signs and symptoms provoked by motion [9]. It was later, in 1881, that Irvin used the term "motion sickness" to characterize the sickness induced by a "motion environment", thus freeing the term from its limited association with sea sickness [10]. The more frequent use of the term "motion sickness" in scientific literature is, however, far more recent, dating from the 1970s [9, 11, 12].

Nowadays, the scientific community widely accepts Reason and Brand's proposed definition of MS as the "sensory rearrangement theory on motion sickness" [12]. Their theory grounds MS as a mismatch between visual stimuli and the vestibular and non-vestibular proprioceptors. This cognitive disagreement may occur between these different systems, or between their cognitive analysis and an anticipation of expectations based on the subject's previous experiences. The two levels of incompatibility identified are either a cognitive visual-vestibular mismatch or an intersensory conflict with otolith accelerometer mismatch. Several levels of incompatibility have been written and classified since then. We will explore three in particular. To begin with, a type 1 conflict corresponds to the majority of MS occurring in the context of a motion environment wherein two cognitive systems report contradictory motion information. A type 2 conflict most often corresponds to simulation sickness, in which case a cognitive system signals a movement in the absence of a corresponding stimuli from another detector. A type 3 conflict, introduced by Guedry, is described as a vestibular-proprioceptor mismatch, termed "cognitive inconsistency" [13]. In 1982, Oman [14] proposed a heuristic mathematical model for predicting cognitive conflict and MS in order to give the scientific community a means to predict MS. By identifying the conflict between a vector representing the available sensory perceived stimuli and another vector representing the expected information, Oman proposes a model close to the intersensory conflicts described by Reason and Brand.

Bles et al. more recently demonstrated that a single conflict could be the cause of MS [15]. This theory represents a considerable advance in the understanding of MS, yet it does not call into question the entire mathematical model of Oman but extends it instead. This model is called the "subjective vertical conflict" or SV conflict. Even though there are several classes of MS from Type 1 to Type 3, Bles et al. observe a common underlying conflict in verticality. Similarly, others have observed that in a centrifuge only head movements on the vertical axis related to the gravity vector cause MS, while sensory shifts induce illusions of movement but do not cause MS [16].

Other studies [17] seem to corroborate that in optokinetic circular motion stimulation only the so-called "pseudo-Coriolis effect" of head tilt cause MS because of the SV conflict effect. This seems to be a simplified model of the classic "sensory rearrangement theory", in which, even in the case of conflict of anticipation between several cognitive systems, only stimuli relating to verticality are likely to produce MS.

14.3 Gender and Motion Sickness

The role of human factors in the onset of MS symptoms is known, to the extent that MS can be induced by the difference between the observed experience and the expected experience prompted by previous experiences. Beyond the merely cognitive, however, previous studies have pointed out a marked difference in who is the most inclined to suffer from MS. Many experiments report that the gender of the individual is key to predicting MS symptoms. In 1977, a survey showed that women seem to be more susceptible to MS than men [18]. In Lentz and Collins' survey of over 4000 college students, men felt they had more muscle coordination, which could have been instrumental in explaining differences in sensitivity. However, several recent studies show that these differences in sensitivity may be due to physiological differences in neuronal and hormonal structures, not necessarily relating to previous experiences, which are substantially the same as those of men before the age of 18 [19, 20].

Since then, a vast literature reported a higher prevalence of MS in women than in men [21–24]. Researchers have tried to explain these differences in sensitivity by the principles of survival and evolution of species. Some research points out that these results are due to social bias: women would more easily report somatic discomfort disorders, whereas men would tend to hide them consciously or unconsciously [25]. On the other hand, several studies struggle to show a difference in susceptibility to MS [26, 27], but this may be due to a self-selection bias. Sensitive women could indeed choose to not participate in this type of experimentation in a process of self-selection. As seen previously, the experimental processes on MS must be very rigorous to avoid bias, particularly regarding gender. It is interesting to note that when the experimental processor does not allow auto-selection, as is the case of a study on the rate of motion sickness among 1350 passengers and crew for 10 days on the ocean [28], their results show that seasickness occurs more often in women than men.

These results seem to be confirmed by studies specifically oriented around VR sickness. In recent research, the probability of inducing VR sickness by two games with the use of an Oculus Rift virtual reality headset was evaluated [29]. Their results show that in games with low vection, the reports of MS do not differ between genders. However, in games with high vection, 56% of participants reported VR sickness effects and the incidence of VIMS among women was greater and more significant (77.78%) than among men (33.33%). This seems to confirm that the gender differences observed in MS are comparable to those of VR sickness. In this case, VR sickness is not induced by a Type 1 MS conflict in which several cognitive systems report contradictory motion information, but by a Type 2 conflict in which the CNS signals movement in the absence of an expected stimulus from biological receptors. It is still relevant to highlight in which ways VR sickness differs from MS, however.

A critical review of the aforementioned studies may, however, reveal some bias not only in their demographic sampling but also a failure to account for behavioral factors beyond those explained by hormonal differences. Indeed, a critique of the studies based on hormonal differences might be relevant to the origins of MS in and of itself. For instance, while Flanagan [25] alleges that the high incidence of MS in females is due to hormonal differences, it is important to note that scientific literature itself can be inconclusive or even contradictory regarding phenomenological differences based on the biology of cis males and females. For instance, Bartley and Fillingim [30] found that women are more likely to report pain than men, whereas Leonel de Nazaré et al. [31] found that women seem to have a higher pain tolerance threshold than men. We posit instead that the high incidence of MS in females in VR might be due to a lack of exposure to the medium itself. According to the Entertainment Software Association, there has been a marked increase in the participation of females in the gaming market from 38% of American gamers being female in 2006 to 48% as of 2022 [32, 33]. Despite this marked increase, it might still be possible that not enough female gamers sampled in studies have had enough exposure to VR to become accustomed to high vection like their male counterparts. The change in demographics also points toward an often-suspected trend: in the early history of video games, there have been fewer female gamers than male gamers. It could be possible that, historically, females have been less exposed to virtual vection than males, thus making them more susceptible to VIMS. It follows that the best approach should be less reliant on gender differences and instead focus more on differences in virtual vection tolerance and gradually increasing exposure to DT environments. Consequently, this chapter shall propose best practices geared to diminish VR sickness triggers and building tolerance for VR vection in the general population.

14.4 VR Sickness

In the context of Web 3.0, Industry 4.0 and DTs, VR represents the key technology for immersive simulation, optimization, and collaborative remote interaction. DTs include processes that allow more flexibility, cost and time reductions that give significant competitive advantages [34]. The combination of Industry 4.0 principles, cloud servers, simulation and big data analysis leads to cyber-physical production systems (CPPS) [35] which produce real benefits for the industry in terms of cost and efficiency [36, 37]. As proposed by Moore et al., CPPS can greatly benefit from the use of VR principles in a "3D virtual engineering approach" (VIR-ENG) [38]. VR is itself defined by authors like P. Fuchs as the necessary combination of immersion and interaction [39], thus VR = I^2 (Immersion + Interaction) [39], which makes it perfectly adapted to help interacting with CPPS. VR is also defined as "a computer-generated digital environment that can be experienced and interacted with as if it were real" [40].

The benefits of VR have been well known for decades. In industrial process design, the use of virtual prototypes saves time and money, and allows iterations faster than classic physical prototyping [41, 42]. In addition to the many cognitive processes at work building in the capacity for immersion and interaction in each user, we must also introduce the necessary willing suspension of disbelief (SoD) [43], which describes the mental process carried out by book readers, movie spectators, or even VR users who agree to put aside their skepticism at the time of the experience. The capacity to fully accept immersion while disregarding the real external environment outside the computer-generated VE, is the fundamental element for the notion of Presence in VR [44] as well as for CPPS DTs. Scientific literature already shows that conscious or unconscious emotional states can promote a high level of presence [45], the SoD introduced consciously by the user being one of these fundamental elements. Thus, a user knowingly refusing suspension of their disbelief will not be able to fully interact with a DT in VR.

Presence in a VE is defined by Lombard and Ditton as "a mediated experience that seems very much like it is not mediated." [46]. "Mediated" refers to a simulated world induced by technology, as in a VR VE [47]. Slater et al. argue that presence is the result of two equal-sided illusions: the place illusion (PI) —the feeling of being there—and plausibility (Psi) — the credulity of VR [48, 49]. In this case, "Presence = Pi + Psi", where all the VE's stimuli must lead to the foundation of a strong SoD. Even tough, from a Spinozian point of view, as Gilbert points out, humans naturally believe the stimuli presented to their cognition and will later analyze the veracity of their surroundings [50]. Gilbert introduces the Spinozian concept of the "natural believer". From a purely cognitive point of view, many studies have shown that cognitive inconsistencies like the one that leads to the onset of sickness simulation symptoms through vection and VIMS [7] generate BIPs [48, 51] which make people skeptical about the plausibility of the VE. These BIPs inevitably lead to the destruction of the SoD model and to disengagement away from the interaction in the VE of the CPPS. The consequences of VR sickness are multiple, and they can potentially lead to an inability to benefit from these VEs. Crowley argued that pilots suffering from simulator sickness end up not using flight simulators, which can have serious consequences in terms of aviation safety [52]. Enrique et al., highlight the difficulties in implementing Industry 4.0 technologies, such as CPPS and DTs, due to the virtual reality sickness inherent to them [53].

By definition, VR sickness may occur when the user is exposed to VR, which is both immersive and interactive, as is the case of CPPS DTs. Symptoms of VR sickness are similar to those of MS: eyestrain, stomach upset, discomfort, nausea, vomiting, paleness, fatigue, drowsiness, listlessness, and more frequently in the case of VR sickness, disorientation [54, 55]. Cybersickness and VR sickness are distinct from MS in that they do not require real self-motion to trigger [56]. They are also distinct from nonvirtual simulator sickness, which is related to oculomotor disturbances, whereas VR sickness is more closely related to disorientation [57]. In the instance of VR in DT interactions, moving visual stimuli are often dominant, even in the case of use of a multimodal immersive model (using several senses). Symptoms of VR sickness may arise if the user repeatedly receives sensory feedback contradicting their cognitive expectations [58], which is in line with VIMS [7]. Several research papers relate VIMS to "vection": the illusory perception of self-motion [5, 6, 59, 60]. Vection can be represented, for instance, by the sensation of motion that a train passenger experiences when their wagon is sitting static at the station while other close wagons are moving on parallel rails. In a real environment, the phenomenon of vection is temporary and does not persist for more than a few seconds. But in a VE, vection can be prolonged indefinitely. In this case, the movement information expected by the vestibular system does not reach the CNS because the VR user (except natural walking models) does not actually move in the DT. The greater the imbalance between information received and information expected, the greater the symptoms of VIMS. VR sickness has been known to be related to intersensory conflict, which strengthens the case for high-fidelity multimodal systems over unimodal systems [61].

Much research has been conducted beyond the importance of DT and into the causes and countermeasures of VIMS detrimental to their adoption. Most of the solutions proposed tend toward the limitation of techniques and content. Some solutions also consider human factors leading to cognitive inconsistencies and adopt cognitive processes that simulate or copy the real world. On the other hand, it is worth acknowledging that there might be certain possible biases in their methodologies.

One must note that even comparative analyses like that of Chang et al. [61] fail to elucidate best practices for the VR industry, since they pay more attention to the technological capabilities of the technology being surveyed, rather than pointing out how their shortcomings induce VIMS. Indeed, most technologies included in these comparative analyses are either technologically obsolete or irrelevant to the study of VR sickness. For instance, one might be pressed to ask how a comparative study between cathode ray displays and a Cave Automatic Virtual Environment (CAVE) could be relevant to contemporary wireless VR headsets. Further, their samples frequently include analyses of technologies not directly related to VR or DTs. It is difficult to find a paper or publication that focuses on the scientific exploration of VIMS exclusive to VR. And to compound to the confusion, scientific papers that do exist on the topic make use of obsolete VR headsets, which inevitably induce

MS due to their technological limitations. We must therefore elucidate our best practices from the partial information still relevant to our field of study.

14.5 VR Sickness Measurement

Let us then address the heart of the problem. There are two types of measurements generally used to determine the level of discomfort and the severity of VR Sickness. The first, and most common, is the post-experimental use of subjective semantic questionnaires where users report the level of discomfort or the severity of the symptoms of VR sickness felt during and after the VR experience. This questionnaire mainly uses a Likert scale which allows a greater degree of nuance than binary information [62]. However, these questionnaires' reliance on users' memory may introduce bias. Indeed, the postponement of past emotions or past suffering varies depending on the individual. Contextual factors when filling the questionnaire can influence their affective state, situational patterns, and cognitive representation [63]. In this case, the use of subjective semantic questionnaires requires the participation of many subjects to compensate for this bias. Subjective semantics questionnaires can also introduce bias if some users report nausea whereas others report discomfort. The implementation of more objective measurement systems tracking physiological data in real time constitutes an alternative or important complement to the use of subjective semantic questionnaires. Postural sway, electroencephalography (EEG), and electrodermal activity (EDA) are frequently used to corroborate physical states related to MS or VR sickness.

That being said, there are many subjective semantic questionnaires that try to correct for their predecessors' relative fragility and potential for bias. For instance, the "Motion Sickness Assessment Questionnaire" (MSAQ) [64] uses four dimensions of VR sickness symptoms, but its reporting process is long. Other rapid and unimodal questionnaires have been developed, such as the "Fast Motion sickness Scale" (FMS) [65] and the "Misery Scale" (MISC) [66], which use a Likert scale from 0 to 20 for FMS and 0 to 10 for MISC. They try to report a single dimension of discomfort every minute of the experiment, making them quasi-real-time questionnaires that avoid memory biases. These subjective semantic questionnaires are seldom used, however. In a study by Chang et al. on VR sickness measurement [61], it is reported that out of 76 studies using subjective semantic questionnaires, 58 of them used the "Simulator Sickness Questionnaire" (SSQ) proposed in 1993 by Kennedy [67]. In multidimensional VR sickness questionnaires realized post experimentation, the SSQ assesses the level of VR sickness in three dimensions: nausea, oculomotor, and disorientation. Included are 16 questions on a 4-point scale from 0 to 3. In 2013, a study by Balk et al. evaluated the use of the SSQ 20 years after its creation for the purpose of modern simulation [68]. The study showed a strong correlation between high estimated SSQ scores (above 30) and user dropout in VR. Even if the SSQ can reveal some biases in the non-crossed methodology of the three dimensions or the post-process data collection based on the memory of the MS, its generalization and ease of use make it an accessible tool whose results are comparable to many previous studies.

As discussed previously, physiological measuring methods are frequently used to compensate for the lack of objectivity in the semantic questionnaires. Some physiological VR sickness measurement methods have been evaluated over time with varying degrees of success. Indeed, VIMS nausea can cause characteristic changes in the CNS, reflected by variations in blood pressure [69], skin temperature (SKT) [70], respiration pneumography (RSP) [71], and photoplethysmography (PPG) [72]. All these methods, however, were used less than three times in 68 of the studies in Chang's report [61], either because their effectiveness was inconclusive, or because their implementation was too complex. The same is true for the following physiological measurement methods, each used less than 6 times on the same 68 studies related to the dynamics of MS: EEG [73], EDA [72], eye-related measurement, and electrogastrography (EGG) [72]. The two most used physiological measurements in studies on VR sickness are ECG and postural sway, known for the longest time as essential markers of the physiological modifications linked to MS-related discomfort [74]. ECG is used up to 10 times in 68 studies, but the most used physiological marker of MS and VR sickness is the postural sway measurement in 22 out of 68 related research publications [61]. Postural sway measures differences in postural instability which predicts the occurrence of MS in the presence of high vection and in VIMS-inducing environments [75, 76]. This method compares pressure differences between the soles in natural conditions versus virtual simulations. The procedure requires initial Center of pressure (CoP) of the foot measurements with eyes open and eyes closed. These measurements are then compared with measurements collected during exposure to vection in a VE. This process is known as recurrence quantification analysis (RQA). As an example, in 2018, research by Palmisano showed a significant correlation between low initial CoP in standard standing and the ratio of strong vection and VIMS, both in optical flow exposure and in VR [75]. Thus, postural sway measurements of spontaneous postural activity through CoP and RQA are relevant for detecting and predicting vection and VIMS in DTs.

Various tools for measuring VR sickness symptoms are essential elements to the design of DTs. Rigorous application of these tools could guarantee optimal use of these paradigms for all users, whether they are extremely sensitive to VIMS or not. If used iteratively during the development process, the SSQ questionnaire accompanied by CoP, RQA, postural sway, and EGG measures bring immediately applicable information that can be used to minimize or suppress VR sickness symptoms.

14.6 Guidelines for VR Digital Twins Development to Avoid VR Sickness

Having reviewed the tools that minimize or prevent VR sickness, we can now elucidate best practices for DT development. Thanks to the aforementioned iterative development models, we can objectively measure the effects and symptoms of VR sickness in order to develop functional and accessible DTs. However, it is also possible, and desirable even, to pre-emptively apply these best practices at a much earlier stage to proactively address the three main causes of VR sickness in DTs, namely: hardware, content, and human factors.

14.7 Hardware Challenges

A significant amount of research has been carried out to determine the combined effects of immersive hardware, the implementation of content, and the uniqueness of users. Unfortunately, the methodologies developed on that front differ extensively and are often not comparable. Even in a sample based on the same MS measurement methodology derived from the SSQ for immersive systems, the contents may differ so much as to make the results next to impossible to compare. For example, some research uses the term "virtual reality experience", defining it as the product of the combination of immersion and interaction [39], but it only uses non-interactive immersive 360° films for their evaluation, which fail to conform to the definition provided [77]. There is also no consensus on the exposure times required for VR sickness experiments. Some experiments only last a few minutes, while others can last between 10 and 30 minutes [78]; and this while a strong correlation has been demonstrated between the length of exposure to VR and the appearance of symptoms of VR Sickness [79–81]. Often the choice of content is also debatable. Some classical examples include simple non-interactive 360° films [77], walking scenic routes [82] stationarily observing scenery [83] and even roller coaster simulations highly likely to induce VR sickness [84]. More contemporary examples include the use of immersive video games of dissimilar generations, from VR Minecraft [85] to Half-Life II [71]. The disparity in the choice of environment and activity makes it very difficult to create a clear correlation between different SSQ results. However, despite these procedural inconsistencies, several studies have compiled information [61, 78, 86] that allow us to cross-examine and conceptualize a pattern of evolution of interactive immersive systems for the optimal use of DTs, thus limiting or even preventing the appearance of VR sickness.

Undeniably, there is an important correlation between cognitive inconsistency mismatches produced by immersive systems and the eliciting of VR sickness.

The evolutionary trend of immersive VR systems tends to reduce or eliminate the cognitive inconsistency arising from prolonged exposure to VEs. For example, while the first-generation "head mounted display" (HMD) had a "field of view" (FOV) limited to 50°, the most recent HMDs offer FOVs of 180°, comparable to a human FOV. Indeed, the gap is gradually being closed between current VR technologies and the optimal resolution perceived by humans, calculated in pixels per degree (PPD). The first Oculus DK1 HMD offered a resolution of 8 PPD. Ten years later, in 2015, Vive's HMD offered 30 PPD. Today, in 2023, some high-end HMDs already offer a resolution of 60 PPD. Paradoxically, the increase in the FOV and the quality of resolution of the HMDs seems to have initially led to worsening VR sickness; with an increase in immersion came an increase in vection and, consequently, in VIMS. In order to reduce the adverse effects of vection, developers introduced "vignetting", inspired by how camera lenses work, artificially limiting the FOV of the user [87–89]. A vignette is a post-processing effect which fades the edges of the screen using the graphics processing unit (GPU) buffer in differed rendering mode. This creates an artificial tunnel effect, effectively reducing VIMS symptoms thanks to users' limited FOV and, therefore, reducing their exposure to an intolerable amount of vection stimuli. Vignetting effects have previously been referenced as a potential tool to reduce VIMS symptoms in VR [87, 89]. However, the increase in VR sickness symptoms reported in SSQs were not simply caused by the increase in resolution and FOV. Rather, these symptoms have been a growing pain in the evolution of HMDs from low-fidelity monoscopic modes to stereoscopic high-fidelity modes [71, 90]. We call this paradox the "VR Dizzy Valley", which is apparent throughout the increase in immersion in the transition from HMDs with low fidelity to higher fidelity models. This paradox can be countered by hardware and software strategies addressing cognitive mismatches. Since most of the aforementioned mismatches arise between incompatibilities in the reports of specific sensory organs, resolving said mismatches requires increasing information coherence in these sensory reports with the use of a multimodal strategy. The Chang report (Figure 14.1) shows how the situation changes when using a multimodal immersive system instead of a unimodal one. As evidenced by the report, the quality increase of the immersive systems corresponds to a drop in VR sickness reports in the SSQ.

Additionally, other factors in the HMD's qualities can impact the severity of VR sickness symptoms, such as latency jitter, which can have a significant impact on VR sickness [91]. This is in addition to other artifacts that may arise when limited hardware is confronted with the strain of rendering a data-heavy 3D environment in real time. These artifacts can generate an intolerable latency between the user's perception and the cognitive feedback necessary for real time interaction with the VE via a human–machine interface (HMI). Indeed, it has been documented that latency above 75 ms can generate severe VR sickness [92]. Conversely, research

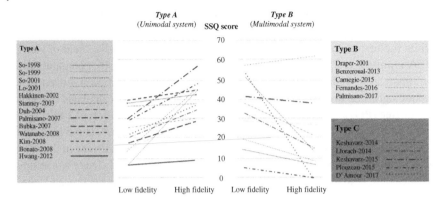

Figure 14.1 The relationship between the level of VR fidelity and VR sickness depends on the type of VR system. Source: Adapted from Chang (2020).

agrees that latency does not cause VR sickness below 20 ms, whether inherently related to HMD jitter or to the communication between the VE, the hardware, and the HMI [39, 86, 92]. Latency can also be related to the global frame rate or frames per second (FPS). The role of a lower frame rate as a cause for VR sickness has been highlighted in several research papers: the "Sensory Conflict Theory", for one, proposes that lower frame rates worsen the cognitive inconsistency between the visual, vestibular, and somatosensory systems [93] [7]. The scientific consensus is that the FPS of a VR experience must reach, at the very least, a 1 : 1 refresh frequency with the HMD (e.g., 90 FPS for a 90 Hz refresh rate). This 1:1 ratio is meant to eliminate latency in the image distribution, which can induce VR sickness. Furthermore, recent research shows that 120 FPS should be taken as an important threshold to prevent VR sickness and make DTs more performant [94]. The same research provided by Wang et al., indicates that 180 FPS and above gives even better performance results in VR. This effectively reduces the cognitive load caused by prediction errors in the path of dynamic objects in the VE. According to the sensory conflict theory, a high framerate can reduce the sensory conflict in the visual-vestibular mismatch [95]. At first glance, a threshold of 180 FPS may appear like a challenge for our current technologies, but research published in 2015 by Davis et al. claims that humans are able to perceive visual anomalies at rates over 500 Hz [96], so the threshold raised by Wang et al. is minimal, at best (Figure 14.1).

However, this is not to say that technology is far from overcoming the Dizzy Valley. One by one, technological hurdles continue falling, leaving us free to manage cognitive inconsistencies within the content of the system, which we will address later on. As it happens, other factors causing VR sickness have already been resolved by advances in technology, such as adapting the headset's interpupillary distance (IPD) to the actual IPD of each user. IPD varies

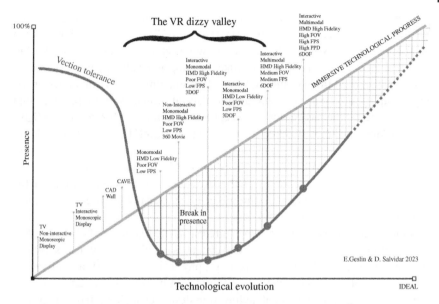

Figure 14.2 The VR Dizzy Valley. Source: E. Geslin and D. Saldivar (2023).

greatly between people, depending on their ethnicity, age, and gender; with an average near-interpupillary distance (NIPD) of 60.91 ± 3.04 mm for men and 59.16 ± 2.55 mm for women between the ages of 26 and 40 [97]. Consequently, an HMD with a fixed IPD is likely to cause VR sickness in stereoscopic systems [98]. In the case of early HMDs, this could have been a likely culprit behind gender differences in VIMS susceptibility. Nowadays most HMDs offer an adjustable IPD, some systems even integrate eye tracking systems for automatic IPD adjustment, which should become a standard in next generation HMDs.

14.8 Content Specificities

Many of the conflicts highlighted in the Sensory Conflict Theory regarding the visual-vestibular mismatch can be reduced or even erased with an appropriate use of current technologies or, eventually, as innovative technologies continue to develop. Yes, immersive environment hardware is an essential element of the DT equation, but other factors come into play as well. Specifically, the elements constituting the content of the VE, the implementation of the "graphic user interface" (GUI), and the "user experience" (UX) model for walking and virtual manipulation.

The principles established by Fuchs wherein VR = I^2 [39] define the intrinsic links between immersion and interaction allowing the establishment of a feeling of high presence [1]. To put things into perspective, even pilots subject to mandatory VR training, may benefit less if their HMD leads to VR sickness and thus become disinterested [55]. Why would voluntary users keep an interest if even the simplest and most essential interactions (like traveling and walking) are not carefully developed to avoid the nuisance of VIMS? Defining unlimited virtual travel walking metaphor protocols corresponding to natural human motility schemes that prevent cybersickness is key for the acceptance of immersive VR in DT paradigms [55, 99]. Most HMI modalities defined as "Magical Travel" [100, 101] have failed to reproduce the movements of walking in a VE without producing cognitive inconsistencies that lead to VIMS [102]. The body-centric natural walking (NW) model considered as a "mundane travel" method by Nilsson [101], seems to immerse the user into the digital wonderland promised by Sutherland [103] even in limited spaces, with limited to no VR sickness.

Some magical travel models garner interest in the realm of DTs because they allow users to move through large virtual distances in tight physical spaces. Others, however, advocate for the use of more ample physical space for virtual exploration. The concept of a 3D maze in VR was introduced by Schnabel in 2001 with the aim of thinking up new ways of architecture design using VR [104]. The principle of a maze in VR allows ample virtual travel and exploration without motion sickness insofar as it uses the NW model, whereas classic interaction models use technical redirection and technical "self-overlapping architectures" [105]. However, DTs have, by definition, an inexorable need for exact virtual representations of real environments making it difficult to implement a VR maze model. However, the concept of the VR maze may inspire enhanced UX models that guarantee more comfortable VEs in DTs.

According to Suma et al., VR mazes can increase the perception of "impossible spaces" from 31% to 56% [105]. Impossible spaces being defined as VE spaces, accessible in layers, yet not accessible in physical natural walking mode. These are mostly used in experiments in spatial cognition [106], in the study of social interaction [107], for entertainment [108] and for police and military training [109]. Although VR mazes are often used as a cognitive research tool thanks to their easy implementation, it is rare to find DT studies interested in the very subject of VR mazes and their contributing immersivity without VR sickness.

It is established that most mundane travel techniques in VR produce little to no cybersickness [55]. Conclusively, there must be no cognitive inconsistency between perceived actions and performed actions, so cybersickness is further mitigated by reducing or eliminating other sensory conflicts [12]. In contemporary HMDs this mitigation can be achieved by producing a Hertzian refresh rate at least equivalent to the application's frame rate, which, in turn, must surpass 90 FPS per eye. Additionally, tracking systems, whether outside in or inside out, must not produce latency in 6DOF. Magical travel techniques, like joysticks or controllers simulating traveling without walking, generally induce more cybersickness symptoms as they are more likely to create situations of cognitive inconsistency. However, magical stationary patterns of interaction and movement such as teleportation are known to produce fewer cybersickness symptoms [110, 111]. Beyond cybersickness, the choice to only use mundane travel techniques in DTs conforms to the desire to favor the immersive aspects of presence [1], which avoids VR usage patterns with non-diegetic user interfaces.

While mundane body-centric mobility techniques are used for 1 : 1 ambulatory movement in a VR-maze, mundane stationary vehicular travel can also be used to simulate movement in elevators and conveyor belts, which gives the user an expected feeling of motion even when stationary, this time benefitting from vection itself [112].

Defining the most appropriate type of GUI for functional tasks in a DT is one of the most important processes during conceptualization. The design and quality of the GUI in VR is decisive for the success of HMIs because it can affect the sense of presence in a positive manner [113]. Traditionally, the quality of the GUI is dependent on the cognitive feedback provided in each interaction. The GUI's success is also reliant on the salience and skeuomorphic quality of the affordances used, whether informative, functional, physical, or sensory [114]. Beyond these aspects, there are two important UX design paths in VR: the GUI can be either non-diegetic — wherein affordances are added on layers above the VE (e.g., a heads-up display with floating name tags)— or it can be diegetic — where the affordances are naturally associated with virtual objects present in the VE (e.g. using a virtual mobile phone to interact in a mundane way with elements in the scene). On the one hand, diegetic VR GUIs generate a deeper sense of presence in experimented users, improving their cognitive involvement and their sense of control [115]. On the other hand, a 2020 study by Ashar Imtiaz shows no significant differences in VR sickness between diegetic and non-diegetic VR GUIs [113, 116]. Thus, the choice of diegetic versus non-diegetic GUIs is left to the discretion of DT designers, with the GUI remaining the most important interface between users and the VE. Indubitably, a diegetic VR GUI is comparatively more immersive and thus more empowering for digital twin users.

14.9 Human Factors

As previously discussed, physiological differences lead to variances in tolerance for VR hardware which in turn may result in symptoms of VR sickness while using DTs. As a remedy, proper applied knowledge of behavioral patterns can guide designers toward frictionless GUIs. Ignoring said patterns could result in unwittingly convoluted GUIs, as unintentional omissions inevitably lead to cognitive inconsistencies that bring forth MS. Therefore, the consideration of human factors in the implementation of DTs becomes fundamental in the prevention of VR sickness.

The first human factor we will explore is the length of exposure to a VE per session. It is well known that VR sickness symptoms can occur well within 10 minutes of entering a VE [71, 117]. Several studies show that the longer the time of exposure, the greater the symptoms of VR sickness. There is also a proven positive correlation between length of exposure to VEs and SSQ scores [59, 80, 118, 119]. Regular breaks can be introduced into the DTs VR UX to avoid worsening VR sickness symptoms, but it is not a functional long-term solution for consistent operation of the Digital Twins, since it interrupts immersion, leading to BIPs that encumber the optimal use of VEs [48]. Breaks do not allow operators to fully interact with VEs, and several studies have shown that passivity and lack of interaction ultimately lead to severe VR sickness [120, 121]. Fortunately, it is possible for users to gradually adapt to the use of VR for DTs and thus become more comfortable with VR. A gradual approach can reduce or eliminate VR sickness for subjects sensitive to VIMS, regardless of their personal differences in sensitivity [122].

Consequently, the time of exposure to VEs should be gradually increased. Studies show a correlation between the time of exposure in a single session and the appearance of VR sickness [123, 124]; fortunately, further studies also report that VIMS decreases with every passing pre-exposure iteration [80]. Increased experience with VR allows the brain to anticipate vection in VEs. These effects on the vestibular system in the context of VR paradigms becomes more and more apparent. An increasing number of quality referential experiences allows, over time, to curb or even eradicate the symptoms of VR sickness. Thus, incremental exposure to VEs is essential for DT users.

14.10 Best Practices

Having reviewed the science behind MS and VIMS and having reviewed the current technological capabilities of HMDs for VR, we propose a series of standards and best practices to minimize or even eliminate the arousal of VIMS in VR.

Table 14.1 presents a summary table proposed to synthesize the information collected previously so that designers of DTs within VEs might make the most out of it. Since we have shown that oftentimes cybersickness in VR is generated by cognitive inconsistencies between human perception and the stimuli produced by VR systems, this table attempts to provide a list of the optimal actions to take and systems to implement; which are expected to prevent MS symptoms by closing the gap between mediated information and their natural perceptual counterparts, as well as reducing the discomfort that may come to novice users.

Table 14.1 HMI minimum standards to fill the gap with human perception for enhancement of digital twins.

Natural human perception	Optimal expected VR HMI and optimization for avoiding VR sickness in digital twins
Natural limitation of perceived resolution in the eye = 60 PPD	Optimal HMD Resolution $\geq$ 60 PPD
Natural FOV = 180	Optimal FOV = 180
Human visual acuity is limited to $\pm 1°$ around the optical axis of the centered fovea	Use of Foveated rendering with eye and gaze tracking
Rate at which Natural Vision is able to perceive flickering visual artifacts > 500 Hz	HMD FPS $\geq$ HMD Hz Frequency Ideally > 120–180 FPS & Hz
Natural latency between the occurrence of physical phenomena and light carrying information into our eyes = close to 0 ms	Optimal feedback latency < 20 ms
Natural cognitive perception = multi sensorial	Optimal feedback system = multimodal
Natural ambulatory speed < 4 km/h	Mundane natural XWY walking mode
Travelling on conveyor belt = Vection present in XW	Mundane stationary XW vehicular travel
Travelling by elevator = Vection present in Z	Mundane stationary Z vehicular travel
Travelling long distance above the limit of the physical time and space limitations = impossible	Teleportation
Human IPD = varies between 60.91 $\pm$ 3.04 and 59.16 $\pm$ 2.55 mm	Automatic or manual IPD adjustment (ideally automatic and with a latency below 20 ms)
Natural interaction = adjusted to human ergonomics	Ideally using a diegetic GUI

Source: E. Geslin and D. Saldivar (2023).

Figure 14.3 proposes a workflow that leads to the production of VR DTs with limited VR sickness effects. The first steps correspond to the gathering of information relating to the expected functionalities of the upcoming DT. For this purpose, we propose designers survey the technical limitations related to the technologies available, as well as audit the available physical space(s) in order to reproduce the qualities and limitations of the real environments in the DT. Indeed, these data are essential to determine a large number of technical choices, as well as the production of the content and the modalities of interaction within the VE.

An optimized design workflow for avoiding **VR-sickness** in digital twins using virtual reality.

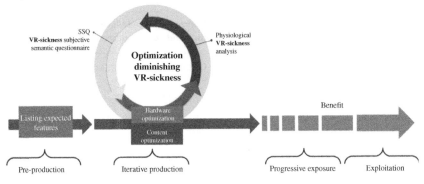

Figure 14.3 caption below:

Figure 14.3 Design process involving digital twin and virtual reality environment without VR Sickness. *Source*: E. Geslin and D. Saldivar (2024).

Afterward, surveys must be carried out on the different iterations of the model in production. These surveys should aim to measure the different levels of observable VR sickness through semantic subjective questionnaires (such as the SSQ) and with additional physiological measurements. Depending on these results, modifications must be iteratively released to the DT model. Eventually, once the optimal DT has been achieved, a strategy for adapting operators and users' cognition must be set up which, over time, will allow users to train tolerance through repeated experiences of vection in VEs, ultimately limiting the arousal of VR sickness symptoms.

14.11 Conclusion

DTs are becoming a major economic and strategic venture for many companies. Coupled with the principles of Metaverses and Web 3.0, their use will probably be a daily routine for many operators. However, technological limitations and uninformed design practices have led to discomfort while using DTs, motion sickness being most prevalent. Motion sickness is a kind of discomfort that occurs when cognitive information attenuated by the CNS does not agree with information from other stimuli. In this case, visual and vestibular discomfort creates nausea. As we have seen in this chapter, motion sickness in DTs or VIMS in the precise case of VR may very likely compromise the future of DTs used in VR immersion unless it is actively diminished, prevented, or even eliminated.

Therefore, an understanding of the fundamental cognitive mechanisms behind VR sickness and the principles that can reduce or cancel VIMS symptoms becomes paramount in the design of DTs. To best illustrate this, we proposed an analysis of the causes and principles of VR sickness specific to immersive use in VEs supporting DTs.

The analysis of the scientific literature allowed us to define several causalities of VR sickness due to vection, VIMS and, more generally, due to a visual-vestibular mismatch according to the Sensory Conflict Theory. A detailed review of the mechanisms of VR sickness and their impact upon VR users during the technological development of DTs led us to visualize a dip in VR tolerance with a Dizzy Valley model. This model allows us to anticipate a possible evolution of systemic and design strategies for DTs that could help VR overcome this gap.

In order to further develop the VR scene into a more user-friendly direction, we proposed a table of best practices to summarily guide motivated designers into the right direction. Our summary table makes it possible to visualize the notable differences between human perceptual systems and the thresholds or techniques expected in virtual reality to limit or eliminate the symptoms of VR sickness in these paradigms.

We also highlight the impact of human factors in the onset of and susceptibility to MS. We show that making the most out of these factors, can inform the creation of a referential background of vection in VE capable of limiting or eliminating, to a certain extent, the symptoms of VR sickness in DTs.

Thanks to these explorations and analysis, we are able to synthetize an iterative design model for a DT environment that takes into consideration the mechanics behind VR sickness issues and the implementation and optimization of alternative strategies using interactive modalities that reduce adverse vection effects.

We hope that this chapter has led the reader to a better understanding and optimization of DT design models using VR. In view of the swiftness of innovations and developments, both in technology and in the procedures analyzing and measuring discomfort linked to cybersickness, we strongly advise designers to update their knowledge of these subjects over time. Likewise, we beseech researchers to keep contributing to the discourse by reviewing technologies and their effects on their users on a regular basis.

References

1 Slater, M. (2002). Presence and the sixth sense. *Presence* 11 (4): 435–439.
2 Sanchez-Vives, M.V. and Slater, M. (2005). From presence to consciousness through virtual reality. *Nature Reviews Neuroscience* 6 (4): 332–339.
3 Martin, J.H. and Jessell, T.M. (1991). Development as a guide to the regional anatomy of the brain. *Principles of Neural Science* 296–308.

4 Riemann, B.L. and Lephart, S.M. (2002). The sensorimotor system, part I: the physiologic basis of functional joint stability. *Journal of Athletic Training* 37 (1): 71.

5 Bonato, F., Bubka, A., Palmisano, S. et al. (2008). Vection change exacerbates simulator sickness in virtual environments. *Presence Teleoperators and Virtual Environments* 17 (3): 283–292.

6 Lubeck, A., Bos, J.E., and Stins, J.F. (2015). Interaction between depth order and density affects Vection and postural sway. *PLoS One* 10 (12): e0144034.

7 Keshavarz, B. (2015). Vection and visually induced motion sickness: how are they related? *Frontiers in Psychology* 6: 472.

8 Nguyen, W. (2020). VR Heaven. VRHeavenNews. https://vrheaven.io/vr-motion-sickness-statistics/ (Accès le 22 02 2023).

9 Money, K.E. (1970). Motion sickness. *Physiological Reviews* 50 (11): 1–39.

10 Münch, W. (1955). Cocculus und seine homöopathische Verwendung. *Allgemeine Homöopathische Zeitung* 200 (01): 27–29.

11 Dichgans, J. and Brandt, T. (1973). Optokinetic motion sickness and pseudo-Coriolis effects induced by moving visual stimuli. *Acta Oto-Laryngologica* 76 (1-6): 339–348.

12 Reason, J.T. and Brand, J.J. (1975). *Motion Sickness*. Academic Press.

13 Guedry, F.E., Rupert, A.R., and Reschke, M.F. (1998). Motion sickness and development of synergy within the spatial orientation system. A hypothetical unifying concepts. *Brain Research Bulletin* 47 (5): 475–480.

14 Oman, C.M. (1982). A heuristic mathematical model for the dynamics of sensory conflict and motion sickness. *Acta Oto-Laryngologica* 94 (392): 4–44.

15 Bles, W., Bos, J.E., De Graaf, B. et al. (1998). Motion sickness: only one provocative conflict? *Brain Research Bulletin* 47 (5): 481–487.

16 Bles, W., DeGraaf, B., and Bos, J.E. (1996). Space adaptation syndrome (SAS) and sickness induced by centrifugation (SIC): vestibular consequences of earth anomalous gravity. *Journal of Vestibular Research* 4 (6): S66.

17 Brandt, T., Wist, E., and Dichgans, J. (1971). Optisch induzierte pseudocoriolis-effekte und circularvektion. *European Archives of Psychiatry and Clinical Neuroscience* 214 (4): 365–389.

18 Lentz, J. and Collins, W. (1977). Motion sickness susceptibility and related behavioral characteristics in men and women. *Aviation, Space, and Environmental Medicine* 48: 316–322.

19 Dobie, T., McBride, D., Dobie, T. et al. (2001). The effects of age and sex on susceptibility to motion sickness. *Aviation, Space, and Environmental Medicine* 72 (1): 13–20.

20 Grunfeld, E.A., Price, C., Goadsby, P.J. et al. (1998). Motion sickness, migraine, and menstruation in mariners. *The Lancet* 351 (9109): 1106.

21 Abe, K., Amatomi, M., and Kajiyama, S. (1970). Genetical and developmental aspects of susceptibility to motion sickness and frostbite. *Human Heredity* 20 (5): 507–516.

22 Bakwin, H. (1971). Carsickness in twins. *Developmental Medicine and Child Neurology* 13 (3): 310–312.

23 Deich, R.F. and Hodges, P.M. (1973). Motion sickness, field dependence, and levels of development. *Perceptual and Motor Skills* 36 (3): 1115–1120.

24 Turner, M. and Griffin, M.J. (1999). Motion sickness in public road transport: the relative importance of motion, vision and individual differences. *British Journal of Psychology* 90 (4): 519–530.

25 Flanagan, M.B. (2005). *Sex, and Virtual Reality: Posture and Motion Sickness*. New Orleans: University of New Orleans.

26 Hu, S., Glaser, K.M., Hoffman, T.S. et al. (1996). Motion sickness susceptibility to optokinetic rotation correlates to past history of motion sickness. *Aviation, Space, and Environmental Medicine* 67 (4): 320–324.

27 Woodman, P.D. and Griffin, M.J. (1997). Effect of direction of head movement on motion sickness caused by Coriolis stimulation. *Aviation, Space, and Environmental Medicine* 68 (2): 93–98.

28 Ooper, C., Dunbar, N., and Mira, M. (1997). Sex and seasickness on the Coral Sea. *The Lancet* 350 (9081): 892.

29 Munafo, J., Diedrick, M., and Stoffregen, T.A. (2017). The virtual reality head-mounted display Oculus Rift induces motion sickness and is sexist in its effects. *Experimental Brain Research* 235: 889–901.

30 Bartley, E.J. and Fillingim, R.B. (2013). Sex differences in pain: a brief review of clinical and experimental findings. *British Journal of Anaesthesia* 111 (1): 52–58.

31 Nazaré, L.d., Soares, M., Silva, G. et al. (2014). Comparison of pain threshold and duration of pain perception in men and women of different ages. *Fisioterapia em Movimento* 27 (1).

32 Entertainment Software Association (2006). *Essential Facts About the Computer and Video Game Industry 2006*. Entertainment Software Association.

33 Entertainment Software Association (2022). *Essential Facts About the Computer and Video Game Industry 2022*. Entertainment Software Association.

34 Rüßmann, M., Lorenz, M., Gerbert, P. et al. (2015). *The Future of Productivity and Growth in Manufacturing Industries*, 9. Boston Consulting Group.

35 Monostori, L. (2014). Cyber-physical production systems: roots, expectations and R&D challenges. *Procedia CIRP* 17: 9–13.

36 Colombo, A.W., Karnouskos, S., and Bangemann, T. (2014). Towards the next generation of industrial cyber-physical systems. In: *Industrial Cloud-Based Cyber-Physical Systems: The IMC-AESOP Approach*, 1–22. Springer.

37 Lee, J., Bagheri, B., and Kao, H.-A. (2015). A cyber-physical systems architecture for industry 4.0-based manufacturing systems. *Manufacturing Letters* 3: 18–23.

38 Moore, P.R., Ng, A.H., Yeo, S.H. et al. (2008). Advanced machine service support using Internet-enabled three-dimensional-based virtual engineering. *International Journal of Production Research* 46 (15): 4215–4235.

39 Fuchs, P. (2006). *Le traité de la réalité virtuelle*. Paris: Presses des MINES.

40 Jerald, J. (2015). *The VR Book: Human-Centered Design for Virtual Reality*. Morgan & Claypool.

41 Falcão, C.S. and Soares, M.M. (2013). Application of virtual reality technologies in consumer product usability, *Design, User Experience, and Usability. Web, Mobile, and Product Design: HCI International 2013*, Las Vegas, NV, USA, pp. 342–351.

42 Zimmermann, P. (2008). Virtual reality aided design. A survey of the use of VR in automotive industry. In: *Product Engineering: Tools and Methods Based on Virtual Reality*, 277–296. Springer.

43 Böcking, S. (2008). Suspension of disbelief. In: *The International Encyclopedia of Communication*.

44 Blake, E., Nunez, D., and Labuschagne, B. (2007). Longitudinal effects on presence: suspension of disbelief or distrust of naive belief? *In Proceedings of the 10th Annual International Workshop on Presence*, pp. 291–295.

45 Geslin, E. (2013). *Processus d'induction d'émotions dans les environnements virtuels et le jeu vidéo*. Paris: Ecole nationale supérieure d'arts et métiers-ENSAM.

46 Lombard, M.T.D. (1997). At the heart of it all: The concept of presence. *Journal of Computer-Mediated Communication* 3 (2): JCMC321.

47 Gorini, A., Capideville, C.S., De Leo, G. et al. (2011). The role of immersion and narrative in mediated presence: the virtual hospital experience. *Cyberpsychology, Behavior and Social Networking* 14 (3): 99–105.

48 Slater, M. (2009). Place illusion and plausibility can lead to realistic behaviour in immersive virtual environments. *Philosophical Transactions of the Royal Society, B: Biological Sciences* 1, 364 (1535): 3549–3557.

49 Slater, M., Banakou, D., Beacco Porres, A. et al. (2022). A separate reality: an update on place illusion and plausibility in virtual reality. *Frontiers in Virtual Reality* 3: 914392.

50 Gilbert, D.T. (1991). How mental systems believe. *American Psychologist* 46 (2): 107.

51 Brogni, A., Slater, M., and Steed, A. (2003). More breaks less presence. *Presence* 1–4.

52 Crowley, J.S. (1987). Simulator sickness: a problem for Army aviation. *Aviation, Space, and Environmental Medicine* 58 (4): 355–357.

53 Enrique, D.V., Druczkoski, J.C.M., Lima, T.M., and Charrua-Santos, F. (2021). Advantages and difficulties of implementing Industry 4.0 technologies for labor flexibility. *Procedia Computer Science* 181: 347–352.

54 Kolasinski, E.M. (1995). Simulator sickness in virtual environments. *U.S Army Research Institute, Technical Report* 1027: 7–23.

55 LaViola, J. and Joseph, J. (2000). A discussion of cybersickness in virtual environments. *ACM SIGCHI Bulletin* 32 (1): 47–56.

56 Kourtesis, P., Linnell, J., Amir, R. et al. (2023). Cybersickness in virtual reality questionnaire (CSQ-VR): A validation and comparison against SSQ and VRSQ. *In Virtual Worlds* 2 (1): 16–35.

57 Stanney, K.M., Kennedy, R.S., and Drexler, J.M. (1997). Cybersickness is not simulator sickness. *Proceedings of the Human Factors and Ergonomics Society Annual Meeting* 41 (2): 334–338.

58 Sherman, C.R. (2002). Motion sickness: review of causes and preventive strategies. *Journal of Travel Medicine* 9 (5): 251–256.

59 C.-L. Liu S.-T. Uang, "A study of sickness induced within a 3D virtual store and combated with fuzzy control in the elderly," *9th International Conference on Fuzzy Systems and Knowledge Discovery*, 2012.

60 So, R.H., Ho, A., and Lo, W.T. (2001). A metric to quantify virtual scene movement for the study of cybersickness: definition, implementation, and verification. *Presence* 10 (2): 193–215.

61 Chang, E., Kim, H.T., and Yoo, B. (2020). Virtual reality sickness: a review of causes and measurements. *International Journal of Human Computer Interaction* 36 (17): 1658–1682.

62 Edmondson, D. (2005). Likert scales: a history. *Proceedings of the Conference on Historical Analysis and Research in Marketing,* vol. 12. 127–133.

63 Fleeson, W. (2001). Toward a structure-and process-integrated view of personality. *Journal of Personality and Social Psychology* 80 (6): 1011.

64 Gianaros, P.J., Muth, E.R., Mordkoff, J.T. et al. (2001). A questionnaire for the assessment of the multiple dimensions of motion sickness. *Aviation, Space, and Environmental Medicine* 72 (2): 115.

65 Keshavarz, B. and Hecht, H. (2011). Validating an efficient method to quantify motion sickness. *Human Factors* 53 (4): 415–426.

66 Bos, J.E., de Vries, S.C., van Emmerik, M.L. et al. (2010). The effect of internal and external fields of view on visually induced motion sickness. *Applied Ergonomics* 41 (4): 516–521.

67 Kennedy, R.S. (1993). Simulator sickness questionnaire: An enhanced method for quantifying simulator sickness. *The International Journal of Aviation Psychology* 3 (3): 203–220.

68 Balk, S., Bertola, D., and Inman, V. (2013). Simulator sickness questionnaire: twenty years later. In: *Driving Assessment Conference*, vol. 7. University of Iowa.

69 Farmer, A.D., Ban, V.F., Ceon, S.J. et al. (2015). Visually induced nausea causes characteristic changes in cerebral, autonomic and endocrine function in humans. *The Journal of Physiology* 593 (5): 1183–1196.

70 Harvey, C. and Howarth, P.A. (2007). The effect of display size on visually induced motion sickness (VIMS) and skin temperature. *Proceedings of the 1st International Symposium on Visually Induced Motion Sickness, Fatigue, and Photosensitive Epileptic Seizures*, Hong Kong, pp. 96–103.

71 Dennison, M.S., Wisti, A.Z., and D'Zmura, M. (2016). Use of physiological signals to predict cybersickness. *Displays* 44: 42–52.

72 Kim, Y.Y., Kim, E.N., Park, M.J. et al. (2008). The application of biosignal feedback for reducing cybersickness from exposure to a virtual environment. *Presence Teleoperators and Virtual Environments* 17 (1): 1–16.

73 Chen, Y.-C., Duann, J.R., Chuang, S.W. et al. (2010). Spatial and temporal EEG dynamics of motion sickness. *NeuroImage* 49 (3): 2862–2870.

74 T. Kiryu, Uchiyama E, Tada G et al. (2007). A time-varying factors model for interpreting visually induced motion sickness. *Proceedings of VIMS2007*, pp. 69–76.

75 Palmisano, S., Arcioni, B., and Stapley, P.J. (2018). Predicting vection and visually induced motion sickness based on spontaneous postural activity. *Experimental Brain Research* 236: 315–329.

76 Apthorp, D., Nagle, F., and Palmisano, S. (2014). Chaos in balance: non-linear measures of postural control predict individual variations in visual illusions of motion. *PLoS One* 9 (12): e113897.

77 Somrak, A., Humar, I., Hossain, M.S. et al. (2019). Estimating VR Sickness and user experience using different HMD technologies: an evaluation study. *Future Generation Computer Systems* 94: 302–316.

78 Saredakis, D., Szpak, A., Birckhead, B. et al. (2020). Factors associated with virtual reality sickness in head-mounted displays: a systematic review and meta-analysis. *Frontiers in Human Neuroscience* 14: 96.

79 Strojny, P. and Strojny, A. (2018). Can simulator sickness be avoided? A review on temporal aspects of simulator sickness. *Frontiers in Psychology*.

80 Stanney, K.M., Hale, K.S., Nahmens, I. et al. (2003). What to expect from immersive virtual environment exposure: influences of gender, body mass index, and past experience. *Human Factors* 45 (3): 504–520.

81 Moss, J.D. and Muth, E.R. (2011). Characteristics of head-mounted displays and their effects on simulator sickness. *Human Factors* 53 (3): 308–319.

82 Kim, H.G., Lim, H.T., Lee, S. et al. (2018). VRSA net: VR sickness assessment considering exceptional motion for 360 VR video. *IEEE Transactions on Image Processing* 28 (4): 1646–1660.

83 Pettijohn, K.A., Geyer, D., Gomez, J. et al. (2018). Postural instability and simulator seasickness. *Aerospace Medicine and Human Performance* 89 (7): 634–641.

84 Gavgani, A.M., Nesbitt, K.V., Blackmore, K.L. et al. (2017). Profiling subjective symptoms and autonomic changes associated with cybersickness. *Autonomic Neuroscience* 203: 41–50.

85 Porter, J. and Robb, A. (2019). An analysis of longitudinal trends in consumer thoughts on presence and simulator sickness in VR games. *Proceedings of the Annual Symposium on Computer-Human Interaction in Play*.

86 Caserman, P., Garcia-Agundez, A., Gámez Zerban, A. et al. (2021). Cybersickness in current-generation virtual reality head-mounted displays: systematic review and outlook. *Virtual Reality* 25 (4): 1153–1170.

87 Teixeira, J. and Palmisano, S. (2021). Effects of dynamic field-of-view restriction on cybersickness and presence in HMD-based virtual reality. *Virtual Reality* 25 (2): 433–445.

88 Bala, P., Dionísio, D., Nisi, V. et al. (2018). Visually induced motion sickness in 360° videos: Comparing and combining visual optimization techniques. In: *2018 IEEE International Symposium on Mixed and Augmented Reality Adjunct (ISMAR-Adjunct)*, 244–249. IEEE.

89 Bala, P., Oakley, I., Nisi, V. et al. (2021). Dynamic field of view restriction in 360 videos: Aligning optical flow and visual slam to mitigate vims. In: *Proceedings of the 2021 CHI Conference on Human Factors in Computing Systems*, 1–18.

90 Naqvi, S.A.A., Badruddin, N., Jatoi, M.A. et al. (2015). EEG based time and frequency dynamics analysis of visually induced motion sickness (VIMS). *Australasian Physical & Engineering Sciences in Medicine* 38: 721–729.

91 Stauffert, J.-P., Korwisi, K., Niebling, F. et al. (2021). Ka-Boom!!! Visually exploring latency measurements for XR. In: *Extended Abstracts of the 2021 CHI Conference on Human Factors in Computing Systems*, 1–9.

92 Caserman, P., Martinussen, M., and Göbel, S. (2019). Effects of end-to-end latency on user experience and performance in immersive virtual reality applications. In: *Entertainment Computing and Serious Games: First IFIP TC 14 Joint International Conference, ICEC-JCSG 2019, Arequipa, Peru, November 11–15, 2019, Proceedings 1*, 57–69. *Springer International Publishing*.

93 Bos, J.E., Bles, W., and Groen, E.L. (2008). A theory on visually induced motion sickness. *Displays* 29 (2): 47–57.

94 Wang, J., Shi, R., Zheng, W. et al. (2023). Effect of frame rate on user experience, performance, and simulator sickness in virtual reality. *IEEE Transactions on Visualization and Computer Graphics* 29 (5): 2478–2488.

95 Nürnberger, M., Klingner, C., Witte, O.W. et al. (2021). Mismatch of visual-vestibular information in virtual reality: is motion sickness part of the brains attempt to reduce the prediction error? *Frontiers in Human Neuroscience* 15: 757735.

96 Davis, J., Hsieh, Y.-H., and Lee, H.-C. (2015). Humans perceive flicker artifacts at 500 Hz. *Scientific Reports* 5 (1): 7861.

97 Evereklioğlu, C., Doğanay, S., Er, H., and Gündüz, A. (1999). Distant and near interpupillary distance in male and female.

98 Fulvio, J. and Rokers, B. (2018). Sensitivity to sensory cues predicts motion sickness in virtual reality. *Journal of Vision* 18 (10): 1066–1066.

99 Ruddle, R.A., Volkova, E., and Bülthoff, H.H. (2013). Learning to walk in virtual reality. *ACM Transactions on Applied Perception (TAP)* 10 (2): 1–17.

100 Nilsson, N.C., Serafin, S., and Nordahl, R. (2016). Walking in place through virtual worlds. In: *Human-Computer Interaction. Interaction Platforms and Techniques: 18th International Conference, HCI International 2016, Toronto, ON, Canada*, Toronto, July 17-22, 2016. Proceedings, Part II 18. Springer International Publishing.

101 Nilsson, N.C., Serafin, S., Steinicke, F. et al. (2018). Natural walking in virtual reality: a review. *Computers in Entertainment (CIE)* 16 (2): 1–22.

102 Wickens, C.D. and Baker, P. (1995). Cognitive issues in virtual reality.

103 Sutherland, I.E. (1965). The ultimate display. *Proceedings of the IFIP Congress* 2: 506–508.

104 Schnabel, M.A. and Kvan, T. (2001). 3D Maze: Getting Lost in Virtual Reality. In 5th Iberoamerican congress of digital graphic SIGRADI 2001, *Conference Proceedings, Conception, Chile*: 145–147.

105 Suma, E., Lipps, Z., Finkelstein, S. et al. (2012). Impossible spaces: maximizing natural walking in virtual environments with self-overlapping architecture. *IEEE Transactions on Visualization and Computer Graphics* 18 (4): 555–564.

106 Morganti, F. (2009). A virtual reality-based tool for the assessment of "survey to route" spatial organization ability in elderly population: preliminary data. *Cognitive Processing* 10: 257–259.

107 Martínez-Reyes, F. and Hern'ndez-Santana, I. (2012). The virtual maze: a game to promote social interaction between children. In: *Eighth International Conference on Intelligent Environments*. IEEE, pp. 331–334.

108 Ubisoft (2020). When physical and virtual worlds blend. https://www.ubisoft.com/en-us/game/vr-maze (Accès le 2023).

109 RAMROD (2021). UTAV - Urban Training Assisted by Virtual Reality, Ramrod XR. https://www.ramrodxr.de/en/utav (Accès le 2023).

110 Loup, G. and Loup-Escande, E. (2019). Effects of travel modes on performances and user comfort: a comparison between Arm Swinger and teleporting. *International Journal of Human Computer Interaction* 35 (14): 1270–1278.

111 Clifton, J. and Palmisano, S. (2020). Effects of steering locomotion and teleporting on cybersickness and presence in HMD-based virtual reality. *Virtual Reality* 24 (3): 453–468.

112 Dichgans, J. and Brandt, T. (1978). Visual-vestibular interaction: Effects on self-motion perception and postural control. In: *Perception*, 755–804. Berlin, Heidelberg: Springer.

113 Salomoni, P., Prandi, C., Roccetti, M. et al. (2016). Assessing the efficacy of a diegetic game interface with Oculus Rift. In: *13th IEEE Annual Consumer Communications & Networking Conference (CCNC)*. IEEE, pp. 387–392.

114 Hartson, R. (2003). Cognitive, physical, sensory, and functional affordances in interaction design. *Behavior & Information Technology* 22 (5): 315–338.

115 Lacovides, I., Cox, A., Kennedy, R. et al. (2015). Removing the HUD: the impact of non-diegetic game elements and expertise on player involvement. In: *Proceedings of the 2015 Annual Symposium on Computer-Human Interaction in Play*, 13–22.

116 Ashar Imtiaz, M. (2020). The effects of different types of HUDs on Cybersickness-Effects of Diegetic and Non-Diegetic displays on Cybersickness in Virtual Reality, Gothenburg, Sweden.

117 Dennison, M. and D'Zmura, M. (2018). Effects of unexpected visual motion on postural sway and motion sickness. *Applied Ergonomics* 71: 9–16.

118 So, R.H.Y. and Lo, W.T. (1999). Cybersickness: an experimental study to isolate the effects of rotational scene oscillations. *Proceedings of the IEEE Virtual Reality* 237–241.

119 So, R.H.Y., Lo, W.T., and Ho, A.T.K. (2001). Effects of navigation speed on motion sickness caused by an immersive virtual environment. *Human Factors* 43 (3): 452–461.

120 Dong, X. and Stoffregen, T.A. (2010). Postural activity and motion sickness among drivers and passengers in a console video game. *Proceedings of the Human Factors and Ergonomics Society Annual Meeting* 54 (18): 1340–1344.

121 Jaeger, B.K. and Mourant, R.R. (2001). Comparison of simulator sickness using static and dynamic walking simulators. *Proceedings of the Human Factors and Ergonomics Society Annual Meeting* 45 (27): 1896–1900.

122 Adhanom,I.B. (2021). Exploring Adaptation-Based Techniques to Create Comfortable Virtual Reality Experiences. Diss., University of Nevada, Reno.

123 Freitag, S., Weyers, B., and Kuhlen, T.W. (2016). Examining rotation gain in CAVE-like virtual environments. *IEEE Transactions on Visualization and Computer Graphics* 22 (4): 1462–1471.

124 Teasdale, N., Lavallière, M., Tremblay, M. et al. (2009). Multiple exposition to a driving simulator reduces simulator symptoms for elderly drivers. In: *Driving Assessment Conference*, vol. 5. University of Iowa, pp. 169–175.

15

Smart Cities Using Digital Twins and Industrial IoT Based Technologies: Tools and Products from the Industry Sector

Aravinda Koithyar¹, Asha Venkataramana², Balasubramanian Prabhu Kavin³, and Gan Hong Seng⁴

¹*Department of Electronics and Communication, New Horizon College of Engineering, Bellandur Post, Bengaluru, India*
²*Department of Master of Computer Applications, New Horizon College of Engineering, Bellandur Post, Bengaluru, India*
³*Department of Data Science and Business Systems, SRM Institute of Science and Technology, Kattankulathur, Chengalpattu, Tamil Nadu, India*
⁴*School of AI and Advanced Computing, XJTLU Entrepreneur College (Taicang), Xi'an Jiaotong – Liverpool University, Suzhou, Jiangsu, P.R. 215400, China*

15.1 Introduction

Despite being envisaged decades ago, digital twin (DT) has gained favor in academic and industry circles lately. Michael Grieves, who first proposed the notion of DT in 2002, said, "At its best, a product's digital twin may provide the same insights as checking the physical version of the product" [1]. By this definition, there are three parts to a DT:

(1) Physical twin: Something existing in the physical or metaphysical world, such as a human being, animal, plant, machine, process, or organization.
(2) DT: The digital duplicate of a physical object that may act just like its analogue in the real world.
(3) Linking mechanism: The automated, bidirectional, real-time data transfer between the two systems.

Academics and businesses may disagree on defining DT, but they cannot disagree on its value. It enhances decision-making, maintenance schedules and operations, remote access, the working/operating environment, safety, and sustainability, while decreasing operational costs and saving time [2]. The usage of DTs across industries has increased in recent years because of their many uses and benefits. The worldwide DT market was valued at US$5.04 billion in 2020

Digital Twins in Industrial Production and Smart Manufacturing:
An Understanding of Principles, Enhancers, and Obstacles, First Edition. Edited by Rajesh Kumar Dhanaraj, Balamurugan Balusamy, Prithi Samuel, Ali Kashif Bashir, and Seifedine Kadry.

and is predicted to reach US$86.09 billion by 2028, expanding at a CAGR of 42.7% between 2021 and 2028 [3]. The recent pandemic epidemic of COVID-19 is one cause for the increased need for DTs. As a result of the pandemic's lockdowns, supply chains were disrupted, workers were laid off, and many businesses moved their operations online or away from their employees. This increased the importance of digitalization and the development of systems requiring minimal human touch [4].

The widespread use of cutting-edge digital technology has allowed cities to improve their infrastructures and services while creating citizen-friendly digital services. These two examples of cutting-edge digital technology cities want to use to improve everyday operations [5]. Cities also seek to improve city security and general monitoring and administration by investing in digital technology. In addition, cities hope to synchronize city operations and entice individuals to engage in urban development projects and activities using cutting-edge digital technology. To that end, cities are taking steps to reduce bureaucracy and increase openness in their operations [6].

Cities have a pivotal role in accepting and implementing sustainability policies, especially in light of climate change and its properties on the environment. Over 75% of the energy and material movements and ingesting occur in urban areas, according to [7]. Understanding of various natural and nonrenewable material fluxes in the metropolitan region, including but not limited to water, energy, food, and trash) is crucial to creating more sustainable urban environments [8]. Smart cities can improve the design and planning of urban metabolism by using technologies like dynamic DTs to test out potential solutions in a simulated setting before committing to them for real. Thus, innovative digital technologies help cities optimize urban metabolism and identify strategies for meeting their climate change and sustainability targets [9].

The IoT has become an integral part of our few short years, largely thanks to the pioneering efforts of the research community with networks. Cyber-physical systems (CPSs) are a key component of smart manufacturing and Industry 4.0 [10]. A new, groundbreaking concept called the DT has emerged as a result of these advancements and the seismic shift toward the integration of machine learning (ML), deep learning (DL), and big data from all things digital, with the aim of increasing the breadth of this sum of applications that make use of it. With CPS as a foundational concept of IT and OT [11], the DT has developed their lifecycle keep the physical and virtual parts permanently connected. DTs may be considered CPS in their own right [12] due to their emphasis on data exchange and integrating physical and digital parts across a wide range of areas. There is no doubt that various development environments suitable for deploying DT-based solutions are available. The platforms make possible city-scale DT scenarios, the most popular of which is urban planning.

The technology provides users and stakeholders with actionable insight into the inner workings of any scheme, the interplay between its many components, and the future behavior of its physical counterpart. Smart cities, metropolitan areas, medicine, engineering, and the automotive sector are just a few examples of the fields that have seen an uptick in DT study and application [13]. This study aims to give a comprehensive picture of the technical problems, constraints, and revision of applications by analyzing the DT concept across IIoT and smart levels. The advantages, future research agenda, and application factors of DT technology are discussed in addition to its technical aspects. In this study, we provide a literature review of many key papers that discuss cutting-edge work in the field of DT. The most important findings from this survey are discussed below:

(1) It provides an impression of the history and theory of DT. It also presents fundamental methods for implementing background.
(2) It emphasizes the current requests for DTs in business, revealing that Population Health Management (PHM) is where DTs are used most, that modeling and simulation are the most significant steps in creating a DT, and that data fusion and transfer are essential.
(3) It discusses the present uses of DTs in smart cities and concludes that these applications are most prevalent in transportation, urban health management, and security.

The respite of the paper is prearranged as shadows: Section 15.2 offers the basic introduction to DT and Section 15.3 presents the DT in industrial IoT. The introduction of DT in smart cities is mentioned in Section 15.4. The application of DT in industrial applications is provided in Section 15.5. A brief discussion of the literature review is mentioned in Section 15.6. The discussion of related works and their challenges is given in Section 15.7. The future direction with open challenges is shown in Section 15.8. In conclusion, the summary of the whole work is as long as in Section 15.9.

15.2 Background Knowledge of Digital Twin

DTs are, at their most fundamental, accurate simulations of real-world objects (Figure 15.1). This material may be a car, a wind turbine, or a factory [14]. Even a whole city like Singapore may be complicated [15]. Understanding DTs requires looking back at where the idea came from and where it has come to be now.

15.2.1 The First DT

In his MirrorWorlds, published in 1993, David Gelernter discussed the potential for software models to reflect a slice of reality. Yet even before that, NASA was among the first to employ intricate spacecraft simulations. During the Apollo 13 mission in 1970, an oxygen tank explosion severely damaged the spacecraft's

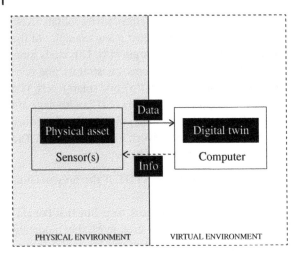

Figure 15.1 A symbolic illustration of the connection between DT knowledge and a physical object.

primary engine and forced it off course by nearly 400 mi/h. The crew's supply was steadily seeping into space, worsening a bad situation. To aid damaged ship. Using a digital duplicate in this way was likely one of the earliest examples of its kind

It is worth noting, though, that the notion of a DT did not exist until the year was 1970. Nonetheless, it was clear that many of the hallmarks of a DT were present in this particular case. For instance, the simulators could detect the spacecraft's actual state and adapt accordingly. Moreover, it assisted the team in addressing what-if circumstances that were not part of the original concept.

15.2.2 Key Properties

As was previously mentioned, DTs are digital copies of actual objects. Let us take a deeper dive into this term and examine some of the crucial topographies of a DT.

First, the twin requires variables to accurately represent the asset and mimic its behavior. To do this, various sensors with specialized data transmission and reception capabilities through the public Internet or a private, encrypted network are required.

Second, the twin must be capable of receiving, storing, and processing massive amounts of data in near-real time. This calls for a massive quantity of data storage, processing power, and computer resources. That is why it has to leverage cutting-edge resources.

Third, the twin needs to understand the massive amounts of data being constantly sent. Since this is beyond the computational capabilities of most humans, it is necessary to employ AI algorithms to sort through data and determine what is actually valuable. This also includes employing AI-powered recommendation and action engines.

In the fourth place, the twin should be able to increase the physical asset's performance by learning from various cause-effect scenarios over time. This necessitates the execution of several what-if simulations, test cases, and alternative scenarios. This amount of intricacy is, once again, incomprehensible to humans. In order for ML algorithms to learn, explore, and evolve the best potential path of action, they must be taught in a controlled environment.

Finally, this information must be accessible to human decision-makers through a digital user interface that encourages interaction. Computers, tablets, and smartphones are all examples of devices that fit this bill (Figure 15.2).

The physical asset, the digital representation and the human decision-makers are the three most important twin implementations. More than that, though, these three things can be in any number of possible states, connected by various communication flows. Most of them will change from job to job based on how comprehensive the DT is expected to be. Table 15.1 summarizes common terminology used in most DT implementations.

Figure 15.2 The flow of information among physical assets.

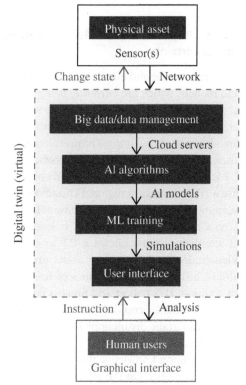

Table 15.1 A summary of key terms used in applying DT.

Description	Key terms
The setting in which the asset is physically located. Only sometimes simple to get to	Physical situation
The situation in which the digital twin's virtual counterpart resides. Simple to gain access to	Virtual simulation
There are several distinct possibilities for states, each representing a change in the physical asset	Sensory states
It is altering the state of an object or its digital duplicate through various transitions	Changes in state
Bringing the state of the physical thing and its digital twin into sync with one another	Twinning
The rate at which synchronization takes place in this process. as near as possible to the real moment	Twinning rate
A number of processes that can result in the asset or twin being in a different state	System processes

15.2.3 Beyond Computer Representations and Dealing with Uncertainty

Costs may be cut and efficiency can be increased by using computer simulations and models, which are currently being used in many fields. DTs can be thought of as computer simulations. Yet there are a number of distinctions. The primary distinction is that computer models are constructed to investigate and foretell a broad variety of situations. The prevalence of coccidiosis, for instance, among farm animals in a certain area may be predicted using a computer model. However, a DT is only the digital counterpart of one physical item. That is to say, you cannot use a DT to forecast a regional coccidiosis outbreak. Instead, it can only assist in keeping tabs on Stacy, the dairy cow at farm 1073 in Kentucky, and her vital signs.

As a result, the focus of a DT is incomplete to a single asset. This narrowing of attention is what allows a DT to excel where other computer models fail. Vicissitudes to the physical assets are reproduced in the digital counterpart almost instantly, with lag times of few milliseconds to a few minutes at most. In addition, DTs collect and process far more data than conventional computer simulations. As a corollary, DTs can create more plausible "what-if" scenarios. These important distinctions (Table 15.2) were discussed in Dr. Smith's latest online resource [16].

It is common knowledge that computer models contain holes in their comprehension of reality. On the other hand, having a comprehensive grasp of the data

Table 15.2 Dr. Matthew Indra acronym.

A DT needs to be	
It must represent an exact thing, e.g., "rather" than a general cow.	Individual
This also implies that the DT should be "always on," meaning that it should be accessible for the same amount of time as its counterpart in the physical world.	Near real-time
A digital measurement of the object that exists in the actual world, such as a soil moisture meter or a frequent satellite observation, must be used in order to bring it up to date.	Data informed
The twin must be an adequately realistic replacement for the real-world thing.	Realistic
The real-world counterpart's input must be capable of inspiring some kind of action.	Actionable

that underpins them helps the designers of the models to foresee and correct any incorrect outcomes that may be generated by the models. They are able to achieve this by utilizing a number of error-correcting algorithms to cut down on the mistakes that are caused by making incorrect environment. In contrast, a DT has the capacity to gather any data for which a sensor is available as the scenario unfolds. As a result, the degrees of uncertainty are significantly reduced when using a DT.

Additionally, in contrast to the majority of generic computer representations, making efficient use of a DT does not require an awareness of all of the practical aspects that went into its production. This is a significant advantage. Users might instead concentrate all of their efforts on gaining a sympathetic of how the DT responds when subjected to certain situations. This is quite similar to what they would do in the real world with any asset, and practically instant response is provided here. When compared to standard computer models, this is a significant benefit.

Several research studies and articles from the business world have previously brought to light the advantages of utilizing Jones et al. conducted a comprehensive literature search and compared the perceived assistances of employing DT to the various articles that detailed it (Table 15.3) [17].

15.2.4 Types of DT

- **ADT instance (DT):** There is a continuous monitoring of the condition of the physical twin, evolution have an influence on the DT. A DT case is characterized as a form of DT that reflects all of its lifespan. In this sense, this idea follows a product or process from the moment it is conceived all the way through its

Table 15.3 Describing the DT: a methodical literature review.

Perceived benefits of digital twins—from characterizing the digital twin research	
Reduces costs	[18]
Reduces risks	[19]
Reduces complexity	[20]
Improves after-sales service	[21]
Improves efficiency	[22]
Improves maintenance decisions	[23]
Improves security	[24]
Improves safety and reliability	[25]
Improves manufacturing processes	[26]
Enhances flexibility and competitiveness	[27]
Fosters innovation	[18]

lifetime, all the while anticipating the behavior of the product or process. It is helpful to validate a product or object's expected behavior and presentation before using the product or using the object.

- **DT prototype (DTP)**: A DT prototype learns and remembers important details about the physical twin that may be used during the product's manufacturing and production processes. CADs, BOMs, and drawings are all examples of data that might be used to connect the manufacturing process to the various parties involved in the production chain [2]. In line with DT features, the DTP can replicate production environments and do validation, evaluation, and quality control tests in advance of real production. By highlighting potential problems with the physical twin before manufacturing begins, this method successfully cuts down on production costs and operating time. In this context, DTPs can also be referred to as experiment table DTs, where a virtual prototype is made available with progressively higher levels of detail as described in [29], and where virtual test consequences provide a sufficiently reliable declaration quality of the project while reducing the sum of typically luxurious hardware prototypes.
- **Performance DT (PDT)**: The PDT can track, gather, and evaluate product data in settings that are more like their physical siblings' in terms of realism and unpredictability. In order to optimize the design, produce maintenance strategies, and draw inferences from a product's performance, the PDT aggregates smart capabilities in order to interpret counterpart [30].

15.2.5   Integration Levels

According to the hierarchy given in [30], DTs are the most combined technology, followed by digital models, which are the least combined.

- **Digital model**: As currently conceived, the digital paradigm will not provide seamless, automated data transfer among the real and virtual worlds, which has no automatic link among any changes must be reflected by making appropriate adjustments in both.
- **Digital shadow**: The digital shadow will have mechanical transfer between the real and virtual worlds. This is most accurately shown by a system in which sensors collect data from the physical perfect and feed signals into the digital counterpart. The incorporation level can be calculated as a digital shadow if data is sent in a polling or interrupted fashion.
- **Digital twin**: An identical twin with seamless bidirectional interaction between the virtual and actual worlds. This means that data is continuously sent across universes without any human intervention. In this scenario, data from the digital realm may be used to modify the actual model or command the actuators to carry out the desired action. On the flip side, data automatically impact the virtual twin such that the virtual twin precisely depicts the present state and future trajectory of the actual counterpart.

15.3   DT in Industrial IoT

DTs were originally mentioned in a 2016 patent filing by General Electric (GE). Based on the ideas presented in the patent, the "Predix" platform was created as a means of producing DTs. Predix is used for data analytics. GE has recently reduced its DT ambitions in favor of returning to their roots as an industrial international rather than a software firm. For its part, Siemens has created the "MindSphere" platform, which is a scheme that links substructure to a DT and so fully embraces the Industrial 4.0 idea. With the help of the billions of linked gadgets and data streams, it aims to revolutionize enterprises and supply them with DT solutions.

The "ThingWorx" platform is an alternate option for creating DTs and AI technologies. PTC's Industrial Innovation Platform is designed to collect data from the IoT and display it to users in an accessible, role-based interface. The platform not only provides the atmosphere for a DT solution but it also makes the development of data analytics easier.

IBM's "Watson IoT Platform" is advertised as a comprehensive IoT data tool for managing large-scale systems in real time using information gathered from millions of IoT devices. Additional cloud services are also available on the platform, all of which suggests that this may serve as the basis for a DT infrastructure [31].

There are two chief initiatives worth mentioning. First, Eclipse's "Ditto" [19] project is a ready-to-use platform for managing DT states, providing access and management for both physical and DTs. The platform plays a supporting function for other connected devices and makes it easier to set up and maintain DTs. Bentley Systems' [32] "imodel.js" is another open-source project that serves as a platform for developing, accessing, and generating DTs.

When used with the IoT, DTs may give global context across several dimensions. This implies it coordinates the different functions of an IoT device. The DT bridges the gap among the physical interface, making it extremely easy to deal with IoT resources. Simply said, a DT is an object's digital version that can communicate with its physical counterpart so that its many features and functions may be accessed and utilized. With DTs, there's no need to worry about consolidating your resources to ensure constant data connection in IoT that make requests to these geographically dispersed IoT resources in the cloud's secure sandbox to create the scenario where IoT and DT share the same workspace.

This sandbox method reduces vulnerabilities since cloud-based DTs do not necessitate moving the corresponding programs. Decreases in development costs have been seen, which bodes well for the rapid creation of IoT applications. As a consequence, DTs hosted in the cloud can pave the way for the outline of new IoT volumes and services that is mentioned as below:

15.3.1 Predict the Future

It is simple to foresee the lifespan of devices and changes in their functions thanks to the data coming from the environment and realized in real time by the IoT sensors in the DT.

15.3.2 Proliferation of Precision

Big data, which is generated by an increase in the number of interconnected devices and software, makes it simple for companies to establish channels of information exchange and gain a more comprehensive sympathetic of the state of their industrial equipment. When machine learning algorithms are combined with information about devices' past behavior, DTs can predict how an item will perform in a certain situation and, thus, prevent unexpected malfunctions.

15.3.3 Producing Intricate Digital Twins

The capabilities of DT allow software requests' real-time connection with remote devices for the purposes of retrieving data from or providing instructions to the device. Real-time data may not be sufficient for full device management, especially if historical and forecasted patterns of behavior are also needed to improve accuracy. The next degree of success may be reached through the twinning of behavioral functions with real-time DT.

15.3.4 Lessen Expenses

Companies always strive to maximize profits while decreasing expenses. Companies in the Internet of Things (IoT) industry, like those in other sectors, analyze their device deployment processes and experiment with novel, cost-effective ways to boost profit margins. Companies may use DTs to perform these tests without having to subject the final product to a wide range of physical changes in real time. Professionally speaking, retailers may save money by employing DT technology to make a digital replica of their shop and monitor customer behavior in relation to purchases through the use of sensors. In fact, DTs aid in keeping a safe distance from flaws while also distinguishing between levels of achievement.

15.3.5 Evading Failure

While avoiding mistakes and failures may seem like common sense in a real-world setting, it really requires significant investment from businesses. With the help of DTs, architects may explore a wide range of possibilities, increasing the confidence with which they can provide praise for an asset's longevity or consistency. Industries that are eager to improve uptime and boost manufacturing can speed things up by using DT.

By shifting from a focus on a particular resource to a more comprehensive resource population, businesses can unlock previously unrealized opportunities for growth. Different DTs of, say, an engine, transmission, and braking mechanism can exist, but they all need to communicate with one another in order to provide the same level of insight as the actual physical engine, transmission, and brake system.

15.4 DT in Smart Cities

Modern governments are launching smart city projects to provide uninterrupted access to all services for their citizens. Following the definition provided previously, DTs produce three-dimensional virtual illustrations of buildings, networks, and other real-world assets, complete with contextual data. It is the accountability of city governments to ensure the maintenance, energy use, space use, and traffic safety of their cities. DTs can help with all of these tasks. Many governments have expressed interest in constructing "smart cities" using DT in conjunction with intelligence.

With the development of technology, a little village in the emergence of Nomoko, a Zurich start-up, the Railways, and other corporations including Swisscom and AMAG are leading the charge to create smart cities in Switzerland. The business makes use of an app platform to create 3D city models. Using this 3D model, city administrators may better plan for and evaluate how to coordinate the many moving parts that make up urban life, from transportation to energy generation and distribution. DT will be used to create the first prototype of a smart city, and the city of Basel will serve as the inspiration for this project.

The new capital may be the first fully formed metropolis born with a DT. At the 2019 World Economic Forum's annual common session in Davos, the preliminary 3D unveiled; it was created by Cityzenith utilizing Smart software. Cityzenith CEO Michael Jansen has detailed how everything that happens in Amaravati would be anticipated in advance to improve results.

15.4.1 The Advantages of Having a DT for Smart Cities

In this context a gather information from sensors, rovers, and other IoT devices and then uses advanced analytics, machine insights into the asset's performance, operation, and profitability in real time. With the goal of addressing key public health, security, and environmental challenges, such knowledge is expected in smart cities across the world.

By linking the gap among the digital and real worlds, we can help shed more light on tough decisions, reduce risks, and increase participation from the general public. The Itron Idea Labs group is collaborating with Azure to use mixed realism to create a virtual depiction of the interconnections between the materials used in construction, the underlying infrastructure, and various types of sensors in a downtown Los Angeles area.

Los Angeles event planners and city experts will use a Microsoft software giant's DT innovation to simulate the installation of smart sensors, the modification of rooftop materials, the modification of activity designs, the planting of trees, and the subsequent effects on individuals, vehicles, schools, and buildings.

The DT will help the city respond to extreme weather events, as well as evaluate an infinite number of other potential situations that might arise in the future. A DT is extremely helpful in situations like the recent unexpectedly high rains and storms. It tells us which structures will be damaged, which charities will be forced to close, and which medical centers may be affected. It quickly shows us which areas are going to be impacted.

The project's creators make it clear that the DT smart city can collaborate with crisis responders to do re-enactments for any location and spot difficulties sooner and simpler.

15.4.1.1 Introducing DT Citizens

While DTs hold great promise for event planners and urban experts, increasing the number of businesses involved in smart city initiatives through the recruitment of new partners is essential to realizing those gains. Together, Cityzenith proposes a DT client ID for each citizen, which would serve as a single sign-on for all governmental information, notifications, services, and apps. Gaining traction in connecting with individuals and businesses may be facilitated by employing DT ID.

Andhra commissioner Sreedhar Cherukuri has said that the city of Amaravati may be a city constructed with the citizens' security as its top priority.

15.5 Industrial Requests of Digital Twins

This unit gives a brief impression of the twin applications explored in the literature. They are highly focused in the industries of creation design, production, and PHM. Figure 15.3 displays numerous examples of how DTs are used in manufacturing.

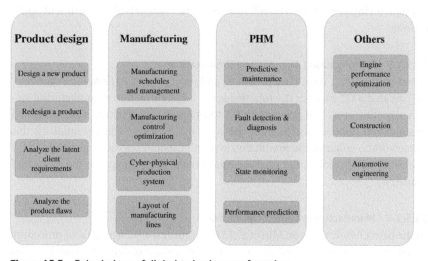

Figure 15.3 Submissions of digital twins in manufacturing.

15.5.1 Product Design

In product development, DTs have found widespread usage for anything from conceptualization to prototyping to analysis of hidden customer needs and product faults.

15.5.1.1 Design a Novel Product

Zheng and Lim [33] detailed a DT perfect that incorporated many apparatuses into the planning and execution of a product line's design and the establishment of its various stages of use. It was employed in the benchmarking and interaction procedures of the product family phase. During the operational phase, it was also applied to the redesign and configuration techniques.

15.5.1.2 Redesign a Creation

Tao et al. [34] created a DT perfect that might aid in the iterative product by connecting the physical model and virtual model.

15.5.1.3 Analyze the Product Flaws

At least six different types of data are incorporated into DT during the product design process to aid in the analysis of potential consumer requests and product defects.

15.5.1.4 Manufacturing

Based on CPS, the DT idea enables a more consistent, adaptable, and predictable manufacturing procedure. The following is a brief synopsis of the pertinent applications.

15.5.1.5 Manufacturing Agendas and Organization

Based on the CPS's resources, products, and process information, Biesinger et al. [35] created a DT idea for the robotic idea of a body-in-white manufacturing system. Agostino et al. [36] presented a DT technique for production scheduling and organization data accessible from the corporeal scheme. The potential for this framework to enhance the production system's capabilities was evaluated using a case study of a business that produces mechanical components for the motorized industry.

15.5.1.6 Manufacturing Control Optimization

In the petrochemical sector, Min et al. [37] introduced a DT agenda and procedure based on the IoT and AI, and they built a realistic cycle of data exchange between the actual DT perfect for optimizing industrial control.

15.5.1.7 Cyber-physical Production System

After evaluating a cyber-physical production system built on automation ML for blisk machining, Zhang et al. [38] first articulated a common cyber-physical manufacturing scheme. In order to guarantee the smooth running of the cyber-physical manufacturing system, Liu et al. [39] combined cyber-physical scheme, DT modeling, and simulation techniques.

15.5.1.8 Layout of Industrial Lines

Guo et al. [40] introduced a DT-enabled honeycomb industrial perfect to address challenges such as illogical line arrangement, uneven skill sets in the production process, inaccurate distribution, and insufficient brainpower in quality control measures.

15.6 Literature Review on Digital Twin Progress in Smart Cities

According to Ref. [41], DTs are the foundation of smart communities and places; smart city and urban planning are two fields where the adoption of the DT idea has recently increased. Smart DT cities are emerging as a new area of research and development, alongside digital replicas of entire neighborhoods for the purposes of maintenance and asset management of smart digital buildings. The goal of implementing DTs on a city-wide scale is to enhance the quality of life, mobility, and facilities available to residents, rather than to maximize economic efficiency. This work, for instance, draws on research [42] of China, the nation with the largest prevalence of smart city DT ideas being explored.

The cities of Bogotá, Helsinki, Brescia, Valencia, Irkutsk, Dublin, Herrenberg, Shenzhen, and Zurich, together with their respective apps, are featured in other publications. Web-based 3D visualization resolutions using joined data for communiqué and teamwork are being used in recent methods and efforts like Bentley's OpenCities and need extensive processing, which may be readily provided via a web border with a robust cloud service in the background. To link assets and surroundings as varied as factories, buildings, and stadiums to glean insights from previously separate gadgets and business schemes, the authors of [43] suggest Microsoft Azure DTs as one strategy. Less focus on the 3D modeling elements and more on operational optimization and refining seems to be the next step that Azure DT is taking. This includes real-time assets across numerous factories or cities across the world. Dassault Systems, on the other hand, are elevating the experience by combining VR with the 3D rendering of the scheme. In addition, the authors in [44] explain why a sophisticated geoportal is necessary for the display of 3D spatial Geographic information system (GIS) models to a wide audience. The authors of [45] discuss DT 3D demonstrating values like Geographic Data Files (GDFs), CityGML, OpenDRIVE, and OGC standards like cities.

There is a noteworthy void due to the fact that several industry leaders are moving toward the competence of such schemes powered by AI. Several methods have been adopted to integrate data from various sources and begin constructing DTs for real-world cities. This method creates a mountain of operational data, which complicates diagnosis and forecasting. Algorithm design, extraction, performance enhancement, and so on will all revolve on big data–based prognosis. Concerning the SRL for this kind of DT, the bottlenecks and a difficulty needing to be overcome are the interests of all players utility.

The authors of [46] examine the fact that modern cities generate data from a wide range of sources, including traffic, transit, electricity production, utility providing, water supply, and trash management, making them smarter overall. In recent years, "exciting testbeds for data mining and ML" have emerged as smart cities, elevating the importance of utilizing ML and DL methods. In Figure 6, the authors present their definition of a six-tiered DT smart city. Information about urban elements including buildings, infrastructure, and transportation is added in successive layers, beginning with the most fundamental (terrain) and culminating in the digital. The last layer, the DT, is automatically connected to the digital layer so that data may be used for simulation and in-depth urban research and studies with machine learning and deep learning. Connectivity among the DT and the city council enables the utilization of data on urban mobility and 3D models, gathered from residents as well as IoT data from sensors. Building catastrophes (flooding), and green space simulations are only some of the situations that may be simulated in this work's innovative and cutting-edge use of DT technology. But in [47], a more in-depth and sophisticated work on DT Earth project is laid bare. The European initiative aims to build. The goal of this data-driven endeavor is to learn more about the Earth system, which encompasses phenomena such as regional and worldwide variations in natural cycles, deep subterranean processes, and the interconnections between human culture and the rest of the natural world (see Figure 15.4).

In [48], the authors explain how DTs can be used. This is just one example of how DTs are being used in smart city applications. This research demonstrates the viability, where data is utilized to draw inferences about crime prevention, traffic consumption, and waste reduction. The difficulties of building simulations for complex dynamic settings, as well as the interpretation of such enormous streams of data, are discussed by the authors as obstacles to a practical implementation of the system. Consequently, progress in multi-paradigm simulation approaches is useful for meeting these obstacles.

Figure 15.4 Smart city cardinal twin layers of addition. Adapted from [47].

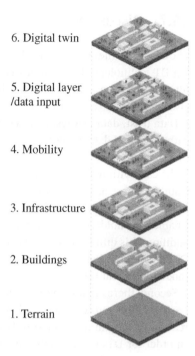

6. Digital twin

5. Digital layer /data input

4. Mobility

3. Infrastructure

2. Buildings

1. Terrain

As on-demand delivery services like Amazon Prime, Deliveroo, Uber Eats and so on continue to grow in popularity, more people will be making impromptu travels inside cities, creating a new and difficult modeling issue. In addition, the use of big data in urban area applications allows for the observation of lifestyle and behavioral trends and their influence on population growth, economic progress, building, and infrastructure. According to [50], which ranks the information datasets based on their importance to present and future applications in smart cities, the top five are as follows:

- Infrastructure data (including transportation, renewable energy, and manufacturing) account for 91%.
- Data from sensors (in homes and on streets) account for 88%.
- Data acquired by smart and linked networks in key utility services like water account for 86% of the Internet data generated by smart cities.
- 86% of the information came from social media platforms (including LinkedIn, Facebook, Twitter, and Pinterest).
- Search engines and other online resources (including YouTube and Google) accounted for the remaining 15%.

15.7   Discussion

The examined literature revealed that the most significant obstacles to implementing DT technology that is discussed in this section. These difficulties serve as a fitting conclusion to the research sub-question SQ1 and a response to it, and they also contribute to answering.

Data trust, data privacy, data security, data convergence and ascendency, data gathering, and data analysis are presents on a huge scale. Designers have a hard time recreating or modeling actions that have no logical basis in numbers. This is the case with regard to ecological stability, social justice, equality, and political stability. These advances in the social and conservation sectors will aim toward lower levels of Societal Readiness Level (SRL), where the possible impact on identifiable stakeholders, the broader society, and the situation is better understood. In addition, this difficulty is associated with levels 3 and 4, where the complexity of DT implementations is significantly hampered by the necessity of supplementing models with real-time.

Second, there are not enough rules and guidelines in place for DT applications. Lack of values and acknowledged interoperability, chiefly in industry, are the main reason for the limited request of DTs, which is explained by the authors of [51]. In order for DTs to be widely adopted, a solid grasp of their benefits, ideas, and structures, as well as the state of the art, must be adopted. In addition, researchers might potentially see the lower level of the Technology Readiness Level (TRL) in surveys and literature reviews, cumulative awareness of fundamental principles and ideas.

Third, the growing number of sensors and processing resources required leads to high implementation costs. Because DT applications are so costly, their use is forced by the resources, which is typically lacking in needy nations. As the number of sensors required rises, so does the difficulty of data transmission and processing, making it harder to get to the third stage of the maturity spectrum. This difficulty also hinders practitioners' ability to allow greater TRLs, such as the demonstration of pilot systems or the incorporation of DTs into a commercial design or full-scale placement.

Fourth, meeting long-term and massive data analysis needs using AI and big data. Big data techniques and the IoT technologies are potent partners that may lend significant assistance to effective DT employments evaluated in DT systems. Moreover, it is difficult to establish uniform norms and benchmarks due to the information that flows from different tiers of indicator systems. This challenge is aimed at maturity spectrum, and if successful, it might pave the way for features such as digital model and asset upkeep stumbling blocks in the path of the communication network. There is a pressing need to develop more rapid and effective communication protocols like 5G. To permit connectivity and working competence for the DT, the authors of [50] note an urgent need to use 5G technology for smart cities due to its many advantages.

15.8 Open Challenges for IIoT and Smart City Development

15.8.1 IoT/IIoT Challenges

IoT face the following difficulties:

15.8.1.1 Data, Confidentiality, and Trust
The proliferation of IoT gadgets in both domestic and commercial settings raises the difficulty of gathering massive capacities of data. The difficulty is in managing the influx of information and making it usable. The problem has grown more severe since the introduction of big data. Big, unorganized datasets are made much bigger by the Internet and organizing data analyses.

Due to the sensitive nature of data, there is a greater risk of it falling into criminal hands. When private consumer information is at stake, the risk to firms increases dramatically. Criminals are progressively resorting to cyber assaults to harm an organization's infrastructure by isolating and knocking out specific systems. There is a likelihood that cybercriminals would target businesses with thousands of linked IoT devices so that they might seize control of the operations. In the case of the Mirai botnet controversy, hackers infiltrated roughly 15 million IoT devices around the world and exploited them to execute a distributed denial of service (DDoS) assault. The proliferation of IoT raises the possibility of DDoSs. The disparity in value placed on privacy and security solutions also leaves us vulnerable to assault. Installing the devices with out-of-date security features and protection leaves a path open for hackers to get access to a wider, networked IoT ecosystem.

15.8.1.2 Infrastructure
The current IT infrastructure is inadequate because of the rapid expansion of IoT technology in comparison to the currently in use schemes. IoT development is aided by the update of aging infrastructure and the incorporation of new technologies. Modernized IoT infrastructure allows businesses to make use of cutting-edge tools and cloud-based services having to invest heavily in replacing their current infrastructure. Integrating legacy hardware into an IoT infrastructure is another area of difficulty. Retrofitting IoT sensors onto older equipment is one strategy for dealing with this problem; doing so helps ensure that data is not lost and that even dated devices may benefit from analytics.

15.8.1.3 Connectivity

Despite this increase in IoT deployments, connection issues continue to be problematic. These occur frequently when real-time monitoring is desired yet difficult to obtain. As the sum of sensors used in a single manufacturing procedure grows, so does the complexity of linking them all together.

This overarching is being hindered by difficulties with qualities like power faults and continual rollout failures. A single disconnected sensor can have a significant impact on a process's output. For instance, IoT devices are a potential source of data for AI procedures, which presents a difficulty in that the algorithms need all the data to function effectively, and any gaps in that data might have a negative impact on how the system functions. One way to guarantee that all data is gathered is to retrofit equipment and gather the data currently being provided by the machine. To guarantee complete connection and to allow the execution of generating substitute values for missing IoT sensor data.

15.8.1.4 Expectations

Organizations and end users struggle with AI's expectations because they do not know what to expect from IoT solutions or how to best use them. The fast expansion of IoT is encouraging because it shows that consumers and businesses alike are beginning to appreciate IoT for what it is and the ways in which a more interconnected world can improve our lives. The assumption that IoT may be used indefinitely without any preparation can be harmful, and its knock-on effects can exacerbate privacy and security worries as well as trust issues. The IoT requires expertise similar to that of AI to be fully utilized.

15.8.2 Open Trials for DT in Smart Cities

This part aims to deliver a high-level overview of this topic by first summarizing the open difficulties of DTs gleaned from the conversation of the request areas described in the preceding units.

15.8.2.1 Datasets and Data Sources

While discussions about markets for smart cities have been going on for some time, the data accessible to city-scale DTs still leaves much to be desired. The obtainability and quality of data is a recurrent subject of technology applications, as many of the recent technical developments upon which DTs have been developed rely on this. Further complicating matters is the fact that the ownership of urban databases is an essential consideration. In many cases, the challenge in designing and implementing a city-scale DT is not in identifying a compelling use case for such an endeavor, but rather in identifying the necessary data and establishing the necessary collaborations and links among multiple stakeholders to feed the DT.

There appears to be less contention over who owns data in particular industrial contexts, which may help to explain why smart manufacturing has been adopting DTs at a considerably faster clip than city-based applications over the past several years. Costs for data and so on may be incurred in urban settings, and data may be held by the city or by other parties in the city, such as universities and private enterprises executing municipal programs.

It is important to remember that DTs in urban areas may have quite different data needs than their counterparts in manufacturing facilities. While several kHz sample rates may be necessary in some industrial settings, several Hertz or even slower rates may be necessary for similar observations in urban areas. This, however, raises the possibility that such values may be generated by an extremely large sum of distributed mechanisms that are prone to error, are expected to run for long periods of time without human intervention or calibration, and so on. When we reflect that we use such data to provide a credible global image of a certain request area on a big scale, concerns of statistics quality and data upkeep become highly essential. Mechanisms for ML-based sensors and data imputation might be useful in this setting, paving the way for reasonably high-quality urban datasets.

15.8.2.2 Data Values and Interoperability

There is a pressing need for speeding up DT rollout, as well as interoperability among them and other existing schemes, such as smart city stages. The fact that the Industrial Digital Twin Association were founded in response to this need is further proof that the industry recognizes the importance of DTs. ISO/IEC JTC 1/SC 41 is also active in this area, showing that international standardization bodies are keen to address this unresolved issue in the DT field. To that end, DTC plans to compile a collection of reference implementations for DTs, which may speed up the process of fixing compatibility problems.

In a more tangible sense, standards influencing parts of the city DTs, such as transportation, are starting to be announced, like in the circumstance of the transportation data requirement. Since it is essential that many parties, typically from the sectors, must cooperate in order to gadget a DT at scale, Centre for Digital Built Britain (CDBB) is an instance of inventiveness to an assistance push onward for the use of DTs standards.

15.8.2.3 Brand Data Expressive and Nearby to City Investors

Following up on the prior open test, it is essential to keep the many organizations involved in city-scale DTs updated about the DT's capabilities and the outcomes it has created in order to maintain their support. Another unanswered question is how to make DT data accessible and more understandable so that its potential and whether or not it truly generates public benefit can be adequately assessed. The bulk of public user interfaces for such DTs so far have relied on 3D maps delivered via specialized programs or websites; in many cases, this component is already accessible from the DT itself as a digital replica of the physical situation. Some prototypes have expanded on this idea by providing virtual reality (VR) perspectives of the DT, such as the DT prototypes in Herrenberg [52] and Helsinki [53].

In many instances, however, we believe the difficulty of interacting with DT end users remains unsolved. The ability to effectively communicate tactic itself is not equally served in all use-case situations by a UI representation, but this is necessary for city DTs to recover ascendency and take well-informed decisions. In fact, we think the sheer size of the DT on its own can be scary and have antagonistic consequences in some circumstances. Given that what-if scenarios are central to the design and philosophy of digital twins (DT), inviting end-users to experiment with these scenarios in urban areas could significantly enhance DT user interfaces, helping local authorities.

15.8.2.4   Privacy, Security, and Ethics

Privacy, along with security and related ethical problems, is an essential component of city-scale DTs. This factor was already essential for smart cities, so it is not shocking that this is an unresolved problem for DTs on a citywide scale. Frameworks like GDPR in Europe provide a strong regulation foundation, upon whose guidelines more specific privacy-protecting implementations can be built.

Responsibility and transparency are central to discussions on the ethics of DTS at the local level. The public benefit should be clear given the widespread participation of communities and stakeholders in the operation of such a DT. It might be met with indifference or anger from these users if they are unsure of its advantages or whether it respects their privacy and security. The DTC's activities can further accelerate DT deployment due to the inherent complexity of privacy-related design and operation challenges in large, complex systems.

15.8.2.5   Cost and Benefits of DT's Application

The hidden costs of deploying a DT across several use cases at scale is an issue that is rarely mentioned in the context of this discussion. Possible causes include a lack of appropriate documentation of the present infrastructure/network setup, the requirement to install brand new sensing infrastructure, or the impossibility to alter the current setup without starting from scratch. Many applications for city-scale DTs have been proposed, but it will be necessary to test them in various settings before the buzz can be translated into reality.

15.8.2.6   5G Technologies and DTs

The rapid development of 5G technologies and their potential to alleviate the aforementioned problems should also be highlighted. Their potential application in smart manufacturing has been a driving force behind their development from the start and a strong argument in favor of their widespread implementation.

The paper explains how 5G networks may facilitate the real-world application of ideas from Industry 4.0. Both DTs and 5G stand to benefit from one another in terms of the difficulties faced by the latter and the former. Denser network installations are needed for 5G networks, at least in metropolitan areas, hence the planning process must be more involved than it was for 4G networks. When applied to city-scale DTs, 5G technologies have the potential to significantly improve upon issues like latency, scale, and complexity while also city landscape, which is crowded with competing solutions all vying for the same market share. To further improve the accuracy of city-scale DTs, private 5G networks may coexist with existing Low-Power Wide-Area Network (LPWAN) systems to provide a more scalable and flexible implementation in tough conditions, such as those seen in metropolitan regions.

15.9 Conclusions

The idea of the twin is gaining traction in discussions about keen cities. Based on the interviews conducted for this study, it is mentioned that a DT platform is vital for the development of smart city services and applications, simulations of urban development, and the presentation of alternative scenarios to city stakeholders such as politicians, decision-makers, and citizens. Despite its usefulness, implementing a DT in a city is complicated by a number of constraints. It takes a manual effort to create quality twins because of the necessity for human data cleansing and preparation. A lot of computation and data processing capacity is also needed for dynamic DTs. Recent advances in technology, however, have sped up the creation of twins, making city models that incorporate DTs easier and more economical to implement. As DTs become more widely used in smart city planning, cities will need to rethink their organizational culture, methods, and structures to ensure that data can move freely across their various computerized infrastructures. Cities will need to take a strategic approach and make a long-term commitment if they are going to succeed in making DTs a standard feature of urban planning. Cities can succeed in this endeavor with the help of investments (both monetary and educational) in their people and their skillsets. This study only includes a small sample of real-world digital context because their adoption is still in its infancy there. What is more, the actual evidence is limited to only one city. But the findings show that DT knowledge in a clever city context is a viable option for addressing these issues. Bottom-up urban development strategies can benefit from the use of a twin in smart city research by facilitating the creation and study of elements such as social interaction. In contrast, the notion of DT knowledge might aid the city in researching urban sustainability elements and generating new inventive solutions, which are all important factors when thinking about cities' urban metabolism and carbon impartiality aims.

References

1 Haag, S. and Anderl, R. (2018). Digital twin–proof of concept. *Manufacturing Letters* 15: 64–66.

2 Singh, M., Fuenmayor, E., Hinchy, E.P. et al. (2021). Digital twin: origin to future. *Applied System Innovation* 4 (2): 36.

3 Wu, J., Yang, Y., Cheng, X.U.N. et al. (2020). The development of digital twin technology review. In: *2020 Chinese Automation Congress (CAC)*, 4901–4906. IEEE.

4 Tao, F., Zhang, H., Liu, A., and Nee, A.Y. (2018). Digital twin in industry: state-of-the-art. *IEEE Transactions on Industrial Informatics* 15 (4): 2405–2415.

5 Zheng, Y., Yang, S., and Cheng, H. (2019). An application framework of digital twin and its case study. *Journal of Ambient Intelligence and Humanized Computing* 10: 1141–1153.

6 Rajab, H. and Cinkelr, T. (2018). IoT based smart cities. In: *2018 International Symposium on Networks, Computers and Communications (ISNCC)*, 1–4. IEEE.

7 Memos, V.A., Psannis, K.E., Ishibashi, Y. et al. (2018). An efficient algorithm for media-based surveillance system (EAMSuS) in IoT smart city framework. *Future Generation Computer Systems* 83: 619–628.

8 Alavi, A.H., Jiao, P., Buttlar, W.G., and Lajnef, N. (2018). Internet of Things-enabled smart cities: State-of-the-art and future trends. *Measurement* 129: 589–606.

9 Bresciani, S., Ferraris, A., and Del Giudice, M. (2018). The management of organizational ambidexterity through alliances in a new context of analysis: Internet of Things (IoT) smart city projects. *Technological Forecasting and Social Change* 136: 331–338.

10 Santos, P.M., Rodrigues, J.G., Cruz, S.B. et al. (2018). PortoLivingLab: an IoT-based sensing platform for smart cities. *IEEE Internet of Things Journal* 5 (2): 523–532.

11 Chatterjee, S., Kar, A.K., and Gupta, M.P. (2018). Success of IoT in smart cities of India: an empirical analysis. *Government Information Quarterly* 35 (3): 349–361.

12 Pawar, L., Bajaj, R., Singh, J., and Yadav, V. (2019). Smart city IoT: smart architectural solution for networking, congestion and heterogeneity. In: *2019 International Conference on Intelligent Computing and Control Systems (ICCS)*, 124–129. IEEE.

13 Andrade, R.O., Yoo, S.G., Tello-Oquendo, L., and Ortiz-Garcés, I. (2020). A comprehensive study of the IoT cybersecurity in smart cities. *IEEE Access* 8: 228922–228941.

14 El Saddik, A. (2018). Digital twins: the convergence of multimedia technologies. *IEEE Multimedia* 25: 87–92.

15 Qi, Q., Tao, F., Hu, T. et al. (2019). Enabling technologies and tools for digital twin. *Journal of Manufacturing Systems* 58: 3–21.

16 10 Things about Digital Twins in Agriculture, Agrimetrics. https://www.agrimetrics.co.uk/marketplace

17 Jones, D., Snider, C., Nassehi, A. et al. (2020). Characterising the digital twin: a systematic literature review. *CIRP Journal of Manufacturing Science and Technology* 29: 36–52.

18 Grieves, M. (2014). Digital twin: manufacturing excellence through virtual factory replication. *White Paper* 1: 1–7.

19 Damjanovic-Behrendt, V. (2018, 2018). A digital twin-based privacy enhancement mechanism for the automotive industry. In: *Proceedings of the 2018 International Conference on Intelligent Systems (IS)*, Funchal-Madeira, Portugal, 272–279. Piscataway, NJ, USA: IEEE.

20 Talkhestani, B.A., Jazdi, N., Schloegl, W., and Weyrich, M. (2018). Consistency check to synchronize the digital twin of manufacturing automation based on anchor points. *Procedia CIRP* 72: 159–164.

21 Rüßmann, M., Lorenz, M., Gerbert, P. et al. (2015). *Industry 4.0: The Future of Productivity and Growth in Manufacturing Industries*, vol. 9, 54–89. Boston Consulting Group.

22 Ayani, M., Ganebäck, M., and Ng, A.H. (2018). Digital Twin: applying emulation for machine reconditioning. *Procedia CIRP* 72: 243–248.

23 Macchi, M., Roda, I., Negri, E., and Fumagalli, L. (2018). Exploring the role of digital twin for asset lifecycle management. *IFAC-PapersOnLine* 51: 790–795.

24 Bitton, R., Gluck, T., Stan, O. et al. (2018). Deriving a cost-effective digital twin of an ICS to facilitate security evaluation. In: *Proceedings of the European Symposium on Research in Computer Security*, Barcelona, Spain, 533–554. Germany: Springer: Berlin/Heidelberg.

25 Söderberg, R., Wärmefjord, K., Carlson, J.S., and Lindkvist, L. (2017). Toward a digital twin for real-time geometry assurance in individualized production. *CIRP Annals* 66: 137–140.

26 Qi, Q., Tao, F., Zuo, Y., and Zhao, D. (2018). Digital twin service towards smart manufacturing. *Procedia CIRP* 72: 237–242.

27 Zheng, Y., Yang, S., and Cheng, H. (2018). An application framework of digital twin and its case study. *Journal of Ambient Intelligence and Humanized Computing* 10: 1–13.

28 Singh, M., Fuenmayor, E., Hinchy, E. P., Qiao, Y., Murray, N., and Devine, D. (2021). Digital twin: Origin to future. *Applied System Innovation* 4 (2): 36.

29 Dahmen, U. and Rossmann, J. (2018). Experimentable digital twins for a modeling and simulation-based engineering approach. In: *Proceedings of the 2018 IEEE International Systems Engineering Symposium (ISSE)*, Rome, Italy.

30 Sharma, A., Kosasih, E., Zhang, J., et al. (2021). Digital Twins: State of the Art Theory and Practice, Challenges, and Open Research Questions. arXiv:2011.02833.

31 Kumar, S. and Jasuja, A. (2017). Air quality monitoring system based on IoT using raspberry pi. In: *International Conference on Communication, Computation, Control and Automation (ICCCA)*, 1341–1346.

32 Eleftheriou, O. T., and Anagnostopoulos, C. N. (2022). Digital twins: a brief overview of applications, challenges and enabling technologies in the last decade. *Digital Twin*, 2, 2.

33 Zheng, P. and Lim, K.Y.H. (2020). Product family design and optimization: a digital twin-enhanced approach. *Procedia CIRP* 93: 246–250.

34 Tao, F., Sui, F., Liu, A. et al. (2018). Digital twin-driven product design framework. *International Journal of Production Research* 57: 3935–3953.

35 Biesinger, F., Meike, D., Kraß, B., and Weyrich, M. (2019). A digital twin for production planning based on cyber-physical systems: a case study for a cyber-physical system-based creation of a digital twin. *Procedia CIRP* 79: 355–360.

36 Agostino, Í.R.S., Broda, E., Frazzon, E.M., and Freitag, M. (2020). Using a c digital twin for production planning and control in industry 4.0. In: *Scheduling in Industry 40 and Cloud Manufacturing*, 39–60. Springer.

37 Min, Q., Lu, Y., Liu, Z. et al. (2019). Machine learning based digital twin framework for production optimization in petrochemical industry. *International Journal of Information Management* 49: 502–519.

38 Zhang, H., Yan, Q., and Wen, Z. (2020). Information modeling for cyber-physical production system based on digital twin and AutomationML. *International Journal of Advanced Manufacturing Technology* 107: 1927–1945.

39 Liu, C., Jiang, P., and Jiang, W. (2020). Web-based digital twin modeling and remote control of cyber-physical production systems. *Robotics and Computer-Integrated Manufacturing* 64: 101956.

40 Guo, H., Chen, M., Mohamed, K. et al. (2020). A digital twin-based flexible cellular manufacturing for optimization of air conditioner line. *Journal of Manufacturing Systems* 58: 65–78.

41 Ivanov, S., Nikolskaya, K., Radchenko, G., Sokolinsky, L., and Zymbler, M. (2020, November). Digital twin of city: Concept overview. In *2020 Global Smart Industry Conference (GloSIC)* (pp. 178–186). IEEE.

42 Ghandar, A., Ahmed, A., Zulfiqar, S. et al. (2021). A decision support system for urban agriculture using digital twin: a case study with aquaponics. *IEEE Access* 9: 35691–35708.

43 Microsoft. What is Azure Digital Twins? https://azure.microsoft.com/en-us/products/digital-twins/

44 Schrotter, G. and Hurzeler, C. (2021). The digital twin of the city of zurich for urban planning. *Journal of Photogrammetry, Remote Sensing and Geoinformation Science* 88: 99–112.

45 Beil, C. and Kolbe, T.H. (2020). Combined modelling of multiple transportation infrastructure within 3D city models and its implementation in CityGML 3.0. *ISPRS Annals of The Photogrammetry, Remote Sensing and Spatial Information Sciences* 6: 29–36.

46 White, G., Zink, A., Codecá, L., and Clarke, S. (2021). A digital twin smart city for citizen feedback. *Cities* 110: 103064.

47 Nativi, S., Mazzetti, P., and Craglia, M. (2021). Digital ecosystems for developing digital twins of the earth: the destination earth case. *Remote Sensing* 13: 2119.

48 Lee, S., Jain, S., Zhang, Y., Liu, J., and Son, Y. J. (2020). A multi-paradigm simulation for the implementation of digital twins in surveillance applications. In *IIE Annual Conference. Proceedings* (pp. 79–84). Institute of Industrial and Systems Engineers (IISE).

49 Marcucci, E., Gatta, V., Le-Pira, M., Hansson, L., Brathen, S. (2020). Digital Twins: A Critical Discussion on Their Potential for Supporting Policy-Making and Planning in Urban Logistics. *Sustainability* 12: 623.

50 Russell, H. (2020). Sustainable urban governance networks: data-driven planning technologies and smart city software systems. *Geopolitics, History, and International Relations* 12: 9–15.

51 Harrison, R., Vera, D., and Ahmad, B. (2021). A connective framework to support the lifecycle of cyber–physical production systems. *Proceedings of the IEEE* 109: 568–581.

52 Dembski, F., Wössner, U., Letzgus, M. et al. (2020). Urban digital twins for smart cities and citizens: the case study of Herrenberg, Germany. *Sustainability* 12 (6): 2307.

53 Helsinki, V. (2021). https://www.myhelsinki.fi/en/see-and-do/sights/virtual-helsinki

16

Digital Twin in Aerospace Industry and Aerospace Transformation Through Industry 4.0 Technologies

Vijay Kandasamy[1], Eashaan Manohar[1], Saiganesh Bhaskar[2], and Ganeshkumar Perumal[3]

[1] Department of Computer Science and Engineering, Rajalakshmi Engineering College, Chennai, Tamil Nadu, India
[2] Department of Electronics and Communication Engineering, Rajalakshmi Engineering College, Chennai, Tamil Nadu, India
[3] Department of Computer Science, Al Imam Mohammad Ibn Saud Islamic University (IMSIU), Riyadh Saudi Arabia

16.1 Introduction

When it comes to using digital twin (DT) technology, the aerospace sector is ahead of the curve. What we refer to as a "digital twin" is essentially a simulation of a real-world system, process, or item in software [1, 2]. It relies on an object's form, function, and behavior being faithfully reflected in a computer simulation. The DT of a system, process, or product is an exact digital replica that can stand in for the actual thing while testing or studying. The use of DTs has been on the rise in several industries, particularly the aerospace industry, in recent years. "Industry 4.0" is shorthand for a new way of doing business in the manufacturing industry that puts an emphasis on adaptability, reactivity, and output. The aviation industry is feeling the effects of Industry 4.0 technology in a major way.

Data from sensors, simulation models, historical records, and real-time monitoring systems are fused together to generate DTs in the aerospace sector. When these data sets are brought together, we get a DT, which is a dynamic representation that can be utilized for analysis, forecasting, and optimization over the course of an aircraft's lifetime [3].

The aerospace sector can benefit from DTs in a number of different ways. DTs let engineers to analyze and simulate several design iterations, test performance, and fine-tune aircraft system configurations before ever manufacturing a real

Digital Twins in Industrial Production and Smart Manufacturing:
An Understanding of Principles, Enhancers, and Obstacles, First Edition. Edited by Rajesh Kumar Dhanaraj, Balamurugan Balusamy, Prithi Samuel, Ali Kashif Bashir, and Seifedine Kadry.

prototype. By taking this route, we may cut back on expenses without sacrificing efficiency in the design process or the quality of the final product.

Predictive maintenance and condition monitoring of airplane equipment rely heavily on DTs. Real-time data from sensors onboard the aircraft is monitored by DTs, allowing for proactive maintenance scheduling, anomaly detection, and component failure prediction. By addressing maintenance concerns before they become serious failures, this predictive strategy reduces aircraft downtime, boosts operational effectiveness, and enhances safety.

DTs also make it easier to optimize the safety and operations of airplanes. The detection of operational inefficiencies, identification of potential risks, and decision assistance for pilots and maintenance crews are all made possible by real-time data integration in conjunction with advanced analytics and artificial intelligence (AI) algorithms in Figure 16.1.

With DT technology, we can test out the application before it goes live, which takes the stress out of automating and implementing the Internet of Things (IoT) [4]. Systems can be studied with DTs to find opportunities, which can then be implemented in the real world.

In manufacturing, DTs have proven useful in minimizing maintenance costs and maximizing output by providing a real-time look at the condition of machinery and other physical assets.

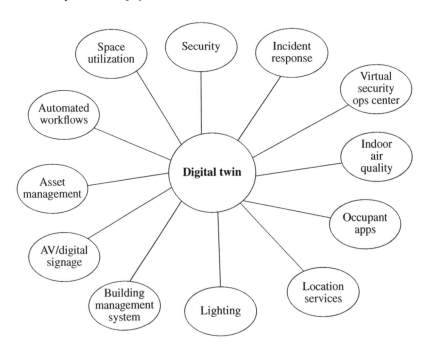

Figure 16.1 Digital twin technology example.

First introduced in 2002, the concept of a DT has only recently become practical and affordable for businesses because to the proliferation of IoT. Taking a step back to examine the vast possibilities of DTs for your company is well worth the time it takes. Once we understand how the DT concept functions, we may move on to its practical applications in industry.

Despite the substantial benefits, there are difficulties in implementing DTs in the aerospace sector. The main challenges that need to be overcome include data integration from diverse sources, data security and privacy issues, and compatibility between various systems. DTs [5], however, have the potential to revolutionize the aerospace sector by enhancing design correctness, operating efficiency, and maintenance effectiveness as technology advances.

16.1.1 Introduction to Industry 4.0 Technologies for Aerospace Transformation

Industrial Revolution 4.0 practices are having a profound effect on the aerospace sector. The purpose of "Industry 4.0," or the incorporation of ICT into manufacturing, is to develop smart, networked, and highly efficient production methods. Just a few examples are the widespread use of Internet-connected gadgets, big data analysis, robots, AI, AM, and fully automated assembly lines.

Industry 4.0 technologies are transforming production procedures in the aircraft industry by fusing digital and physical systems [6]. IoT makes it possible for different parts, devices, and systems to communicate with one another, enabling real-time data collection and analysis. Combining this data with cutting-edge analytics and AI algorithms enables proactive decision-making, manufacturing process optimization, and quality control.

Another industry-transforming technology in the aerospace sector is additive manufacturing, sometimes known as 3D printing. With less waste and shorter lead times, it enables the manufacturing of complex geometries, lightweight components, and customized parts. Aerospace businesses can get better flexibility, cost savings, and design freedom by utilizing additive manufacturing [7].

The manufacturing of aerospace products is also utilizing more robotics and automation technology. Robots are used in material handling, assembly, and inspection operations to increase productivity, accuracy, and worker safety [8]. Material movement and logistics efficiency can be increased through the use of automation technologies such as automated guided trucks and automated storage and retrieval systems in factories.

A further factor pushing innovation in aeronautical operations and services is Industry 4.0 technology. Aerial performance is improved, downtime is decreased, and condition-based maintenance is made possible via real-time data analytics and predictive maintenance. Drones and autonomous systems are being used for cargo transportation, surveillance, and inspection. Digital platforms and connectivity are enabling on-demand services, which are revolutionizing the logistics and supply chain for the aerospace industry.

However, the aircraft industry has challenges when attempting to adopt Industry 4.0 technologies. Considerations such as cyber security, data privacy, the availability of a trained workforce, and the necessity to combine old and new technologies all must be made.

The overall potential for enhanced productivity, efficiency, creativity, and safety offered by the transformation of the aerospace industry through Industry 4.0 technologies is enormous. The aerospace sector can maintain its competitiveness, respond to changing customer needs, and promote sustainable growth in the digital era by embracing these innovations.

The phrase "Industry 4.0" refers to a new paradigm that encompasses several different types of enabling technologies that have the potential to completely revolutionize the way conventional factories operate. There is a wealth of literature discussing the concept and the nature of these enabling technologies. The DT, AR/VR, cloud computing, IoT, autonomous/collaborative robotics, big data, and the cloud are all covered in this book [9–11].

Figure 16.2 depicts the progress of the aero structure in Industry 4.0 [12]. The process of designing aerostructures is going increasingly digital. It is widely believed that businesses will reap the greatest strategic benefits from investing in the future of product lifecycle management, industrial automation, and composite materials workflows.

1. Highly specialized analysis, design, and production technologies that completely support and comprehend the creation of composite materials.
2. The use of a DT to facilitate cross-disciplinary teamwork and problem-solving throughout development.

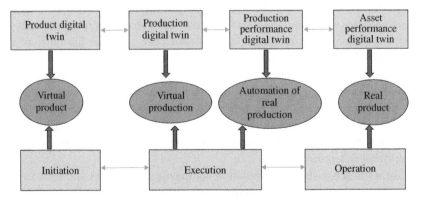

Figure 16.2 Industry 4.0 for composite aerostructure development.

16.2 The Concept, Components, and Applications of Digital Twins

16.2.1 Concept

Because DT concepts are continually evolving and the academic world has so many different points of view, the very definition of DTs is ever expanding. This section attempts a fresh method of approaching the definitions of DTs in order to help readers tell them apart from other related ideas. At first, DTs are recognized as a type of informational system. The goal of developing DTs is to give industries dependent on physical assets a way to visualize asset data across the whole span of that asset's existence. As a result, the phrase "digital twin" to characterize the fusion of digital and physical data has gained popularity. Whoever coined the term "digital twin" [5] also offered the following definition:

"When we talk about something having a 'Digital Twin,' we're referring to a collection of digital creations that, when put together, accurately reflect every facet of the original concept or physical object [1]. At its best, a manufactured good's DT can provide the same details as would be gleaned from inspecting a real-world prototype."

Figure 16.3 uses a hybrid of the previous concepts to illustrate the positioning of DTs from a number of angles. Simply said, a DT is an accurate and current depiction of a fully operational physical asset, complete with all necessary context.

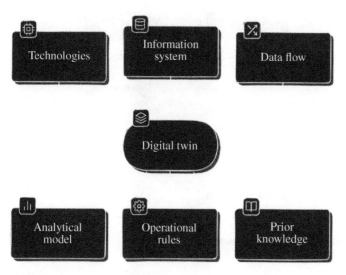

Figure 16.3 The Position of DT.

DTs can be used to do more than just assess an asset's current state; they can also be used to forecast how it will act, enhance management, and enhance efficiency.

16.2.2 Major Component of DTs

16.2.2.1 Physical Side: Infrastructure with Working Sensors
Computing and data capture are not possible without hardware. Data transmission technology and sensors, as shown in Figure 16.4, are required for any DT to achieve real-time data collection and synchronization. Text, audio, hyperspectral images, video, temperature, and pressure sensors, among others, help paint a more complete picture. There has been a shift in the focus of hardware firms toward providing a wide variety of physical solutions as the enabling technologies have become more widely available. IBM, General Electric, Siemens, and Oracle are just some examples of companies that offer DT hardware solutions. The SAP cloud platform is compatible with Amazon Web Services (AWS), Alibaba Cloud, and Microsoft Azure's IoT suites. Large amounts of data are no problem for any of the three cloud providers.

16.2.2.2 Virtual Side: Analytical Models and AI
Digital side's primary functions are data collection, processing, and analysis. Libraries full of models for diverse purposes can be found all over the digital world. Unlike data-driven modeling, traditional modeling relied on physics in its early practical uses of DTs. Data-driven model development, which fits well

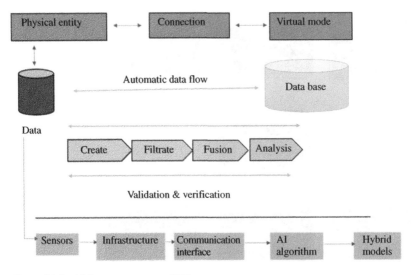

Figure 16.4 Major components of DT.

within the big data and IoT revolutions, became a realistic possibility when processing power began to quickly advance in the 2010s [4]. One of the most appealing features of data-driven models is their adaptability. Engineers can now model in a fair period of time using modern machine learning (ML) techniques without resorting to time-consuming human effort in the form of considerable experimentation or specialized knowledge.

16.2.3 Digital Twins and Their Industrial Use Case Studies

DTs have gained traction in a number of industries over the past decade. In businesses like auto repair and jet engine production, the capacity to anticipate and manage assets is crucial [13]. The DT significantly improves the asset's tractability and predictability by offering a high-fidelity model and utilizing monitoring and prediction capabilities. The need for skilled data and asset managers is widespread across sectors, from manufacturing to urban planning. The unified model and in-built data processing and analysis capabilities of DTs make them perfect for such uses. Although DTs offer a wide range of potential applications, recent research has concentrated on three particularly fruitful areas: "smart cities," "smart healthcare," and "smart manufacturing."

16.3 DTs Roadmap: A Path Toward the Future

This section outlines the limitations of the traditional DT and briefly highlights three areas where further study could yield more effective DTs.

16.3.1 Basic DT

In terms of usage, the most common form of DT is also the simplest. The term "digital twin" is used here to describe a virtual representation of a physical object that permits real-time data collection and synchronization [14]. One-function models and simulation models (often non-3D models) are the building blocks of a DT, with the latter used for essential tasks like monitoring and analysis. Tetrad-driven technology integration shares several hallmarks with conventional DT. It shares some philosophical ground with DT but is less detailed in its implementation guidelines.

16.3.2 Interactive DT

The importance of DT that can operate as a standalone system has been disregarded in favor of constructing data flow and analytical data models. The potential advantages of the digital equivalent can be further explored with the use of interactive DT. Figure 16.5 shows the traditional DT paradigm in three dimensions; [1, 5] by adding the fifth dimension of services and DTs information, this model is expanded to five. The two additional dimensions center on the connection between the digital and physical components of a purchase. To illustrate the importance of human-machine interactions in DT applications, Figure 16.5 shows how functional models are used for detection, diagnosis, and prediction; high-fidelity simulation is achieved through visualization technologies; and consumer services facilitate access to and analysis of historical, real-time product data.

16.3.3 Standardized DT

Standardization in the development, maintenance, and deployment of DT models is what these guidelines are all about. It aims to provide the user with actionable instructions for dealing with problems such as, "How can we build a DT?" and "Where can we start with digital twins?" A consistent DT architecture and tools are required, especially in light of the various formats, protocols, and standards in enabling technologies. The direct stakeholder in the standardizing of DT will be the business community. It will significantly cut down on DTs' research and development (R&D) time and uncertainty. Customers do not have to be concerned about an inconsistent information environment or which models are superior. DT standardization may increase the commercial worth of DTs.

16.3.4 Intelligent DT

Creating a high-fidelity twin and collecting synchronization data are two main areas of study in DT. The amount of data gathering, learning, and reasoning necessary to carry out more complex tasks calls for more than just twinning if DT is to materialize [1].

16.3.5 Industrial Engineering in the Age of Industry 4.0: Defining the Future Role of Digital Twins

Research into DT in Cyber-Physical Systems (CPS)-based manufacturing facilities is growing. Maya recommends utilizing digital continuity and continual real-world synchronization to increase productivity while decreasing production times [15, 16]. This is accomplished through the use of interdisciplinary integrated modeling and forecasting methodologies. The ultimate goal of this endeavor is to develop a plant DT that can aid in anything from initial conceptualization to powering down the facility. Artists working in Maya make things with the intention of merging the digital and physical worlds Modeling the CPS attributes as semantic metadata is necessary for improving the DT representation and

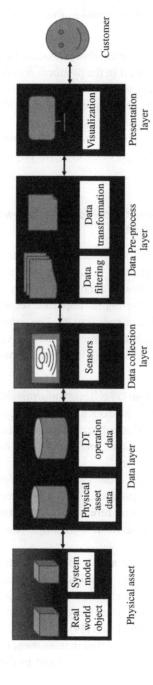

Figure 16.5 Five-dimensional DT structure.

simulation of the actual manufacturing lifecycle [17]. The centralized support infrastructure is the infrastructure that makes this feasible by helping with the following:

- The semantic metadata model, which structures data for easy digital access to information generated at any point in the production system's life cycle.
- The simulation framework [18], which integrates different simulation methods and tools to produce a transdisciplinary representation of the underlying physical system.
- Massive amounts of field data may be synchronized and updated in real time because of the communication layer connecting the physical CPS and the digital world.

16.3.6 The Digital Twin: From Its Core Ideas to Real-world Applications in High-tech Production

Making judgments and performing predictive maintenance based on observed data. has been getting a lot of attention in the high-tech manufacturing industry. Developing a virtual representation of a physical system, procedure, or item is what this term refers to [13]. The digital duplicate records the live dynamics and physical traits of its analogue counterpart. The term "digital twin" is an abbreviation for a more comprehensive notion that has various applications in contemporary manufacturing. Let us examine some of these foundational ideas and uses:

- **Sensing and Data Acquisition**: DT relies on information gathered in real time from various sources including sensors and IoT gadgets. Information on the physical system's condition, performance, and surrounding environment can be gleaned from these datasets. Accurate and reliable data collection is crucial for creating a faithful digital depiction.
- The process of virtual modeling entails building an electronic replica of the real-world system, process, or product. This model accurately represents the physical counterpart's geometrical, functional, and behavioral characteristics. Simple static representations to intricate simulations based on physical laws are all possible.
- **Fusion and Integration of Data**: DT combines and integrates data from several sources, such as sensors, manufacturing equipment, and business systems. To merge and reconcile the many data sources, data fusion techniques are used, guaranteeing correctness and consistency. The virtual representation is constructed using this combined data [19].
- A DT permits continuous monitoring and management of a physical system via its DT in real time. Comparing the virtual model to the physical world on a

regular basis will reveal any discrepancies. As a result, preventative measures can be taken for system upkeep and enhancement.

- Predictive analytics can be performed using the data that the DT has acquired. One can predict the behavior of a physical system by using statistical models and ML methods. This enables proactive decision-making, failure or performance prediction, and maintenance schedule optimization.
- Simulated experimentation and optimization are made possible by DTs. Manufacturers can evaluate the effects of design changes, process alterations, or operational strategies before putting them into practice in the actual world by simulating various scenarios using the digital model. This lowers risks, lowers expenses, and boosts overall effectiveness.
- **Lifecycle Management**: A product or system's DT is used throughout the whole lifecycle. It aids in the creation of products, the scheduling of maintenance, the production of them, and even their disposal. It offers a comprehensive assessment of the system's performance, enabling ongoing development and wise choices at every turn.
- Collaboration and communication are encouraged among the various stakeholders engaged in the production process thanks to DT. It enables information sharing, the visualization of complex data, and problem-solving collaboration amongst engineers, operators, and managers. Better coordination, knowledge exchange, and decision alignment are encouraged as a result.

The automotive, energy, healthcare, and aerospace industries are just a few of the many that can benefit from DT technology in today's industry. It enables a wide variety of uses, including predictive maintenance, enhanced production planning, quality assurance, virtual commissioning, and supply chain optimization. Virtual prototyping and testing of vehicle designs is possible using DTs, minimizing the requirement for physical prototypes. By simulating manufacturing operations and analyzing real-time data, manufacturers may optimize production processes and quality control. Predictive maintenance that uses real-time data from vehicles' sensors can reduce downtime and boost customer satisfaction.

Supply chain optimization: By tracking the flow of parts and components in real time, DTs can optimize supply chain logistics.

16.4 Industry 4.0

The aerospace and military industry as a whole may be affected by Industry 4.0-inspired technology, from the largest original equipment manufacturers to the smallest wholesalers. However, it seems that not all companies are capitalizing on these innovations to expand their client base and enhance their bottom line.

In a recent poll of aerospace and defense (A&D) industry leaders, 40% said that creating new business or delivery models was the most difficult part of their company's digital transformation efforts. This finding suggests that creating innovative products and services still presents significant challenges for the majority of A&D companies.

Despite the broad adoption of Industry 4.0 technologies in fields like manufacturing and logistics, businesses in the A&D sector have been hesitant to adopt comprehensive strategies for digital transformation across the organization. The sector's problems can be traced back to the fact that many local businesses do not consider Industry 4.0 a top priority, preferring instead to implement new technology on their own. Digital transformation is essential for success in today's business environment, but A&D companies risk falling farther behind if they limit their digital strategy to only a subset of company functions. All types of enterprises will need to learn how to take advantage of new technologies if they want to benefit from the Industry 4.0 revolution.

The benefits of these technologies can be realized by A&D businesses of any size, but especially those of medium or small scale, by beginning with a pilot project and then rolling it out to the rest of the business. A&D managers need to reevaluate their function inside the company, rather than seeing new technologies as an add-on to old ways of doing things. The purpose of this research was to analyze the successes and failures of A&D companies as they made the transition to digital transformation and to provide recommendations for how these companies can thrive in the age of Industry 4.0.

16.4.1 What is Defense and Aerospace 4.0?

The use of modern manufacturing processes, such as additive manufacturing, as well as the incorporation of sensors and connection into existing items all contribute to the development of new, more affordable products and services. The A&D industry is modernized with the help of Industry 4.0 technology.

16.4.2 Describe Industry 4.0

"Industry 4.0," sometimes known as the "Fourth Industrial Revolution," describes a trend toward teamwork, information exchange, and innovative problem resolution in the field of digital manufacturing. In particular, it supports the iterative cycle of planning, acting, and reflecting depicted in Figure 16.6.

The term "Industry 4.0," also known as the "Fourth Industrial Revolution," is commonly used in the context of digital production to indicate a change toward a more collaborative, information-sharing, and innovative approach to solving long-standing challenges. It is the lifeblood of the PDR cycle (Figure 16.6) of planning, doing, and reflecting.

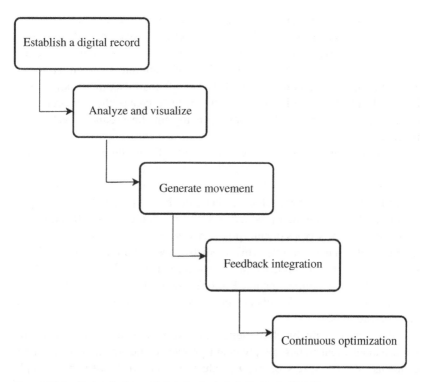

Figure 16.6 Technologies and the physical-digital-physical cycle.

Some A&D firms are already feeling the game-changing consequences of these technologies. Cost reduction, supply chain restructuring, faster delivery, and ubiquitous product and device connectivity are only some of the topics of interest.

16.4.3 Transformational Shifts in Technology Use, With an Emphasis on Aerospace and Defense Priorities 4.0

A&D 4.0 technologies and their effects on the economy as a whole lower the bar to enter and shake up the status quo in established markets. Threatening new entrants to the market are those who look beyond the manufacture of traditional physical products to the exploitation of consumer data and product performance in order to create new revenue streams [20]. Players that are quick on their feet and adept with a number of new technologies may eventually grow to dominate their field. Our research suggests the following four areas are most important for A&D companies to focus on during their digital transformation:

- Businesses will need adaptive production and predictive quality controls as A&D consumers increase their demands for delivery schedules and customization. To keep airplane manufacturers busy until 2026 at the current average yearly production pace, some 14,700 commercial aircrafts were on backorder as of the end of September 2018 [15]. Aircraft manufacturers may benefit from Industry 4.0 since it eliminates the need to create physical prototypes of certain components and systems, speeding up the product development process.
- Make use of creative methods: The vast volumes of data generated by the interconnected systems of Industry 4.0 can be put to good use. The sensors and gadgets used in today's A&D products can number in the tens of thousands, each of which produces enormous datasets. Therefore, in addition to developing, producing, and operating their products, A&D firms should also use this information to develop new types of enterprises. Access to performance data might be used to establish new services, while predictive applications and advanced analytics could be used to enhance Maintenance, Repair, and Operations (MRO) processes [9, 17]. Examples include Boeing's establishment of AnalytX, a central location for the distribution of data analytics software and services to the company's customers.
- Get the supply chain involved right away. Regular roadblocks include program management issues, supply chain problems, and unanticipated vendor concerns. Manufacturers of A&D products should work with their suppliers to create best practices in digital manufacturing in light of consumers' requests for higher capacity and on-time delivery. The fields of digital design, simulation, and integration can help A&D firms achieve this goal by facilitating the conceptualization and digital construction of a simulated prototype or process. This has the potential to increase transparency in the supply chain and enable digital integration throughout the entire manufacturing process.
- Recognize and adopt the new cybersecurity paradigm: Data proliferation and connected system adoption have increased cybersecurity threats for A&D organizations. In fact, according to a poll of A&D executives, the most prevalent technology difficulties that organizations confront as they work to advance their digital transformation program are cyber risks and data ownership. A&D businesses place a high priority on controlling vulnerabilities that arise with greater digitization and are particularly concerned about security. The National Institute of Standards and Technology (NIST) guidelines and other national and international legislations controlling cybersecurity are well within their knowledge base.

16.4.4 Getting on the Right Track for Industry 4.0 Success

The size of an A&D company, its location in the supply chain, the type of the company's activities within the supply chain, and most importantly, the company's commercial goals all play a role in determining the extent to which Industry 4.0 technologies will affect the company [4]. Successful enterprise-level implementation of Industry 4.0 requires A&D companies to undergo a digital transformation across customer engagement, new product/service and platform development, and intelligent assets, as shown in Figure 16.7. Begin by establishing specific goals for implementing DT technology in your aviation operations. What are the main concerns or challenges you hope to overcome? When it comes to enhancing aircraft design, optimizing maintenance, or increasing safety, and having a well-defined goal is critical. Data is at the heart of DT technology. Ascertain that you have access to high-quality data from a range of sources, including aircraft sensors, manufacturing processes, and supply chain operations. Combine these data into a unified platform for analysis. Choose DT technologies and solutions that are suitable for your objectives [18]. This includes choosing software, hardware, and communication protocols that can handle the intricacies of aircraft systems.

Create precise DTs of your aerospace systems and operations. These models must be capable of simulating real-world scenarios and reacting to changes in real time [11]. Use DTs to implement predictive maintenance solutions. Predictive analytics are used to schedule maintenance on aircraft components before they fail, reducing downtime and boosting safety. Use DT technologies to model and optimize aircraft designs. This can help to save development time, improve fuel efficiency, and enhance safety features.

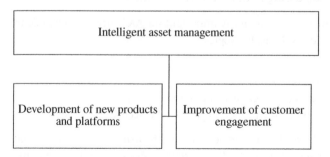

Figure 16.7 Aerospace and defense enterprises undergo digital transformation at three levels.

- **Improving customer involvement:**
  Companies in the A&D industry can improve their standing in the market by using Industry 4.0 tools to provide superior service and encourage greater client participation. By leveraging state-of-the-art technologies [19], A&D businesses may better understand their clients, respond to their needs, and meet their evolving tastes and demands. Raytheon's goal is to raise customer satisfaction through the use of front- and back-end measurement technologies such product data management, manufacturing systems, and supplier assessment tools.

- **Developing new platforms and goods/services:**
  The A&D business is notoriously slow to bring new products to market because of the lengthy R&D and commercialization processes involved. Digital technology can create new avenues for innovation in the A&D sector by speeding up the design, development, and commercialization of new goods as well as facilitating feedback from many stakeholders during the development process.

- **Creating intelligent assets:**
  Industry 4.0 technologies, when applied to A&D firms' assets, have the potential to boost efficiency and productivity by making those assets smarter and more self-aware. Engineers may monitor the performance of an IoT-connected jet engine and its interactions with other plane components from any location. The advantages of this openness include better engine design, performance tracking, and maintenance prediction, as well as a better customer value and experience [13]. Lockheed Martin uses information gathered from networked systems to create brand new parts, products, and even entire programs. Eventually, these data strands weave together to form a "Digital Tapestry," which helps to forge links between seemingly unrelated organizations.

16.4.5 An All-encompassing Plan for Digital Innovation

The following four areas are particularly important for A&D firms to advance in their utilization of Industry 4.0 technologies:

1. Increasing agility
2. Creating novel business models
3. Fostering close collaboration between suppliers
4. Securing data and intellectual property against cybersecurity threats.

There are many paths to digital transformation for businesses, but those that leverage intelligent assets to provide differentiated offerings, ground-breaking platforms, and heightened consumer engagement have the greatest chance of success.

In the eyes of business moguls, Industry 4.0 is a long-term bet on innovation, efficiency, and productivity [6]. They need to adopt a comprehensive approach to the problem of online misbehavior rather than only addressing individual incidents as they arise. Because of its ability to outlive economic cycles, digital transformation should be a priority for A&D executives. In the "disrupt or be disrupted" advertising and design industry, making the most of cutting-edge, market-shifting ideas may be a question of survival.

16.4.6 A Quickening Rate of Change

In order to break into new markets, the aerospace industry must hasten its digital modernization and restructuring. Because of the last pandemic's effect on passenger numbers and demand for new aircraft orders, A&D manufacturers were driven to streamline their production methods and shift their attention to new technologies and developing markets. The European Union (EU) has joined the international race to deploy low Earth orbit satellite broadband networks in response to the rising need to eliminate Internet blackspots in rural areas and provide a secure, resilient backup for terrestrial broadband [12, 21]. Demand for space technology like hypersonic weapons and reusable rockets will increase as a result of the militarization of space and the arms race between China, Russia, and the United States to bolster their interplanetary military presence. The annual growth rate for launch services into space is projected to be 15.7%.

16.4.7 Transforming a Traditional Industry

Recently, McKinsey conducted a survey which found that only 16% of aircraft manufacturers were using digital manufacturing initiatives on a large scale, while 44% were only beginning to explore the possibilities [22]. Adoption has been sluggish and patchy, and it has mostly been utilized for specialized purposes. Aviation manufacturing needs a full-scale digital revolution so that new technologies may be developed and brought to market more quickly. Aviation manufacturers need to employ unified systems that bring together design, engineering, and production to get rid of design iterations, overengineering, and manufacturing waste and to allow for new innovations that can be fine-tuned to specifications in real time.

With end-to-end production process monitoring and real-time data collecting, manufacturing processes may be continually optimized. The manufacturing process becomes more intelligent as a result of this feedback loop. The vertical integration of manufacturing, in which all aspects of production, from design to engineering, are digitally integrated, is made possible by the advent of the IoT [11, 23]. Digitally aggregating data across manufacturing lines to power real-time in-line quality control and machine calibration paves the way for more efficient production and end-to-end traceability across products.

It is critical for A&D firms to rapidly develop interoperable industrial execution systems that are integrated with new tools and technologies in preparation for Industry 4.0. [6]. This is crucial for the speedy adoption of digital methods in manufacturing. A third-party solution is automatically checked by new factory planning and execution systems to make sure tools are calibrated accurately in real-time and operating at peak efficiency. Complete component traceability is made possible by open, interoperable manufacturing execution systems.

Manufacturing processes can be quickly updated and spread when they are digitally integrated, making production more flexible and agile. This promotes a manufacturing mindset that is open to innovation, experimentation, and change [2]. The production of airplanes must use adaptable manufacturing techniques that support rapid innovation and quicken time to market.

16.4.8 Aerospace Digital Twin Navigation

With the help of DTs, physical prototypes might be eliminated, designs could be improved, and goods could become more intelligent, networked, and environmentally friendly [10]. They have the potential to be a single source of truth throughout the whole product lifetime, from design to manufacture to in-service [16]. It is a technology that is revolutionizing the A&D sectors and accelerating innovation. However, implementing DTs can be difficult. The foundational elements of DTs, simulation, high-performance computing (HPC), and data, require the disruption of process silos, convergence of processes, and real-time asset-based cooperation.

Enterprises also require seamless integration between cloud and on-premise computing, with security issues and data flow properly managed [19]. It makes sense that corporations have hired VPs and C-level digital transformation executives to oversee the adoption and use of DTs.

It also performs the following tasks:

- Differentiating DTs
- Silos and convergence management
- How important vendor flexibility is
- Selecting appropriate DT components for integration
- Picking the ideal Cloud and on-premises combination
- The promise of predictive maintenance.

16.4.9 Why Are Aerospace and Defense Undergoing Digital Transformation?

Increasing the digital maturity of the aerospace, aviation, and defense value stream could generate US$20 billion in annual EBITDA, according to a recent McKinsey study conducted in collaboration with the Aerospace Industries Association (AIA).

Several reasons are causing this transformation:

- Automation innovation
- Increased need for efficiency and speed
- Uncertainty in the supply chain necessitates more efficient procedures
- Digital capabilities at the cutting edge as a competitive advantage.

Here are a few of the more particular issues that digital transformation in the aerospace industry is resolving [10].

- Lack of efficiency in operations
- Interruptions in the supply chain
- Problems with upkeep
- Lengthy periods between product iterations
- Exorbitant running expenses
- Difficulty with legal compliance
- Low levels of consumer involvement and contentment
- Inefficient handling and application of data

16.5   The Twins Are Driven by Technology

- **IoT and Big Data**: Organizations [6] can acquire insights into real-time operational performance thanks to the proliferation of smart sensors on assets or components linked to connected systems;
- **Advanced analytics**: Data can be utilized to model and anticipate the future state of assets through ML;
- **Processing power**: The cost and accessibility of the processing power needed to run large-scale DT models have significantly improved thanks to cloud-based technology [22];
- **Accessibility**: Previously, a DT would have been imprisoned in the organization's or factory's control room, but now information can be accessed from mobile devices anywhere.

16.6   Conclusion

Aero-DTs are becoming increasingly important in the aerospace sector; thus, more organizations and institutions are funding their study and development [20]. However, many experts are unsure of where to seek or how to get started learning about the various DT enabling technologies and their benefits and drawbacks. To help shed light on the topic, this research classifies, introduces, and analyses key enabling technologies for aero-DTs, providing a comprehensive overview of the field. There is no a perfect solution for aero-DTs yet; for example, research reveals that DT developers often use ML models without considering the physics [19]. Developers of DTs must therefore make aero-DTs from the most ideal DT components.

This study will examine the function of the DT in factories that have embraced sector 4.0 technology in order to weigh the pros and cons of bringing a concept developed for the aerospace industry into the manufacturing sector. Manufacturing benefits from DTs' ability to stand in for physical objects [7, 24]. A semantic data model allows for prognostic studies to be performed on these digital representations during the design phase (static perspective), and for the virtual representation of the object to be continuously updated in real time to reflect observed data. Accurately depicting the system's status enables optimization, judgment, and condition-based predictive maintenance in real time.

DT is an essential aspect of the progress in the aerospace sector for the implementation of sector 4.0 technologies. In order to foster innovation, efficiency, and better decision-making across the product lifecycle, the aerospace industry has embraced the DT. Real-time monitoring, preventative maintenance, and performance improvement are made possible by DT by building virtual duplicates of aircraft, engines, and other aerospace systems. It makes it easier to combine data from avionics, operational systems, and sensors, giving you a complete picture of the physical assets. With greater reliability and operational efficiency as a result, this enables aerospace businesses to proactively identify and address issues, minimize downtime, and optimize maintenance schedules.

The virtual simulation and testing of aircraft designs, production methods, and operational scenarios is made possible by DT. This lessens the need for actual prototypes and enables quick iteration and improvement [8, 13]. Aerospace producers can make use of simulation tools to assess the effects of design modifications, assess various operational plans, and enhance performance and fuel economy.

The aviation industry may be able to fully fulfil the potential of DT by utilizing industry 4.0 technologies such as the IoT, big data analytics, AI, and the cloud [11]. Improved analytics and ML techniques, simplified integration of data from many sources, and safe, scalable data processing and storage are all made possible by these developments.

DT enables the virtual simulation and testing of aircraft designs, manufacturing processes, and operating scenarios. As a result, fewer real prototypes are required, and iteration and improvement can happen more quickly. Simulation tools can be used by aerospace manufacturers to evaluate the consequences of design changes, compare different operational scenarios, and improve performance and fuel efficiency.

Using cutting-edge sector 4.0 technology like the IoT, big data analytics, AI, and cloud computing, DT could help the aviation industry grow into new markets. These methods not only facilitate cutting-edge analytics and ML algorithms, but also make it simple to combine information from many sources, all while keeping it safe and scalable.

In conclusion, the advent of DT made possible by Industry 4.0 technologies has fueled transformation and innovation in the aerospace industry. It transforms the processes involved in designing, building, and maintaining aircraft, increases operational effectiveness, and improves decision-making. Due to DT and Industry 4.0 technologies, the aerospace sector is now at the forefront of technological progress, guaranteeing safer, more reliable, and more environmentally friendly air travel in the future.

References

1 Talkhestani, B.A., Jung, T., Lindemann, B. et al. (2019). An architecture of an intelligent digital twin in a cyber-physical production system. *at-Automatisierungstechnik* 67 (9): 762–782.

2 Aldoseri, Abdulaziz, Khalifa Al-Khalifa, and Abdelmagid Hamouda. A roadmap for integrating automation with process optimization for AI-powered digital transformation. (2023).

3 Grieves, M. (2014). Digital twin: manufacturing excellence through virtual factory replication. *White paper* 1: 1–7.

4 Tao, F., Zhang, M., Cheng, J., and Qi, Q. (2017). Digital twin workshop: a new paradigm for future workshop. *Computer Integrated Manufacturing Systems* 23 (1): 1–9.

5 Grieves, Michael. Digital twin: manufacturing excellence through virtual factory replication. White paper 1, no. 2014 (2014): 1–7.

6 Xiong, M. and Wang, H. (2022). Digital twin applications in aviation industry: a review. *The International Journal of Advanced Manufacturing Technology* 121 (9-10): 5677–5692.

7 Fang, X., Wang, H., Liu, G. et al. (2022). Industry application of digital twin: from concept to implementation. *The International Journal of Advanced Manufacturing Technology* 121 (7-8): 4289–4312.

8 Corallo, A., Buccoliero, F.O., Crespino, A.M. et al. (2022). Internet of Things and shop-floor digital twin: an aerospace case study. In: *2022 7th International Conference on Smart and Sustainable Technologies (SpliTech)*, Split/Bol, Croatia, 1–6. https://doi.org/10.23919/SpliTech55088.2022.9854314.

9 Carou, D. and Carou, D. (2021). Aerospace transformation through Industry 4.0 technologies. In: *Aerospace and Digitalization: A Transformation Through Key Industry 4.0 Technologies*, 17–46. Springer.

10 Can, O. and Turkmen, A. (2023). Digital twin and manufacturing. *Digital Twin Driven Intelligent Systems and Emerging Metaverse* 175–194.

11 Zhang, C., Zhou, G., Qingfeng, X. et al. (2023). A digital twin defined autonomous milling process towards the online optimal control of milling deformation for thin-walled parts. *The International Journal of Advanced Manufacturing Technology* 124 (7-8): 2847–2861.

12 Aldoseri, Abdulaziz, Khalifa Al-Khalifa, and Abdelmagid Hamouda. A roadmap for integrating automation with process optimization for AI-powered digital transformation. (2023).

13 Negri, E., Fumagalli, L., and Macchi, M. (2017). A review of the roles of digital twin in CPS-based production systems. *Procedia Manufacturing* 11: 939–948. ISSN 2351-9789, https://doi.org/10.1016/j.promfg.2017.07.198.

14 Pal, S.K., Mishra, D., Pal, A. et al. (2022). *Digital Twin – Fundamental Concepts to Applications in Advanced Manufacturing*. Springer ISBN: 978-3-030-81814-2.

15 Zhong, Dong, Zhelei Xia, Yian Zhu, and Junhua Duan (2023). Overview of predictive maintenance based on digital twin technology. Heliyon 9 (4): e14534.

16 Kaur, Harshpreet, and Munish Bhatia. Scientometric Insights of Digital Twins in Industry 4.0. (2024).

17 Lee, J., Bagheri, B., and Kao, H. (2015). A cyber-physical systems architecture for Industry 4.0-based manufacturing systems. *Manufacturing Letters* 3: 18–23. https://doi.org/10.1016/j.mfglet.2014.12.001.

18 Ashton, K. (2009). That 'Internet of Things' thing. *RFID Journal* 22: 97–114.

19 Zhang, R., Wang, F., Cai, J. et al. (2022). Digital twin and its applications: a survey. *The International Journal of Advanced Manufacturing Technology* 123 (11–12): 4123–4136.

20 Sarma, S., Brock, D.L., and Ashton, K. (2000). *The Networked Physical World*. Springer.

21 Zhou, Baoru, and Li Zheng. "Technology-pushed, market-pulled, or government-driven? The adoption of industry 4.0 technologies in a developing economy." Journal of Manufacturing Technology Management 34, no. 9 (2023): 115–138.

22 Yao, J.-F., Yang, Y., Wang, X.-C., and Zhang, X.-P. (2023). Systematic review of digital twin technology and applications. *Visual Computing for Industry, Biomedicine, and Art* 6 (1): 10.

23 Can, O. and Turkmen, A. (2023). *Digital Twin and Manufacturing.* Springer https://doi.org/10.1007/978-981-99-0252-1_8.

24 Bisanti, Giovanni Marco, Luca Mainetti, Teodoro Montanaro, Luigi Patrono, and Ilaria Sergi. Digital twins for aircraft maintenance and operation: A systematic literature review and an IoT-enabled modular architecture. Internet of Things (2023): 100991.

Index

Digital Twins in Industrial Production and Smart Manufacturing:
An Understanding of Principles, Enhancers, and Obstacles, First Edition. Edited by Rajesh Kumar Dhanaraj,
Balamurugan Balusamy, Prithi Samuel, Ali Kashif Bashir, and Seifedine Kadry.
© 2024 The Institute of Electrical and Electronics Engineers, Inc. Published 2024 by John Wiley & Sons, Inc.